Is That a Bat?

Is That a Bat?

A GUIDE TO NON-BAT SOUNDS ENCOUNTERED DURING BAT SURVEYS

Neil Middleton

PELAGIC PUBLISHING

Published by Pelagic Publishing
PO Box 874
Exeter
EX3 9BR
UK

www.pelagicpublishing.com

Is That a Bat? A Guide to Non-Bat Sounds Encountered During Bat Surveys

ISBN 978-1-78427-197-8 *Paperback*
ISBN 978-1-78427-198-5 *ePub*
ISBN 978-1-78427-199-2 *PDF*

A CIP record for this book is available from the British Library

Cover: *Pipistrellus* sp. (CreativeNature_nl/iStockphoto); harvest mouse, *Micromys minutus* (Greg Newman/Pixabay); Roesel's bush-cricket, *Metrioptera roeselii* (Charlie Jackson/Flickr); and barn owl, *Tyto alba* (dmbaker/iStockphoto).

Printed and bound in India by Replika Press Pvt. Ltd.

Contents

About the author

Neil Middleton is the owner of **BatAbility Courses & Tuition**, a training organisation that delivers bat-related skills development to customers throughout the UK and beyond. He has a constant appetite for self-development, as well as seeking to develop those around him, and to this end he has designed and delivered in excess of 200 training events covering a wide range of business and ecology-related subjects. Neil has had a strong interest in the natural world since childhood, particularly in relation to birds and mammals. He has studied bats for over 25 years, with a particular focus on their acoustic behaviour (echolocation and social calls). In 2014 he was the lead author of an important and well-received book about bat social calls, entitled *Social Calls of the Bats of Britain and Ireland*. Two years later he wrote *The Effective Ecologist*, which tackles the challenges facing ecologists, at all levels, as they endeavour to perform to the highest standard within their working environment.

neil.middleton@batability.co.uk
www.batability.co.uk and www.timefor.co.uk

Also available by the same author:

Social Calls of the Bats of Britain and Ireland. Neil Middleton, Andrew Froud and Keith French. Pelagic Publishing, Exeter, 2014. ISBN: 978-1-907807-97-8.

The Effective Ecologist – Succeed in the Office Environment. Neil Middleton. Pelagic Publishing, Exeter, 2016. ISBN: 978-1-78427-083-4.

About the illustrator

Joan Punteney studied illustration and graphic design at Edinburgh College of Art. She works as a freelance artist on a wide variety of projects and subjects, and her artistic talent is always in demand. Her main work, and passion, has always been painting animals. She has undertaken countless commissions for life-like paintings of horses, pets and wildlife. Joan is also an excellent cartoonist, as demonstrated by the work produced for this book, as well as by her much-loved and humorous illustrations in Neil Middleton's *The Effective Ecologist*. Joan is available to produce high-quality commissioned artwork.

jlpunteney@yahoo.co.uk

Key contributors and supporters

The author is especially grateful to the following people who have consistently shown their support for this work. In no way, however, should it be construed that any opinions or statements made by the author concur with the views of those shown below. The author is solely responsible for any thoughts expressed within this book, especially anything contentious, and the dodgy attempts at humour.

David Darrell-Lambert is an accomplished ornithologist and the owner of Bird Brain UK Ltd (www.birdbrainuk.com), based in London, operating throughout the UK and further afield. David's knowledge and passion for his subject is both immense and inspiring, and he has delivered numerous educational events, as well as making regular media appearances. He has previously been the chair of the Ornithological Section of the London Natural History Society. As well as birds he has expanded his knowledge to include butterflies and dragonflies. In 2018 he wrote an excellent book about birding within the London area, entitled *Birdwatching London* (Safe Haven Books). (Picture Credit: Mark Gash, Turnstone Ecology Ltd, 2018).

Stuart Newson is a Senior Research Ecologist at the British Trust for Ornithology, where he is involved in survey design and data analysis from national 'citizen-science' surveys. While birds have been the core of his work, he also has an interest in bats and acoustic monitoring – in particular, how technology can deliver opportunities for conservation and provide ways to engage with large audiences. Stuart set up the Norfolk Bat Survey in 2013, a novel citizen-science approach enabling unprecedented large-scale bat recording using static acoustic detectors, an approach which has since been extended to other parts of the UK. While working on bats, he became interested in bush-crickets, which are recorded as bycatch, and more recently in the sound identification of other mammals. Stuart is also a member of Natural England's Bat Expert Panel.

Matt Binstead is Head Keeper at the British Wildlife Centre in Surrey (www.britishwildlifecentre.co.uk), which is home to over 40 species of wild animal found living in Britain today. The Centre also plays an important role in carrying out captive breeding and release programmes. Everything, from harvest mouse to red deer, is housed in large natural enclosures, providing a 'real-life' natural history lesson for the 10,000 people who visit each year. *'It is a privilege to share my passion and knowledge of British mammals with our visitors. Often overlooked for the more exotic species found abroad, it gives me a real buzz to see people light up when engaging with our own wonderful wildlife.'*

Keith French has lived in the countryside throughout his working life and has always had a keen interest in all flora and fauna. Working firstly with cattle and sheep, then as a gamekeeper on a small country estate, gave him daily access to the natural world. It was when he moved to Epping Forest, in 2001, that his interest in bats was born, and while attending a bat training course he met the author of this book and Andy Froud. This led to them working together on many bat-related studies, culminating in 2014 with the book *Social Calls of the Bats of Britain and Ireland*. Now retired, Keith continues to pursue his interest in bats, as well as being a keen photographer of wildlife, in particular butterflies, dragonflies and other insects.

Preface

During the period that I have been working on this idea, I have described the project to myself in many ways. The original goal was 'a book for bat workers and ecologists describing many of the non-bat-related sounds that could be recorded or heard during bat surveys', and hopefully this is the end result. I also considered a number of other titles, including some sensible ones such as *Forensics for the Bat Call Analyst*, and during some lighter moments, *Guess Who's Squeaking*. At one point, as I fell badly behind schedule, the abbreviation of the final title, ITAB?, announced *Is This A Book?*, especially when I felt there was just too much data coming from so many directions. To try and fit all of this information into the space available was definitely a challenge, as many of the chapters would warrant a book in their own right.

The final result is intended to be a practical approach, with serious intent, which at times doesn't take itself too seriously. Included within, there are some light-hearted thoughts that many bat workers and those interested in wildlife acoustics should be able to relate to. Hopefully, this means that what may at first appear to be quite a technical book invites the reader to progress through the material in an engaging manner. There will be points, no doubt, when I am not going to sit precariously balanced on the proverbial barbed-wire fence. I will definitely say what I really think, when I feel it is needed. I do this, not to be argumentative or to offend anyone, but to give different perspectives on what it is we all do.

When I first had the idea of writing this book (early in 2017) I have to confess that I was not fully aware of what was already known and published regarding sound created by species groups other than bats and birds. As the project progressed I quickly came to the conclusion that I had been operating in a batty silo for the past 25 years, immersed so completely in Chiroptera (hmm, a bath-foam idea for the Bat Conservation Trust, perhaps) that I hadn't taken much time, or until now much interest, in anything other than bats. Now it would be unfair to paint all bat workers with the same broad *'all we want to talk about is bats'* brush, but my guess is that a good proportion of us view the world in a similar manner. We are only really interested in finding out about what we are really interested in. I am sure this holds true for many ecologists, researchers and hobbyists, and not just in relation to natural history, but for almost everything in life. I digress, ever so slightly, to get the point across. This book started off as me trying to come up with explanations about non-bat-related noise in order to make bat workers' lives easier, and it still, hopefully, achieves this initial aim. However, very quickly, as I learned more about these other noises, I became more and more intrigued, fascinated and educated about creatures I never knew much about before I started.

There is definitely a lot more going on that potentially impacts upon bat work than what may initially meet the eye, or indeed the ear. Just knowing what bat-related noise looks and sounds like is not enough. We need to know what other things look and sound like so that they can also be considered during our field excursions, analysis and interpretation. This means, within reason, knowing what else can potentially make a noise within the same soundscape as a bat. In our part of the world we don't have as many bat species, small terrestrial mammals, insects or nocturnal birds as elsewhere. Yet despite our relatively restricted diversity in these respects, I have experienced some fairly jaw-dropping moments while doing the research for this project. I am now wondering how much more there is to discover about sound that could be mistaken for bats within the British Isles. I suspect, in some respects at least, that what I have managed to gather so far is only scratching the surface. And what would happen if we started doing similar exercises in other countries and on other continents, in places where there are considerably more non-bat species emitting more sounds, as well as many more species of bat?

So what am I hoping this book and the accompanying Sound Library will achieve? First of all, I want to help bat workers to be more confident and effective when they encounter sounds that they are unsure of, and to become more aware of the potential for non-bat-related noise to impact upon their time and accuracy during analysis and interpretation. I also hope that introducing bat workers to other species groups may in some cases have spin-off benefits for conservation and studies beyond bats. Second, I want to introduce non-bat workers with an interest in natural acoustic studies to a perspective from another group of people, as well as other sounds that they themselves might encounter while involved in their own specialist area. Those studying acoustics in other taxa may also need to ask themselves similar questions, for example '*is that a mouse?*' or '*is that a bird?*' And finally, I aim simply to highlight many of the sounds produced by wildlife that anyone might hear while out at night. This is something that may be appreciated by a wide range of readers interested in natural history.

This book is not, and was never intended to be, a comprehensive account or an academic research project. There are many omissions and areas that I and those contributing were not able to record or were not able to establish full information about. As such, the examples given in this book by no means represent the full repertoire for any of the subjects discussed, and the material included will not give all of the definitive answers relating to the subjects covered. I have tried to be accurate and thorough when recording subjects myself and accepting recordings from others, but these are complex areas of study, which still, relatively speaking, require a greater level of understanding.

The objective is to give the reader an insight into what other possibilities exist when he or she is hearing sounds in the field or carrying out software analysis. It should be thought of as more of an eye-opener (or an ear-opener), as opposed to a full appraisal of everything that ever squeaked or squawked during the hours of darkness. It isn't about fully understanding what all these noises mean to any great degree of thoroughness. It is about trying to make people aware of what other things look and sound like, irrespective of an animal's communicative intentions or emotional state. All of this, so that we can move closer to becoming more effective and more accurate in our assessments, and spend less time trying to work out what it is we are looking at or listening to.

So, in conclusion, please do not feel that you should read this book with a view to knowing everything there is to know by the end of it. In many respects there are still far more questions than answers (not an uncommon concept for those familiar with bats!). For the time being, please regard this as a source of reference, blended with my own experiences and thoughts, with the aspiration of creating a better appreciation of the subjects covered. This may possibly (hopefully!) inspire others to carry out further research, in order to help improve our understanding, as well as our ability to accurately interpret the encounters we have with the natural world.

Acknowledgements

There are not enough words, even for me (which might surprise some people), to express how much I appreciate the encouragement and support I have received during the process of writing this book. There were moments when it was a bit of a struggle to keep the momentum going, but the interest and encouragement of those close to me spurred me on. So let me introduce the people and organisations who have come together to make this book and the Sound Library possible. As you read and listen to what has been produced, please do so with everyone's involvement in mind. If you know, or ever meet, any of the following people, please say thanks to them for their contributions, and pass on my sincere gratitude.

First of all I would like to thank the 'production team', those who physically made the finished product. Joan Punteney, for yet again producing the excellent illustrations accompanying each chapter. Then we have Aileen Hendry, my partner in life, for her support and feedback (lots of feedback, in fact!) during the entire process, as well as the essential proofreading stage. Why have one excellent proofreader, when you can have two? Massive thanks are due to Laura Carter-Davis (Managing Director, Echoes Ecology Ltd) for her support, feedback, proofreading and critical appraisal of the material. And then it all goes to a professional copy-editor for some much-valued input. In this respect I owe great thanks to Hugh Brazier for his excellent skill and attention to detail at that critical stage of the process. Next, Lars Pettersson (Pettersson Elektronik AB) for being on board, once again, in allowing the extensive use of BatSound software to create the numerous figures that have been included. Thanks also to Chris Corben (www.hoarybat.com) for the use of the AnaLookW software, as well as for some very useful insights along the way. Finally, a big thank-you to Nigel Massen of Pelagic Publishing Ltd, for his faith that I might, once again, have something worth saying, and for his continued support and professional expertise.

Then we have those people who went well beyond what anyone in my position would ever dare to hope for, and helped in the gathering of lots of data, sharing experiences and supporting the cause in so many essential ways. So, for really going 'above and beyond', special thanks are due to David Darrell-Lambert (Bird Brain UK Ltd), Matt Binstead (British Wildlife Centre), Stuart Newson (British Trust for Ornithology) and Keith French, all of whom provided, in different ways, considerable support and encouragement. Special thanks also to Amy Ashe, Kari Bettoney (Mid Devon Bat Rescue Centre), Fiona Cargill (Wood PLC), Melanie Findlay (Findlay Ecology Services), Andy Froud, Leif Gjerde (www.flaggermus.no), Lorna Griffiths (Nottinghamshire Dormouse Group), Andrew Hargreaves (Wildlife Provençale), Nick Hull (twoowlsbirding.blogspot.com), Valentina Iesari, Harry Lehto (University of Turku), Aaron Middleton (WSP), Chris Nason, Erik Paterson (Clyde Amphibian and Reptile Group) and Raimund Specht (Avisoft Bioacoustics).

For providing assistance in so many ways, including unearthing research papers, making recordings available, arranging site access, suggesting ideas, or for general support and encouragement of the project, I would also like to acknowledge all of the following: Mingaile Anderson (Echoes Ecology Ltd), Arnold Andreasson (cloudedbats.org), Henry Andrews, Avisoft Bioacoustics (www.avisoft.com), Sarah Barry, Bat Conservation Trust (in particular Lisa Worledge and Naomi Webster), Rob Bell, Adrian Bicker, Claudia Bieber (University of Camerino), Simon Bowers, Ian Brady (Wildlife Sound Recording Society), Heather Campbell (Echoes Ecology Ltd), Andrea Catorci (University of Camerino), Tom Clarke, Clyde Muirshiel Regional Park, Keith Cohen, Mark Cubitt (National Moth Recording Scheme Recorder, East and West Lothian), Ann Deary Francis, Kate Denton, Adrian Dexter, ECOSA Ltd, Elekon AG, Andrena Ellis (WSP), Pat Emslie, Sophie Ewert, Falkirk Community Trust, James Faulconbridge (Landscape Science Consultancy Ltd), Hazel Forrest, Anders Forsman, Leonida Fusani (University of Camerino), David 'Izzatabat' Galbraith, Mark Gash (Turnstone Ecology Ltd), Sheila Gundry (Froglife), Lorraine Hamilton (Cairn Ecology Ltd), Jodi Handley Bell, John Harrison-Bryant (HB Bat Surveys), Alexander Hatton, Les Hatton (ERM), Waldemar Heise, Philip Higginson, Rosanna Hignett (Echoes Ecology Ltd), Amelia Hodnett (ERM), Michael Hoit, Jon Horn (Nurture Ecology Ltd), HS2, Morgan Hughes, Laurence Jarvis (Froglife), Rogan Jones, Inger Kaergaard, Russell Keen (Echoes Ecology Ltd), David King (Batbox Ltd), Erik Korsten, Roy Leverton (National Moth Recording Scheme Recorder, Banffshire), Liza Lipscombe (British Wildlife Centre), Jochen Lueg, Craig Macadam (Buglife), Craig Macdonald (CSM Ecology and Environmental Assessment Ltd), Jennifer Anne MacIsaac, Lindsay MacKinlay (Parnassus Ecology), Koenraad Mandonx, Dwayne Martin, Tony Martin (E3 Ecology Ltd), Ben McLean, Emily Millhouse (Froglife), Emiliano Mori, Kirsty Morrison, Helen Muir-Howie, National Trust for Scotland (Threave Estate), Nottingham Trent University, Johnny Novy (The Amphibian and Reptile Conservation Trust), Martin O'Neill (The Amphibian and Reptile Conservation Trust), Mark Osborne (OS Ecology Ltd), Simon Parker (Turnstone Ecology Ltd), Claire Parnwell (Greenwillows Associates Ltd), Hannah Paxton, Liz Petchell (Astell Associates), Nick Pinder, Robin Priestley, Sarah Proctor (WSP), Jo Richmond, RSPB Mereshead, Marius Ruchon, Paola Scocco (University of Camerino), Martin Scott (HiDef Aerial Surveying Ltd), Scottish Natural Heritage (Licensing Team), Graham Sennhauser (Tetrix Ecology), Kieran Shaw, Helen Simmons, Heather Simpson (Echoes Ecology Ltd), Greg Slack, Bill Slater, Rune Sørås (Norwegian University of Science and Technology), Sandie Sowler (Bat Training Partnership), Laura Spence, Logan Steele, Joe Szewczak (www.sonobat.com), Natalie Todman, Chris Toop, Angel Torrent, Laura Torrent, Goedele Verbeylen (Natuurpunt), Michael Walker (Nottinghamshire Bat Group), Cecilia Wide, Dan Wildsmith, Will Woodrow and Kathy Wormald (Froglife).

Finally, close family are the ones who are impacted upon the most when someone with my intensity of focus and downright stubbornness sticks to the project deadlines in order to drive it through to the end. Personal and heartfelt thanks go to my son Aaron Middleton, my daughter Emily Simpson, my mother Audrey Middleton, and finally, most importantly, my tolerant, understanding, gorgeous, intelligent wife, Aileen Hendry, who, especially carefully, proofread this final sentence, and made a couple of last-minute, *minor* edits.

The Sound Library

For a full appreciation of the bat and non-bat sounds discussed in this book, the reader also needs to be a listener. To accompany the text I have therefore provided a series of downloadable sound files, available in a *Sound Library* via the following link:

www.pelagicpublishing.com/pages/itab-sound-library

The files are mostly in .wav format and therefore can be opened with most sound analysis software, as well as with systems such as Windows Media Player. The files are grouped into folders which tie in with the chapters where they are discussed.

In the book, when reference is made to a file in the Sound Library, the symbol 🔊 is inserted in the text. Where the 🔊 symbol is attached to a figure (such as a spectrogram), the figure number doubles as the Sound Library file number. Note, however, that when you open any such sound file, the file itself is often larger than the portion that is shown in a figure. The sound file thus has the potential to show you more on your own computer screen than what is provided in the text. As well as the sound files that tie in with the text or figures, the Sound Library contains many additional recordings not referred to in the book.

Each track has an abbreviation included in its title, enabling you to know the type of recording that you are about to download or listen to. The following abbreviations are used:

AU	Audible recording. Usually representing what you would hear with your ears if you were in the field without a bat detector.
FD	Frequency division recording. Note that frequencies on the y-axis of a spectrogram are actually 10 times higher than shown (e.g. 5 kHz is actually 50 kHz).
FS	Full spectrum, real-time recording. Usually ultrasonic, meaning that you won't normally hear anything unless you convert the call to time expansion.
HE	Heterodyne recording, as recorded by a heterodyne bat detector.
TE	Time-expanded recording. Slowed down by a factor of 10, meaning that what you hear is 10 times slower in terms of time, and 10 times lower in terms of frequency.
ZC	Zero crossing analysis file. The format used by AnaLookW software, for example.

Clicking on the track will open it with your default audio software (e.g. Windows Media Player). Alternatively, if you first open the sound analysis software of your choice, and then use the software browse facility to open the track from within its folder, this should open the file in that software. Note that AnaLookW will not recognise .wav files, and .wav utilising software will not usually recognise the .zca file format used by AnaLookW.

In addition to sound recordings, within the Sound Library there are some documents that you may find useful:

- **Anabat/ZCA examples**: some additional spectrogram examples using AnaLookW software.
- **Listening to your software**: a pdf document giving a screengrab tutorial as to how to listen to files on various bat analysis software systems and Windows Media Player.

Test yourself

In the Sound Library there is also a folder entitled *Test Yourself*. This folder contains the following resources:

- Twenty full spectrum recordings related to the subject matter of this book. Note that most of the calls are in addition to those contained elsewhere in the Sound Library.
- A *Question Sheet*, for you to print off, which also contains space within which you (or someone in your team) can write your answers to the questions.
- An *Answer Sheet* giving the answers to the questions posed on the question sheet.

If you are feeling confident, open the calls within the *Test Yourself* folder, and using your favoured full spectrum analysis software, have a bash at answering the questions. Good luck, and no advance peeking at the answers!

CHAPTER 1

Well, what on earth could it be?

We only know what we know. Therefore, we don't know what we don't know, and how much there is still to know.

What on earth could it be? Well it's definitely a book, but what's it all about? At first it looks as if it's all about bats, but skimming through the chapters, apart from this one, there doesn't seem to be much mention of bats. Correct. It is more about the other animals, and other noises, that bat workers (and others) might hear while conducting bat surveys, or encounter during the occasionally boring, sometimes brain-melting analysis thereafter when you end up with all those query sounds. Are any of those bats, and if they are bats, what species? And what about all these other 'noise' files? Surely it can't all be meaningless and uninteresting?

This book is also, at least in part, about bat workers ourselves, and the other people who go out to study and record acoustics in the natural world. Yes, we know you other people exist. As bat workers, however, we will rarely see you, or rather you will rarely see us. We operate during the hours of darkness, and sometimes we are not even there when we are making the recordings. We use automated bat recording systems (static bat detectors) to do it all while we are elsewhere, transecting over some dark, remote, windswept hillside being chased by bullocks, in an attempt to save the planet, one bat at a time. So, in effect our apparent laziness achieves nothing more than freeing up our time to strenuously and tirelessly carry out night-time manoeuvres elsewhere. All meaning, of course, that we have even more sound analysis to plough through, in order to hit deadlines for reports that were probably due before we had even been appointed in the first place.

How bat workers interface with the natural world, and the associated processes, is as important a consideration as the bats we seek, and anything else that squeaks at night, or gets inadvertently recorded with the potential to cause confusion or clog up our disc space. Our knowledge, our behaviours and our desire to know everything can at times conflict with our misplaced embarrassment to admit that '*I just don't know*', when we are supposed to be, in the eyes of others, the expert. Will this book answer all of these '*I'm really not sure*' queries? I doubt it, as frankly I ain't that good myself, and I certainly don't have enough intelligence, time or space to commit to the study of every sound ever made during the dark hours. But it will at least give you some clues, ideas and practical guidance relating to the subjects covered. And indeed, it may also develop your interest in things other than just bats. To be honest, when I started writing this book, it *was* all about the bats. Now I just want to go and learn even more about all the other stuff that's out there. Apart from the pure intrigue, I have demonstrated to myself that knowing about all of this other stuff isn't just interesting or fun – although it is both of those. In fact, it's essential. The next stage of the process is to demonstrate this to you, and there are certainly things covered within these pages that may surprise many readers (and perhaps not always in a good way).

So, in conclusion, yes, this is a book for bat workers (and others recording and researching natural sounds) that takes account of bats, as well as other species, human behaviour and the need to be more knowledgeable about what we are talking about, with an aim that better interpretation and even higher levels of professionalism can be achieved. I will leave you, at this point, with a snippet from a conversation I had with a colleague some years ago:

'Neil, is that a bat?'

'No, it isn't.'

'So what is it then?'

'Hmm, to be honest, I don't know.'

'Well if you don't know what it is, how can you be certain that it isn't a bat?'

'Please, just accept what I am saying. I can't explain it, but I know that it isn't a bat.'

Probably not a conversation that I should admit to, but too late now! On reflection, however, there is one thing that I definitely do know. I was wrong to suppress that person's desire to be better informed, more confident in her interpretation, and to develop herself further. I should be punished accordingly.

So the eighteen months that it has taken me to write this book, in order to answer what I now know was a relatively straightforward and worthwhile question, has been my penance.

It's not a bat – so why does it matter?

I have worked alongside bat surveyors of all shapes and sizes for over 25 years (in fact, I've been a few different shapes and sizes myself over that time), and the number of times we have had conversations about things other than bats that we have heard while out in the field is countless. Occasionally, someone has been able to say, 'that's definitely a …' Quite often, however, we move on without knowing precisely what it was we just heard. Does it matter? I guess quite often it doesn't. But if you don't know, then how can you confidently say 'it doesn't matter?'

While carrying out field surveys (e.g. emergence/re-entry surveys or transects) we are potentially one of only a small number of people who are visiting that area during the hours of darkness. Occasionally there may be an ornithologist out at this time, and those carrying out the great crested newt surveys may be around for part of a season. Bat workers, however, are more likely to be on site, in numbers, during the spring, summer and autumn periods. This is a considerable amount of resource (eyes, ears, recording equipment) that if better informed (even slightly) could potentially unearth important data relating to a site, if not from a specific development's point of view, then from a wider conservation perspective. When we hear something, we have two choices. We can choose to ignore it (after all, it's not a bat), or we can choose to at least attempt to identify the source of the noise. It could be important. It could impact upon the reason why we are there in the first place. A rare bush-cricket, a Schedule 1 bird species (e.g. barn owl or kingfisher), a European Protected Species (e.g. otter). Or it may just be something of local interest that someone, somewhere, may thank you for. And as you tell people about these findings, they may reciprocate, and the knock-on benefits can be substantial.

Therefore, as bat workers we have the potential to gather vast amounts of non-bat-related data, and at least some of these data could be put to good use and contribute towards the wider knowledge of the species concerned, as well as for conservation purposes. An excellent example of a bat-related study coinciding with other species groups has been demonstrated by the British Trust for Ornithology (BTO). Their citizen-science bat project (Norfolk Bat Survey) has not only gathered huge amounts of bat-related data, but also as 'bycatch' lots of additional data relating to bush-cricket species. Not only has this extra information contributed to further the understanding of bush-cricket distribution with the potential for better-informed conservation decisions, but those involved were also able to see additional positive offshoots, for example the development of an automated classifier for bush-cricket identification (Newson *et al.* 2017).

In a more specific set of circumstances, but equally valid, imagine that you saw a barn owl while carrying out a bat survey. You know that the breeding bird survey has already been concluded, but not whether that team had picked up on the presence of this species. I bet that even if you were only a half-decent ecologist, when you got back to the office tomorrow morning you would report this information to the ornithology team. Now imagine that you had only heard a screeching sound that might have come from a barn owl.🔊 You say to yourself, 'Hmm, I think it's just a bird. It's probably not that important.' Come the following morning it would not be the first thing on your mind. In fact, it may not enter your mind ever again. Just another sound, made by something, heard in the dark. Really!

Now let's move swiftly along towards the end of a bat survey process, to the sound analysis that gets carried out on the PC, while we are sitting comfortably at our desk. When considering bat-related sound, we should already have a good idea of what we are looking for. However, there could be other unexpected, unexplained noises that distracts, confuses or intrigues us. If we were able to be more confident with our interpretation of all this 'uninteresting' non-bat noise, then this surely helps us to focus on the bats we are interested in, as well as other things that may be unusual or worth our effort in investigating further. Arguably, this could be of benefit even if it is purely from a self-interest or self-development point of view. And here we go, the first of many challenging and potentially controversial thoughts. If you are sitting there thinking, '*I know what a bat looks like and I am not interested in anything else*', in an '*it's not my job*', or '*it's not where my specialism lies or the direction I wish it to develop*' kind of way, then look no further than Chapter 2 (*Terrestrial mammals*). Go on, have a look now. Look at the spectrograms. Ask yourself truthfully, is there stuff there that surprises you? Is there stuff there that actually looks like bat noise, but isn't?

Personally, I am in a different place to where I was prior to starting on this work. I am not quite so sure about a number of things any more. To be honest, there was a lot that I wasn't that sure of previously in any case. When people send me calls now, asking for my opinion, I really have to take a step back and consider not only bat possibilities, but every other darn thing that potentially goes 'chirp' or 'squeak'. I now feel that during my bat working life I have spent a lot of time listening and learning from brilliant bat people, but I should also, perhaps, have spent some time with people who knew about other stuff. However, please be aware that if you do begin to develop an interest in bioacoustics in other species groups, in some instances there is even less widespread knowledge about these other groups than what we know, or think we know, about bats. But at the very least this does demonstrate that anything which contributes towards the knowledge of acoustics in less-studied groups is going to add value.

Pulling all of the above together, and in order to complete a case in favour of the value of considering what else is out there, Figure 1.1 summarises many of the key points I have mentioned so far, as well as adding a few others.

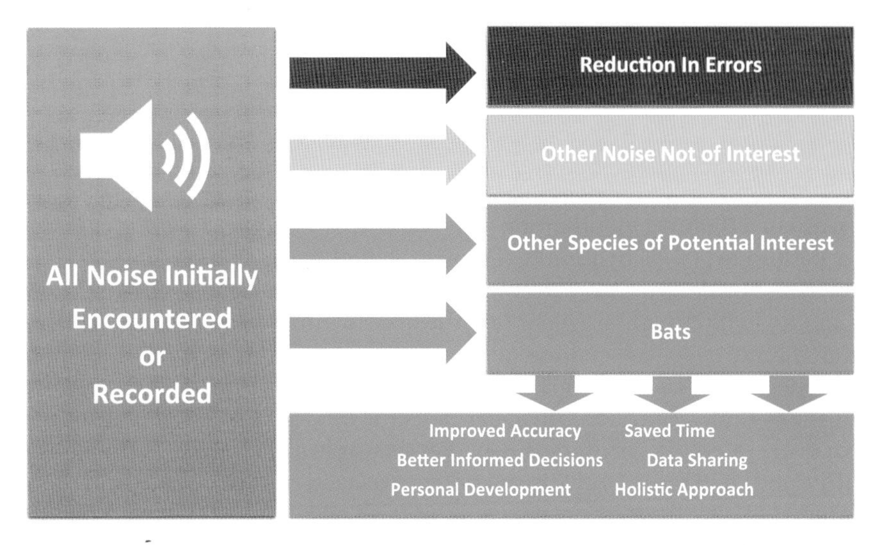

Figure 1.1 Why does it matter? The benefits of knowing more about all noise encountered or recorded.

Spectrograms in Chapter 1

Spectrograms in this chapter use the following scales, unless otherwise indicated in the figure legend:

Time (x-axis): 0.5 seconds (500 ms)
Frequency (y-axis): 0–110 kHz

When a figure legend includes the 🔊 symbol, this means that the figure has been created from a file in the Sound Library. The figure number matches the file number there. For more information about how to access and download files from the Sound Library please see pages xv–xvi.

Use the equipment you already have

We hear stuff all the time. Goodness, most of the time, in the dark, we actually hear considerably more than we are able to see. *'What was that noise?' 'There it goes again.' 'What is it?' 'Where is it?' 'Is it on the ground, in amongst vegetation, or is it above our heads?' 'A UFO, or a "big cat" perhaps. Or someone else doing a bat survey, just like us.'* There we are, in the moment, trying to work out what is making all that racket. We are ecologists, and natural sounds should therefore, at the very least, intrigue us. So, instead of all this thinking about it, record it, and quickly, before the silence returns. It might be important. Even a recording of the sound using the video mode of a mobile phone camera could be good enough for someone else back in the office to tell what it was, and establish whether or not it was of any significance. And having gone through this process, there is a very good chance that you will remember it, so that next time you will recognise it.

For some of us, depending on the equipment we have, it is even easier than getting the phone out. Now, I have tested what I am about to discuss with over a decade's worth of audiences on bat survey methods courses. Some people reading this will have been present at one of these events and will be able to verify that what I am about to say is surprisingly true. Not that many bat workers use (or in some cases even know) what that little white 'Ref' button on the front of a Batbox Duet (Batbox Ltd, West Sussex), or the black 'Com' button on a Pettersson D230 or D240X (Pettersson Elektronik AB, Sweden) actually does. These buttons are an extremely useful feature on those detectors, in that they allow for commentary, enabling the user to integrate real-time voice notes within (i.e. not saved as separate sound files) any recordings that are being made by the detector. Have a listen to an example of a voice note I have provided with the Sound Library. 🔊 Other bat detector models also allow you to make voice notes, albeit in a different way. For example a Batlogger M (Elekon AG, Switzerland) will record voice notes fairly effortlessly, and all hand-held models of Anabat (Titley Scientific, Australia), as well as Echo Meter Touch (Wildlife Acoustics, USA) also have specific voice-note options. Now let's go one step further. If these features allow the recording of speech, then there is a pretty good chance that, using the same process, they will pick up any non-bat-related audible sound. OK, so how good are these recordings? Well, have a listen to this call of a robin 🔊 that I recorded using the 'Ref' button on a Batbox Duet, and then this one of a blackbird 🔊 recorded with a Batlogger M, in standard mode, just pointing the detector in the direction of the bird and listening to the sound in its original, real-time format on BatExplorer software. Reasonably good, I am sure you would agree. OK, perhaps not the quality you would achieve using equipment specifically aimed at the job, but good enough to let you play it to a colleague for a second opinion.

So, sometimes we do actually have all the gear, but no idea how to maximise its use. I do want to stress, however, that this isn't necessarily the fault of the user, for example if no-one has ever explained these functions or this approach to that user in the first place. They do only know what they know, after all.

There are many people doing vast amounts of bat sound analysis every year, and the technology that is available today compared to when I was a lad is amazing. In fact, if you had said to us 25 years ago that it would be possible to record in real time, without losing any data from the field, and then look at the call on our PC, investigating it to the *n*th degree in whatever way we wished (e.g. real time, time expansion, heterodyne, frequency division, zero crossing analysis), most of us would have said, back then, that it just wouldn't be possible. In fact so much so that we would not even have been talking about it as a wish. It was not so much 'impossible', but more 'inconceivable'. With the technology we all have on our desks today (as well as in the field), it makes sense to use it to its fullest. Why, then, do many of us either not consider, or choose to ignore, some of the functionality that we have and can easily access with our fingertips?

I regularly come across people doing bat call analysis who, more often than not, don't take advantage of any of the audio functions readily available on the software they are using. In other words they are carrying out the analysis based purely on what the visual representation is showing them. There could be lots of reasons for this, ranging from lack of training to pressure of time, or just not having thought about it as being of any value. Granted, on some systems (e.g. ZCA files on AnaLookW) there is no audio option, and if it's not there then obviously it can't be accessed. So someone who has been using AnaLookW all of their working life has not had the option, and therefore possibly not felt any need to use sound in their assessments when subsequently using other systems. And anyway, the bulk of the time it's a pipistrelle. You can see it's a pip, and it just wastes time to listen to it, even if you could. I can't argue with that, and have no desire to suggest that anyone should do otherwise. However, what about when you are not sure what it is that you are looking at? For example, it looks like a bat social call, and there were swarming bats going back into their roost that morning, and '*wow, look at this spectacular spectrogram.*' '*Let's check out one of the reference books on the subject* (Russ 2012, Middleton *et al.* 2014, Barataud 2015) *and see if we can find a match. There is nothing like this in there. It's got to be something of interest, surely.*' OK, so let us have a look, here and now (Figure 1.2).

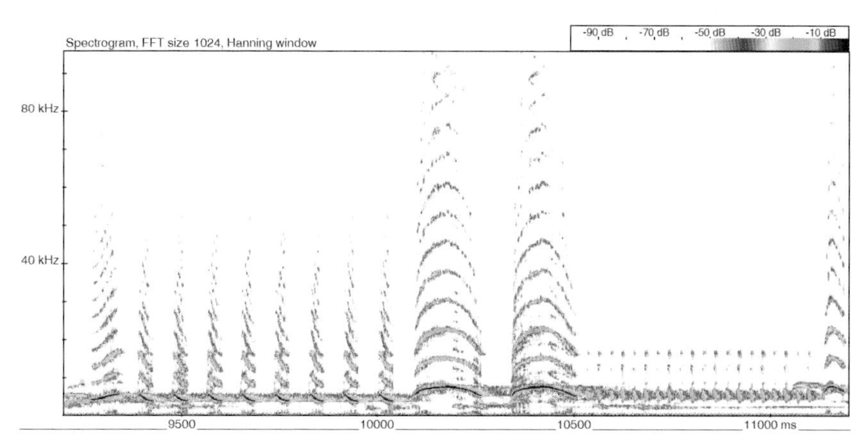

Figure 1.2 🔊 Recorded during a dawn survey while bats were swarming (A Dexter, 2018; frame width 2000 ms, frequency scale 0–96 kHz).

I am sure you would agree. Figure 1.2 does look pretty amazing. OK, I now want you to do something for me, and go through the following sequence of tasks. It's going to be a really valuable personal development exercise if you do, and it will stand you in good stead for other things that follow, as you progress through this book. First of all open up the Sound Library and go into the folder entitled *Chapter 1*. Now click on and listen to the heterodyne file, reference '1.2: HE', through an audio (i.e. audible) system (e.g. Windows Media Player). Now do the same again with file reference '1.2: FD' (frequency division), and then reference '1.2: TE' (time expansion x 10). Now open up the audible, real-time track, file reference '1.2: AU', and once again listen to it in your audio player.

I have been through this process countless times. Very often when I do, it is well worth it. I remember the first time very clearly. I had to say to the bat surveyor that he had wasted the best part of 30 minutes trying to work out what it was visually, when if he had only listened to it in audible, real time in the first place, he would've known within seconds what it was. Thankfully, I knew that surveyor really well. It was me. I just held my head in my hands, thinking, '*Middleton, you plonker.*' As with so many things in life that experienced people hand down to others, the only reason they know what they know, isn't because they are any more intelligent than you. It is merely because they have already made that same mistake that you are about to make. The only difference being that when they were about to do it there was no-one experienced there to point out what was about to occur, and in some cases repeated regularly over a long period of time. Personally, I have a vast amount of experience, all owing to the numerous mistakes I have made throughout my life. By the way, in case you are still wondering what Figure 1.2 actually is, it's a bird. A wren in fact. And to be clear, Adrian Dexter knew what it was when he sent it to me, initially as an example to be used within the bird chapter.

So yes, I do find it odd that bearing in mind it is sound that is being studied in this context, more people do not take the opportunity to also engage their ears as part of the analysis process. Imagine someone was describing a picture to you as a list of line angles and colour variables, with a whole load of adjectives and nouns thrown in. Would it not be easier to improve upon the image in your imagination if you just looked at the picture? Likewise, yes in theory there is nothing you can hear that you shouldn't be able to see on a spectrogram, but, by heck, especially if you are not sure what it is that you are looking at, listening to the sound, if available, can at times have a big impact upon your conclusion.

There is now another example I would ask you to work through. As per last time, there are benefits to doing this, as it will help you to be more aware of things to think about as you continue reading. Look at Figure 1.3. It was recorded during darkness, in Norway (August 2018). It got us all very excited when we saw it on a PC projector screen the following morning. If it wasn't a bat, then perhaps a bird. Take another look at Figure 1.2. Figure 1.3 looks similar to the part of the spectrogram between 10,000 and 10,500 on the x-axis of Figure 1.2. A lot bigger and more spaced out, perhaps. Hold on a minute – the frame width in Figure 1.2 is 2000 ms, but in Figure 1.3 it's zoomed in to 500 ms (the chapter default), meaning that everything is going to look closer, and therefore bigger and more spaced apart. Yes, it's probably going to be a bird. Go on, have a listen.🔊 It's in the Sound Library (track reference 1.3: AU).

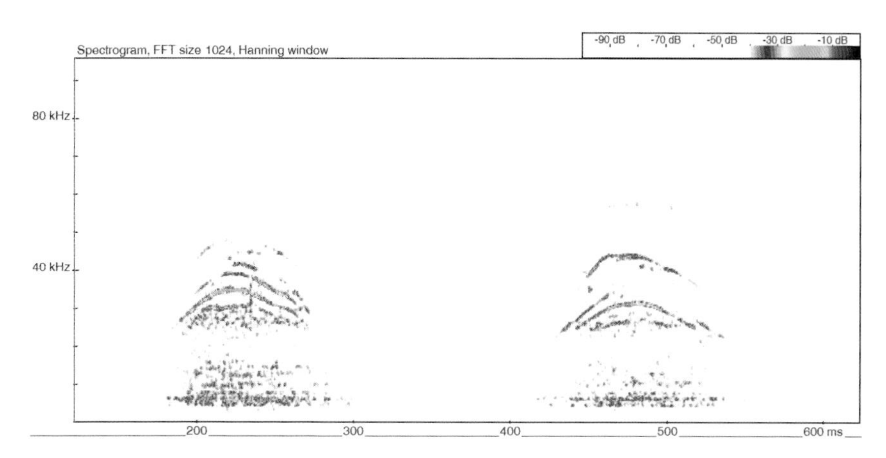

Figure 1.3 🔊 Recorded during a bat survey (K Mandonx, 2018; frequency scale 0–96 kHz).

Yes, you are probably now in no different a place to where we were when we all 'sniffed out' what it actually was. Just to be clear, it was me sniffing during a conversation I was having with a co-surveyor (Laura Torrent, Catalonia). We were standing next to a gentleman who had his bat detector switched on (Koenraad Mandonx, Belgium) searching for bats (which is probably what Laura and I should have been doing). No-one knew that it had happened at the time. Why would anyone consider that such a normal activity could have a potential impact on our bat survey interpretation? And it didn't at the time. But come the following morning, when the analysis was being carried out by Koenraad, it caused great confusion when he found this 'call' in amongst a load of bat-related files. And well done that man, as, after a bit of a debate in the room as to what it might be, he was the first to actually listen to it.

Yes, we definitely need to be using our ears more often, especially for things that don't look like the regular bat passes we are accustomed to seeing. This is very often possible with many of the recordings made in today's world, where sound is retained within the file being analysed. Full spectrum recordings can, for example, be listened to in real time (useful if it's not a bat), or quite easily converted into time-expanded versions for listening purposes (if it is an ultrasonic noise). Software programmes such as BatExplorer, Kaleidoscope, SonoBat and Anabat Insight give the user numerous listening options, for example, not only real time and time expansion, but also heterodyne and frequency division. Knowing what bats, and other things, sound like on any, or all, of these options can be extremely helpful, especially since some recorded sound can, otherwise, take a lot of effort to analyse.

As you read through this book you will encounter lots of examples of sound created by things other than bats, that can look, and in some cases sound, very much like a bat (e.g. small terrestrial mammals). As well as these examples, we will also cover other sounds that don't look bat-like at all. Knowing what these other sound sources look like, and in some cases sound like, means that the bat worker can usually rule these out fairly quickly during any analysis process, as well as identify other potential areas of interest, and/or adaptations required in the survey methods. Within each of the chapters, as appropriate, I am going to provide some guidance about things to consider when there is the potential for these other non-bat encounters to occur. This is summarised in Table 1.1.

Table 1.1 A summary of the most likely areas where there is potential for non-bat-related noise to impact upon the deliverables of a bat survey programme.

Potential for ...	Small terrestrial mammals Chapter 2 Table 2.9	Birds Chapter 3 Table 3.4	Amphibians Chapter 4	Insects Chapter 5 Table 5.9	Electrical & mechanical Chapter 6 Table 6.1	Weather & human Chapter 7 Table 7.1
Equipment damage	Yes	Yes	No	No	Yes	Yes
Inadvertent recording	Yes	Yes	No	Yes	Yes	Yes
Noise filter errors	Yes	Yes	No	Yes	Yes	No
Bat classifier errors	Yes	Yes	No	Yes	Yes	No
Inexperienced human error	Yes	Yes	No	Yes	Yes	No

It is a bat – thank goodness!

My guess is that many people reading this book will be fairly clear as to what bats sound like and look like on bat detectors and sound analysis software. At the very least when they are echolocating, and to an extent, perhaps, when social calling. It does, however, make sense to discuss bat-related sound to a point within this chapter, as some people may be reading this with less knowledge about bats, for example, from wider perspective, or with an interest in a non-bat species group. Apart from that, even if you know quite a bit about bats, what follows does dip into some areas that are not always covered to the same extent elsewhere. However, do bear in mind that this book is not designed to cover what other bat books and resources have already achieved extremely well, and in considerably more detail (see Appendix 3).

Audible versus ultrasonic

It does no harm to remind ourselves that echolocation is not always ultrasonic (e.g. European free-tailed bat, *Tadarida teniotis*), and ultrasound does not always equate to an animal echolocating (e.g. brown rat). Also, the term 'ultrasound' itself is not something easily definable in nature, it is merely a human perspective on what is happening naturally.

If something is described as 'audible', that normally means audible to the human ear. Now, as far as I am aware, every sound deliberately created by any animal is audible to another animal of the same species and/or, in some cases, of another species. Conversely, something described as 'ultrasonic' is not audible to the human ear. But what is ultrasonic to a human may be audible to another animal. At an individual human level, we could even go so far as to say that what's ultrasonic to one person may be audible to another. So a person of my age is encountering (or rather isn't) much more ultrasound than someone in their late teens. In fairness, too many Motörhead gigs probably haven't helped me in this respect. My good friend Andy Froud, however, despite being well gigged out and in his forties, has excellent high-frequency hearing, picking up noctule echolocation and pipistrelle social calls without a bat detector. He *is* a human bat detector, and a very useful chap to have around, albeit making the rest of us feel inadequate.

There is, thankfully, a scientific definition for the term 'ultrasound', describing it as a sound beyond the normal upper limits of human hearing, medically established as being 20,000 Hz (or 20 kHz) in a healthy young adult.

Echolocation

Echolocation is a process described as 'the analysis by an animal of the echoes of its own emitted sound waves' (Altringham 1996). It is used by the majority of bat species (there are exceptions) in order to navigate successfully, in darkness, around their habitat. In many cases it is also used for foraging purposes, to 'home in' on food. It is worth noting that although all of the bat species occurring in the British Isles use echolocation for commuting and foraging, in some parts of the world there are species that use it purely for navigational purposes. To be effective, bearing in mind that bats usually operate in darkness and travel quickly through the air, it needs to be repeated constantly, usually many times per second. Also, it tends to be at higher (i.e. ultrasonic) frequencies for a number of reasons, including attenuation (see Appendix 1, Figure A1.2) and target discrimination, all as outlined in Table 1.2.

Table 1.2 Echolocation – the benefits of high-frequency sound.

Feature of high-frequency sound	Examples of benefits
Attenuation Higher frequencies attenuate more quickly (i.e. travel less distance) through air than lower frequencies. For example, 30 kHz may reach a distance of 30 m, whereas 100 kHz, may only reach 10 m	May help avoid interfering with other bats Allows close approach to prey Less likely to be picked up by predators
Target discrimination for foraging **Greater detail** for obstacle avoidance	Higher frequencies are good for reflecting off (i.e. echo creation) and therefore locating small objects. The higher the frequency the smaller the object it can detect (e.g. small insect prey)
Fewer natural sounds present at higher frequencies	A bat can more easily focus on its own produced sounds, with less interference evident

Given the purpose for which a bat would echolocate, a single pulse of high-frequency sound is not going to tell it very much. The bat needs a constant and fast throughput of data in order for it to assess what's going on. It is therefore unlikely that you are going to record a single echolocation pulse. So, usually, if you are looking at echolocation on a spectrogram (Figure 1.4) you will see a series of repeated calls, or call sequences (such as the alternating calls produced by barbastelle, *Barbastella barbastellus*) that will look similar to each other. Either that, or a gradually changing sequence of echolocation pulses over a period of time.

Bat echolocation is a useful subject to understand at a scientific level, as it can provide information in order for us to narrow down the identification of the bat species, or species group, responsible for its production. In some cases establishing the exact species can be very reliable, depending on which other species occur within the same habitat type or distribution range. For example, identifying soprano pipistrelle (*Pipistrellus pygmaeus*) in the British Isles is usually fairly straightforward – but if you travel further south within mainland Europe it becomes more problematic owing to the presence of Schreiber's bent-winged bat (*Miniopterus schreibersii*), an entirely different species that has similar echolocation. A similar situation can occur when trying to separate Nathusius' pipistrelle (*Pipistrellus nathusii*) from Kuhl's pipistrelle (*P. kuhlii*) where they both occur. In cases such as these, echolocation alone may not always provide 100% certainty as to an individual bat's identification.

And then there are species that are problematic considerably more often. In our part of the world, the *Myotis* species fall firmly within this group. Yes, in certain circumstances, with certain *Myotis* species, we can be more confident in identification. However, quite often, the 100% degree of confidence (i.e. without any shadow of doubt whatsoever) at species level is, if we are completely honest with ourselves, just not there. This may be due to a number of factors, including the quality of the recording, the habitat in which the bat was recorded, the surveyor not having any visual references. But to be fair, all of these things, plus others, could potentially have an impact on our ability to identify many bat calls. So, irrespective of these other factors, the bats within the *Myotis* genus create challenges in identification due to them producing similarly structured, short-duration, wide-bandwidth, frequency modulation (FM) calls (Figure 1.5). They have all, in our part of the world anyway, evolved to be 'closed' habitat specialists, and the structure of their echolocation calls is ideal for gathering information in this setting. This means that, as a group, they are using the same type of acoustic 'torchlight' in order to be effective. Also, their call structure doesn't go through the same level of alteration, in different habitat settings, as seen in other, often more easily separable species such as pipistrelles (Figure 1.4). *Myotis* bats can alter their call structure and call sequences to a point (e.g. start frequency, end frequency, repetition rate), but within this genus are many areas of overlap between individual species when considering call parameters, added to which are the degrees of variability displayed by individual bats in different settings. All in all, these factors, along with many other things, make the creation of diagnostic identification keys problematic, especially when we would want to know the definitive answer for each bat pass. Quite often at this stage of our understanding, it just can't always be done to a safe level, and we may need to accept that in some instances it just won't ever be possible to achieve diagnostic identification of every bat species, on every occasion, using purely their echolocation.

Considering the fact that echolocation is an acoustic tool that bats share within their environment, then bats displaying similar habitat preferences, while within those habitats, will choose, within their physical capabilities, the same 'best tool for the job'. This will have been driven by the evolutionary history of any particular species, in that evolution would have played its part in aspects such as an animal's physical size, wing shape and acoustic capabilities developing over time, in conjunction with habitat preferences and prey selection. Hence, it is easy to see why two different species of bat of similar size, physical structure and habitat preference would evolve to use echolocation in a similar manner. Which, in a nutshell, adds to the earlier statement – that it may not always be possible for us to use echolocation for confident identification purposes. The bats themselves are not emitting it with this in mind, although some studies have shown that in certain species it may be possible, as a by-product of echolocation, that a bat's physical condition, sex or age, for example, may be determinable by conspecifics (Kazial & Masters 2004, Siemers *et al.* 2005, Andrews *et al.* 2011). So, when studying bats, it is useful to remind ourselves that echolocation is not performing a function like that of bird song, and within the echolocating behaviour of an individual bat species there can be considerable variation driven by many factors, first of all relating to the bat itself and where it is, and secondly relating to the quality of the recording being investigated (Russo *et al.* 2017).

Figures 1.4 to 1.7 show some examples of spectrograms demonstrating the diversity of echolocation found in bat species occurring within western Europe. It is not the intention to be thorough here, but to give an essence of what bat echolocation looks like on a spectrogram, and also what it sounds like (via the

Sound Library). As bat workers we know that echolocation is not as simple as what is shown in these figures. As mentioned earlier, there is variation in call structure and call sequences, as well as overlap in measurable parameters, all of which are influenced by habitat and behaviour (e.g. commuting or foraging). This can occur at an individual bat level, as well as at species level, and more broadly across a range of species within a genus or a wider group, where each can perform echolocation in a similar manner.

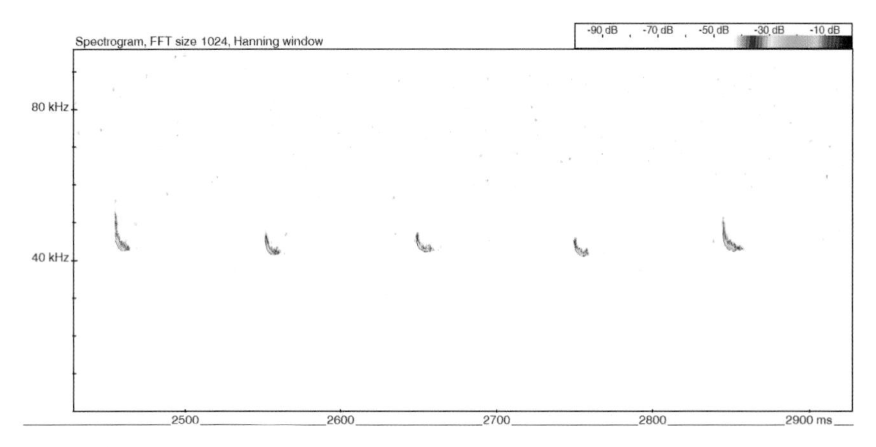

Figure 1.4 🔊 Common pipistrelle – echolocation (frequency scale 0–96 kHz).

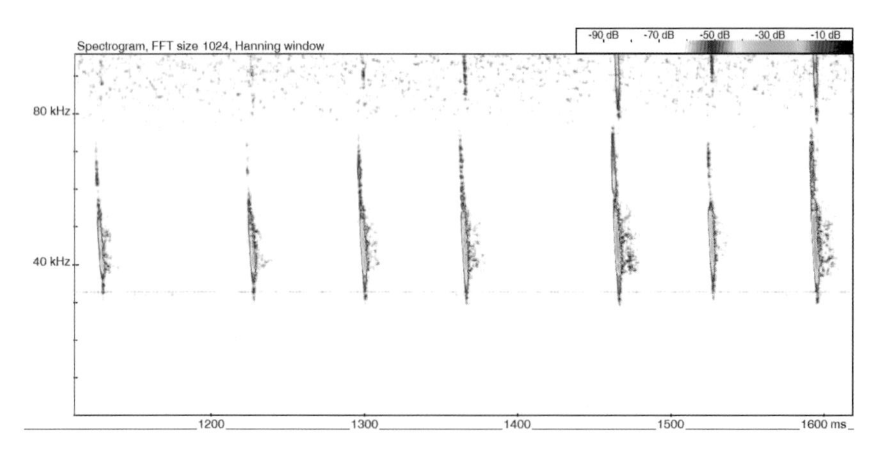

Figure 1.5 🔊 Daubenton's bat – echolocation (frequency scale 0–96 kHz).

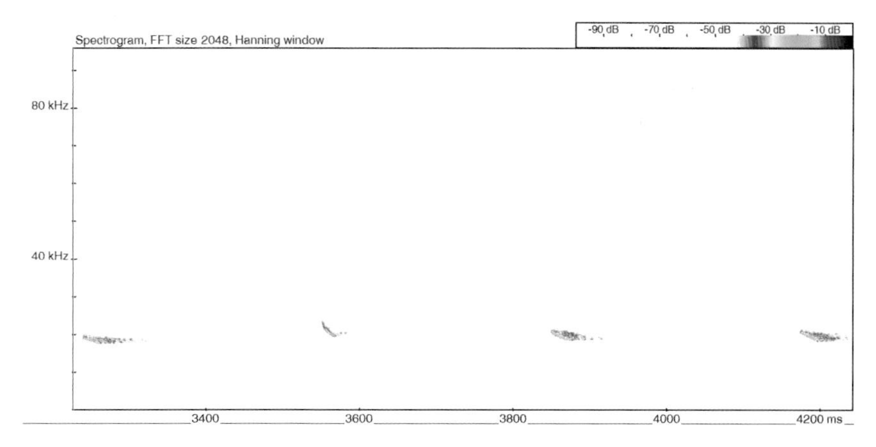

Figure 1.6 Noctule – echolocation (frame width 1000 ms, frequency scale 0–96 kHz).

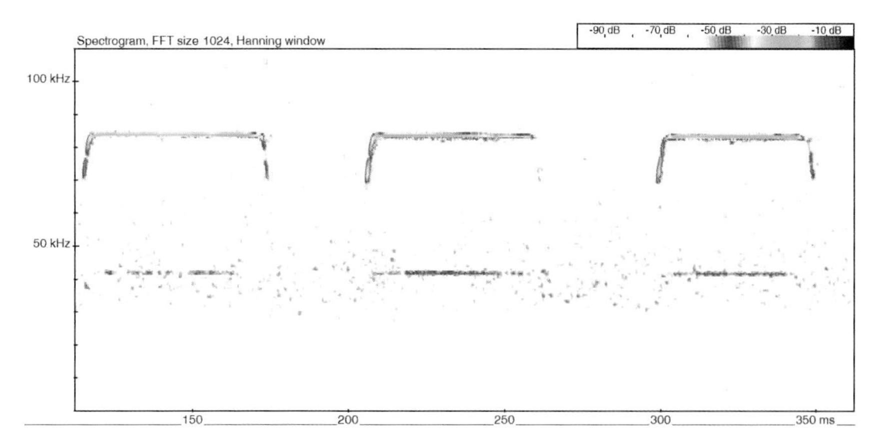

Figure 1.7 Greater horseshoe – echolocation (frame width 250 ms).

Social calls

As well as echolocation, bats also produce sound in a social context in order to communicate with other bats. A wide range of social calls has been documented and described for many species in the British Isles (Middleton *et al.* 2014). In some cases these calls have been shown to be species-specific, and therefore diagnostic, making them useful to us for species identification. In fact, in some circumstances, for example Nathusius' pipistrelle and parti-coloured bat (*Vespertilio murinus*), the social calls produced in certain contexts (e.g. type D, male advertisement calls) can be far more reliable than echolocation for identification purposes. Unlike echolocation, these 'advertisement' calls *can* be considered in a similar way to bird song.

Social calls or call sequences can be fairly complex or simple. For the bat species occurring within Europe, many of the known social calls have been classified into behavioural and/or structural categories, as in the case of the Vespertilionidae (vesper) species (Pfalzer & Kusch 2003), and some Rhinolophidae (horseshoe) species (Andrews & Andrews 2003, Andrews *et al.* 2006, 2017). Therefore, as well as social calls being useful, in some instances, in order to arrive at a correct identification, they can also, in certain circumstances, provide information as to the behaviour of the bat involved (e.g. male advertisement, distress, threat, group cohesion, contact and isolation). Figures 1.8 to 1.11 provide a small range of species examples, each falling within a different behavioural context.

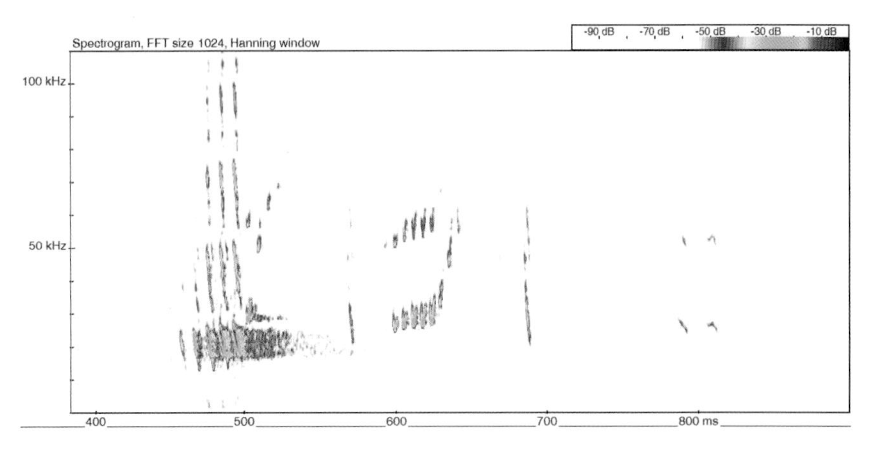

Figure 1.8 Nathusius' pipistrelle – 'male advertisement' type D social call (Middleton *et al.* 2014).

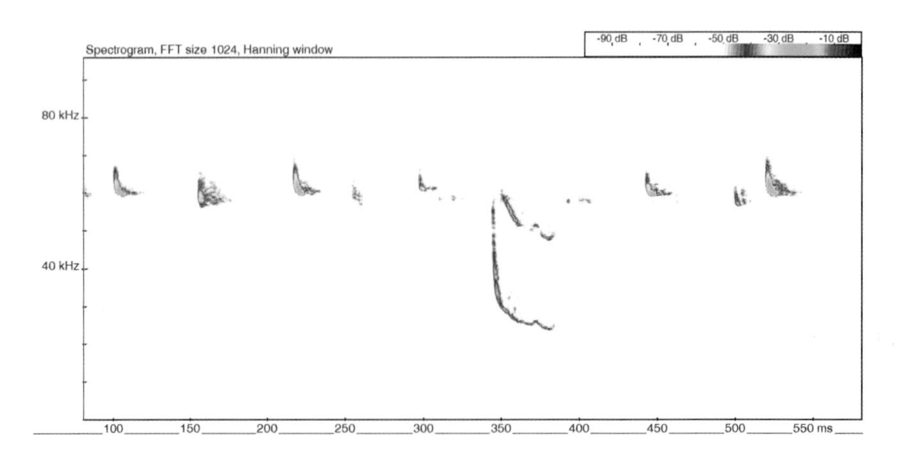

Figure 1.9 Soprano pipistrelle – 'contact' type C social call (W Woodrow, 2014).

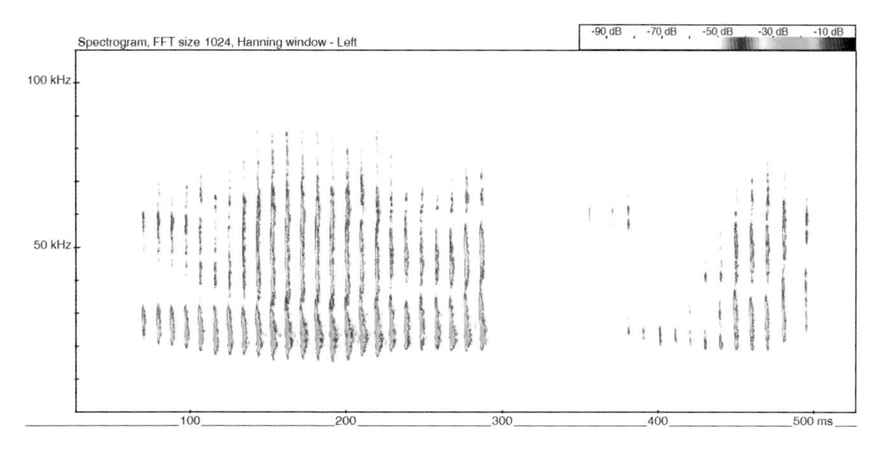

Figure 1.10 Brown long-eared bat – 'distress' type B social call (Middleton *et al.* 2014).

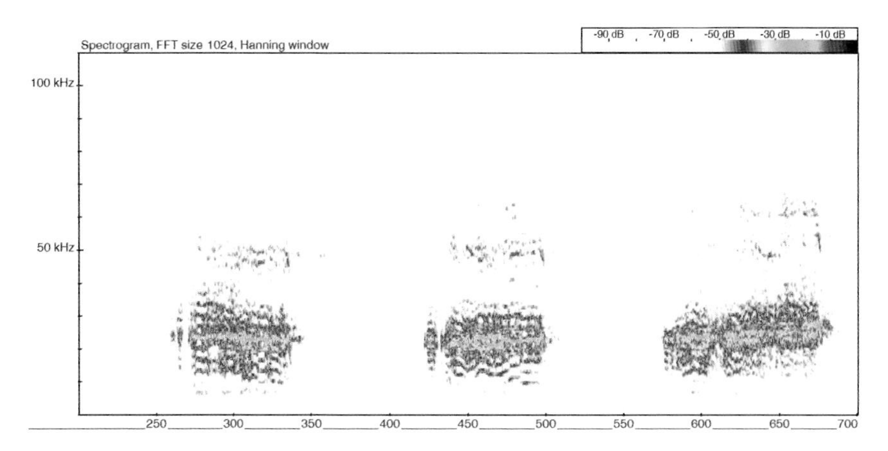

Figure 1.11 Daubenton's bat – 'threat' type A social call (Middleton *et al.* 2014).

Fieldcraft

How a bat worker performs in the field, including how he or she reacts to different scenarios, can impact positively or negatively on the specific survey objectives set. Having good technical knowledge, with the benefit of valuable experience and carrying out the survey using appropriate techniques can all be helpful in identifying individual bats to species or group level, as well as, on occasions, behaviour encountered. Table 1.3 briefly highlights certain aspects of fieldcraft that may be relevant when collecting bat-related sound data, taking some relatively common occurrences and discussing points to bear in mind.

Table 1.3 Bat call recording – considerations regarding fieldcraft.

Query/context	Plausible suggestions
You hear an audible sound that may be a bat (i.e. with your ears)	If it is a bat it is more likely going to be a lower-frequency calling species (if you have good hearing), or social calls (e.g. roost chatter), which can be audible to most of us. Switch the bat detector on and scan through at lower frequencies, pointing the microphone in different directions. Can you isolate the sound source, either by frequency or by direction? If so, this may help answer the question.
You hear a sound on your bat detector that may be a bat, but you are not sure	Provided it's a naturally occurring sound, then quite possibly it will be a bat. The bat detector can, however, also pick up other naturally occurring sounds, as well as non-natural occurrences. So remove your headphones and listen with your ears. Can you hear it now? Many of these other things may be audible to you, and knowing where the sound is coming from (e.g. stationary from a bush) may help. Also remember that your bat detector microphone is directional. When/where is the sound louder? Is it fading in and out, as if it is something quick, moving across the sky (i.e. a bat)?
You can see it's a bat (e.g. it's just flown past in front of you), but it's not making a noise	It may very well be a quiet/silent long-eared bat, or a horseshoe species where echolocation is very directional (i.e. if the bat wasn't close, flying directly towards you, you may not have picked it up). It may also be that the detector did pick it up, but you didn't hear it at the time, due to a combination of the call being quiet, background noise, or the frequency that the detector was tuned to. Definitely worth taking a note of the time and file reference, to then check it out later using software.*
You are picking up lots of feeding buzzes from a pipistrelle, but it doesn't appear to be echolocating	One of the more difficult things that a new bat worker needs to learn in the field is to tell the difference between a feeding buzz and a type D pipistrelle social call. Here are some tips: • Feeding buzzes always occur with echolocation. In the case of social calls, echolocation may not be present, or may not be noticed (if bat was distant). • The texture of the two sounds differs, with the buzz being preceded by faster echolocation as it builds towards the final phase. Social calls are far more random, sound harsher than a feeding buzz, and occur unexpectedly to an experienced ear.
You pick up quite a few *Nyctalus* type calls, with the bat detector tuned in the low 20s kHz area. You haven't had any visuals, and you were not expecting *Nyctalus* species	A common issue with inexperienced bat workers, who may not be familiar with either *Nyctalus* echolocation or type D social calls produced by *Pipistrellus*. The confusion can occur because: • The repetition rate can be similar, and if a pipistrelle was distant or not echolocating, it may not be considered as a sound source. • Pipistrelle social calls typically occur in a similar frequency range to where *Nyctalus* could be encountered echolocating. • An AutoID system on a bat detector may misidentify pipistrelle social calls as a *Nyctalus*, convincing users that they are right, when they are actually misguided.

* See main text for guidance.

Before moving away from Table 1.3, there is an analysis tool present in many software packages that sometimes gets ignored by bat workers, especially in the UK. The oscillogram tool (see Appendix 1, Figure A1.1) measures amplitude against time, and is useful for assessing the quality of a recording, taking accurate time measurements (e.g. call duration and inter-pulse intervals), as well as, in certain circumstances, being a useful aid to identification (Barataud 2015). In some other species groups (e.g. frogs, toads, bush-crickets) it is used for identification purposes considerably more often than we tend to do with bats. The reason I am mentioning it now is to develop further the point made within Table 1.3 (marked with an *). This is where someone thinks they have seen a bat, but it did not appear to have been echolocating as it passed by. Assuming the surveyor had the bat detector switched on and recording, the easiest way to establish whether or not something was echolocating, after the event, is to open up the file, go to the point when the bat was seen but not heard (good field notes are essential), and see if anything is visible on the spectrogram, or indeed audible, albeit quietly so. Depending on the quality of the recording and the software settings (e.g. threshold) you may find yourself faced with something similar to what is shown in Figure 1.12.

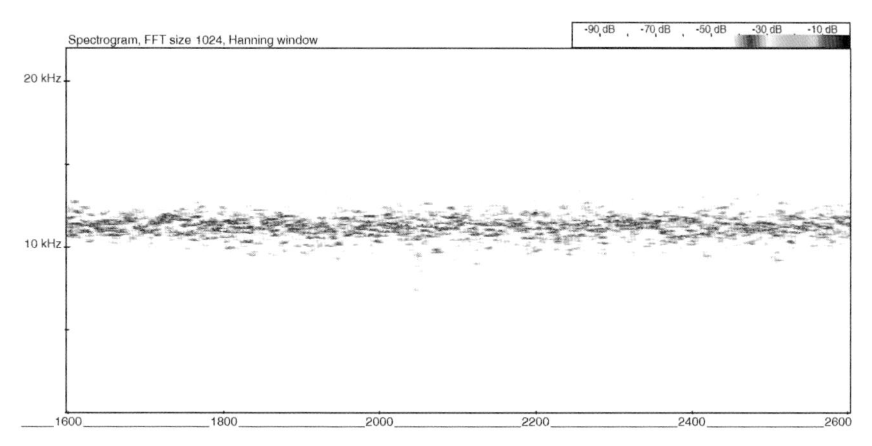

Figure 1.12 Spectrogram showing no bat echolocation. Note that frequency scale needs to be corrected (i.e. actually 0–220 kHz) as this is a frequency division recording.

As perhaps anticipated in the scenario presented, nothing obvious is seen, so maybe the bat wasn't echolocating after all. Many bat workers would go no further than this, but I would like us to be absolutely certain. Let's now look at an oscillogram of the same sequence (Figure 1.13). What are those consistently occurring peaks, and why are they different to the rest of the background noise? Yes, those peaks are demonstrating where the bat echolocation pulses are. They are very weak and not obvious at all on the spectrogram. Due to the proximity of the bat to the surveyor (who was close enough to see it in the dark), it is typical of what a quiet echolocating bat, such as a brown long-eared bat (*Plecotus auritus*) would look like when carrying out sound analysis. The actual recording in the Sound Library comes from a bat that was present inside a loft and flying about 2 metres in front of the surveyor. Have a listen to the call in frequency division.🔊 We can hear the annoying background hissing sound, and in amongst it the faint echolocation pulses. And these instances happen far more often than many would think. Brown long-eared bats are one of our most common species, after all (Mathews *et al.* 2018). On this occasion the surveyor alerted us to its presence, highlighting the value of having not only your ears, but also your eyes fully focused on the job. But what if the surveyor hadn't seen it because it was too dark, or it had been recorded by a static detector? It is very useful, especially when surveying structures or while in habitat where 'quiet' bats may be present, to use the oscillogram tool to look for patterns of regular peaks occurring at a repetition rate similar to a bat. It is often easier to find them this way, as opposed to regularly altering threshold settings, or listening to long recordings without knowing if there was anything there to hear in the first place. Taking this approach, it's not unusual to come across stuff missed by a surveyor, or picked up by a static detector, and not immediately obvious during analysis. This process should be considered part of a standard sound analysis procedure, to help ensure that any calls which may be relevant haven't been missed.

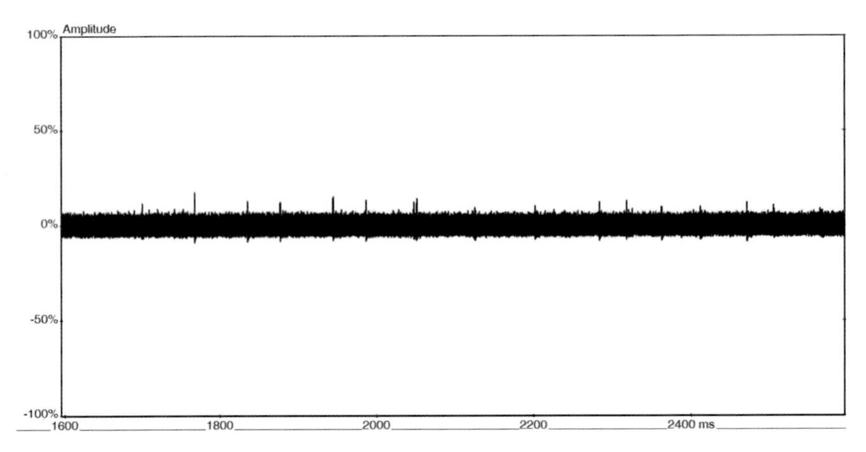

Figure 1.13 🔊 Oscillogram showing regularly occurring bat echolocation pulses (brown long-eared bat).

Embroiled within the subject of 'fieldcraft' is how the surveyor integrates with the equipment in use, and in what way this impacts upon the job, both positively and negatively. There are field-related behaviours that have been considered as beneficial for quite some time now, irrespective of the technology being used (Table 1.4). As well as the points outlined in Table 1.4 there are other behaviours or occurrences, specific to the more advanced technology that has been introduced into the bat world fairly recently. These could also impact upon the quality or accuracy of the recordings being made, and this is an area we are going to consider next.

Table 1.4 Bat detector surveys – positive surveyor behaviours with rationale.

Behaviour	Benefits
Wear headphones	• Surveyor can hear bat-related sound more clearly. • Surveyor can hear what is being recorded more clearly.* • Less chance that bat passes will occur unnoticed. • External noise, which could be a distraction, is reduced. • Speaker noise is not recorded by the same or a different bat detector. • Surveyor is more likely to notice recording issues caused by a faulty lead, equipment failure, battery dying etc.* • Saves detector battery life, as the external speaker is disengaged.*
Limited use of torchlight	• Causes less disturbance to bats. • Less likely to impact upon bat behaviour. • Eyes are not constantly readjusting to differing light levels, with the risk that certain events or behaviours are being missed.
Use voice-note facility on bat detector, or another recording device, to make field notes	• Notes tend to be more thorough and descriptively more accurate as surveyor has time to talk while continuing to watch for activity. • When added directly within the bat recorded sound file the verbal note can integrate precisely with a bat encounter of interest. Written notes, using time references that do not tie in exactly with the time on a recording device, often cause problems in matching observations up with recorded data, especially when there is other bat activity during the same period. • The process of writing notes, which may be necessary if voice recording is not available, takes the surveyor's eyes away from watching for activity (i.e. the surveyor is no longer doing the job!). • Conversations (e.g. someone else talking via a walkie-talkie) can be recorded directly onto the sound file, removing the risk of not remembering accurately what was said by another person. • Audio notes can be stored digitally.
Use walkie-talkies between each surveyor location	• Surveyors can discuss any unusual scenarios they are witnessing, or bat behaviour occurring, while it's happening, with everyone being made aware simultaneously. • Health and safety benefits, should there be a potential issue occurring on site (e.g. the presence of third parties).
Regarding all of the above, health and safety considerations should always be borne in mind. *See Appendix 1, Figure A1.3. If using a separate recording device to the bat detector, ensure that headphones are connected from the recorder, as opposed to from the detector.	

Thank goodness for today's technology

Generally speaking, as humans develop into new areas of scientific understanding, we need accurate knowledge to be certain of the facts we thereafter consider to be true. Acquiring this knowledge can take a lot of hard work, and new technology often presents us with solutions to the problems or barriers we are seeking to overcome. As such, the technology needs to arrive before the factual knowledge can be sought. In order to fly to the moon, to dispel the myth that it's made of cheese, we needed a rocket (i.e. new technology). No rocket would have meant that no-one could go and prove what the moon was made of.

In recent years the advances made in bat detecting equipment and the associated software have been substantial. There is definitely a viewpoint, however, that some of the technology we are using has perhaps appeared a little too early, ahead of the level of knowledge (across a broad range of users) that is required to fully understand and appreciate the technology, its benefits and its shortcomings. And, of course, this same new technology is being used by the 'best of the best' in order to find scientific answers to the questions that the rest of us are grappling with on a daily basis. There are undoubtedly benefits with much of what we now have available to us, but there are also drawbacks. Referring to Figure 1.14, my intention is to do no more, and no less, than make us think through what it is that we are using, and to be aware of the potential negatives that come hand in hand with all of the valid positive reasons that were evident when the product was purchased in the first place.

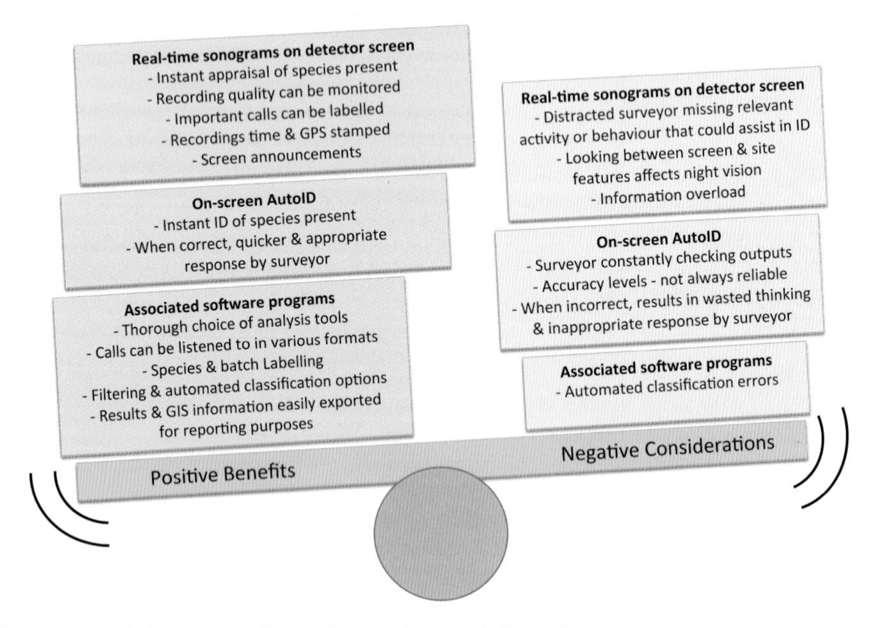

Figure 1.14 Today's more advanced technology – a balanced perspective.

Considering further Figure 1.14, including the negative aspects, we definitely do not want to throw all of this marvellous technology out of the window and go back to what we had 25 years ago. Despite a number of potential drawbacks that could impact upon good fieldcraft, without any shadow of a doubt, once we get the data onto our PC, and start doing the analysis phase of the job, the power we have at our fingertips is immense and extremely impressive, especially to us bat workers of a 'slightly' older generation who remember what it was like when bat-related technology was in its infancy. The technology is amazing, and should be welcomed. Most of the products that are finally produced (once any teething problems have been resolved) are fabulous. But I don't think the science (we still have so much to learn) and therefore the wider levels of knowledge being used on a practical basis, day by day, is close to knowing enough about everything. We have definitely built a rocket, and we are counting down to lift-off, but we haven't yet reached the moon. I sometimes feel that it would be beneficial to stop the technology now and give ourselves some time to catch up in confirming the accuracy of what we already know, what we think we might know and what we have still to learn about these animals. When this improved knowledge and the technology eventually do come together, then we will be in a far stronger position to more fully understand these amazing mammals, which are so in need of our consideration and protection.

Automated bat classifiers – useful, but not the Holy Grail

An *automated classifier* (López-Baucells *et al.* 2019), also referred to as *automated bat identification software* (Brabant *et al.* 2018), and AutoID for short, is a piece of analytical software which has been developed, using trained algorithms, to quickly identify a bat species, or range of species, based on a recording. It works by comparing a call from an unknown species (i.e. the recorded call) against calls from known 'library' species. The call of unknown origin passes through a series of 'rules', to then be assigned to the library species whose calls are the best fit (Russo & Voigt 2016). As well as assisting a user at an individual bat pass level, these programs have become increasingly popular as a means of processing large datasets very quickly, thus saving time and money (López-Baucells *et al.* 2019).

But why am I about to dedicate so much time to this subject matter, when it may appear to be a little off-piste? In a nutshell, it is not only humans that struggle with whether it is a bat or not, or which species of bat it is. Classifiers can also struggle to come up with the right answer. When this happens there is a lot of blame focused on the classifier as it produces a result that is clearly wrong, at least to an experienced user. There are a few things, however, that need to be considered quite seriously in amongst all of this. In the following paragraphs I am going to sound quite negative about these systems. I must stress that if it creates that impression, it is merely to get a point across, in order to help users understand to a better level what they are engaging with when they use a classifier.

First of all, an admission of guilt. I do use classifiers myself. I use them in conjunction with my own experience, and the experience of others, to provide another opinion on something we might be struggling with. At times this has proven to be very valuable. Secondly, any classifier system can only give you an answer that has already been factored in during its development. If the classifier only has a handful of bats to refer to, then any bat call that it investigates can only be assigned to something within its call library, or alternatively labelled as 'other noise' or 'no ID' for example. Finally, the manner in which classifiers are applied, and how the results are then interpreted, falls firmly and only upon the user. So, ultimately, the user decides. In my opinion, there is not enough balanced guidance available to the standard user as to how to apply such systems and interpret their outputs. Many of us focus too much on needing to obtain an accurate identification for every one of our individual calls, without, for example, consideration for looking at patterns of outputs, weightings of suggestions and removing substandard recordings, non-search-phase sequences (see Appendix 1, Figure A1.4) and calls with low levels of confidence from the process.

There are other important considerations that also need to be accounted for. First of all, a classifier is only as good as the data that have been used to populate its call library, and, as previously mentioned, the rules created by those developing its decision-making process. Secondly, for some types of bat-related work, degrees of potential error may not be that big a deal, provided those using and interpreting the data are open and transparent about how they have used the classifier, and what thought processes and manual systems they used when accepting or rejecting outputs generated by the software.

Those using classifiers need far better guidance from others who fully understand how they operate, as well as coaching on how to gather and interpret outputs. There needs to be far more 'ground-level' explanation about the strengths and the weaknesses of any such system, with this all balanced against the purpose for which the user is gathering the information in the first place. For example, the benefits and pitfalls of such a system will differ between a job that is attempting to create an accurate inventory of species present within a specific site, on the one hand, and the relative activity, by habitat, of all bat species collectively across a large area (e.g. county level). In the first example species identification has to be completely accurate, while in the second example accurate data would be great, but inaccuracies would not necessarily impact upon the overall message to be drawn from the study. So in the latter example the report authors should attach degrees of caution at species level, as appropriate, so as not to misinform an inexperienced reader about what they should take from the results.

Following on from what has been said above, many of us need to wake up to the potential, as well as the reasons, for mistakes, and the need for proper training and tempered expectations when using these systems. There is certainly a huge amount of power at our disposal, and it would be crazy to deny ourselves the benefits which are undoubtedly available. Conversely, and especially allowing for some of the issues that are raised elsewhere in this book, you need to be fully aware of the potential for misinterpretation when using such systems.

I am going to continue with a non-bat example. A classifier, to many, is a bit like having a pair of binoculars with a built-in identification system, which through the lens might deliver a message announcing 'willow warbler'. An observer who didn't know any better might say 'yes, it looks like a willow warbler, let's move on.' But if that observer waited patiently for the bird to sing, it might reveal itself to be a chiffchaff. A more experienced observer would know that willow warbler and chiffchaff are very similar to look at. It would have been obvious, with experience, not just to take the equipment's word for it without gathering more information. So extra experience, knowledge and fieldcraft are all needed in order to audit what this expensive pair of binoculars is saying. But we so want these binoculars to be right. They cost us a lot of money, and save us loads of time. If they were just right, all of the time, life would be so much easier and fast-paced as we skipped from one bird to another, possibly not learning that much about how to actually identify them ourselves. Then, years down the line, against a history of numerous 'questionable' field records that no-one has actually queried, no-one knows how to do the basics. No-one is left who has the ability to question results, and no-one has the experience to tell anyone otherwise. All in all, not really a robust scientific approach.

Which brings me on nicely to another useful point from the world of ornithology. When I was lad, there was (and still is, I hope) a rule. You were not allowed to claim that you had seen a species of bird unless you were able to confidently identify the bird yourself. There is a very good reason as to why an approach like this is useful – essential in fact! It is how we learn our craft, and how we turn knowledge into experience, the two of which are entirely different things. Upon becoming experienced we then have the ability and credentials to investigate matters arising, thoroughly and accurately, and question what others (perhaps with contradictory opinions or motives) are telling us.

Most experienced echologists (have I invented a new word?) will tell you many stories about these automated classifiers producing inaccurate results. However, to be fair, none of the developers of these systems claim that the results are always going to be accurate. If you read the small print that comes with these systems (and to be fair it isn't usually that small), it says quite clearly words to the effect that 'there is no substitute for experience on the part of the user.' To me, this translates as, *'if you are not experienced, then you shouldn't be going anywhere near this automated process until you are experienced enough to cross-examine any of its interpretations.'*

Anyway, there is an acceptance on all sides that an issue lies in the accuracy of the results produced in certain circumstances (Rydell *et al.* 2017, Brabant *et al.* 2018, López-Baucells *et al.* 2019), which can be variable, depending on the process used to collect accurate sample bat calls (the library calls) that are used to develop the software, the reliability of these samples being an accurate reflection of typical wild behaviour, and the species or species group in question being able, in the first place, to be reliably separated from other similar echolocating species or groups. It should also be borne in mind that a big part of being able to more reliably separate and identify many bat species from echolocation alone is ideally to consider only what are known as 'search-phase' calls (Murray *et al.* 2001). These are more likely to show consistent structure and repetition rate during a sequence (see Appendix 1, Figure A1.4), and better still if the call sequence was emitted in a relatively open habitat for the species, as opposed to an edge or cluttered environment (Ahlén 1990). When bats are behaving in this way, then their calls, in some cases, can become more safely separable, and a good classifier would be expected to include these types of calls in its library. When the query call relates to what are known as 'approach-phase' calls and/or 'feeding buzzes' (see Appendix 1, Figure A1.4), these examples can become considerably more problematic, if not impossible, for an automated classification system to adequately cope with, depending upon the species involved.

In some respects, with some species, very high levels of classifier accuracy (> 90%) are achievable (López-Baucells *et al.* 2019), but with far lower levels for other species using the same data-collection methods and automated classifier process. In a study comparing two automated classifiers it was demonstrated that there was a sizeable difference in performance, with much greater accuracy for species with structurally distinct echolocation calls (e.g. soprano pipistrelle) than for species with less characteristic calls (e.g. *Myotis*) (Rydell *et al.* 2017). Another study used a dataset of good-quality known calls, recorded in what would be regarded as ideal circumstances (Brabant *et al.* 2018). These calls were fed through four commercially available classifiers, with the overall accuracy rate for these programs ranging from 31% at worst to 77% at best. That study went on to say that 'while the tested programmes may be considered valuable tools to detect bat calls from recordings, a trained bat expert needs to cross-check the automated species identifications to avoid erroneous conclusions.'

Irrespective of how accurate any specific classifier is, we also need to consider the quality of the recording(s) being processed, along with other factors that may affect the analysis (Figure 1.15). All of this could have an impact upon accuracy, as well as what potentially gets treated as 'other noise' due to the application of any filtering process to remove what may initially be perceived as unimportant frequency bandwidths, when in fact 'interesting' data may be present. Guess what? The classifier can't control what the user chooses to put through its process. If the data gathered are not great, for whatever reason, then the results may, at least in part, be a reflection of this. So, here we have an example of a classifier being blamed for something that is beyond its control.

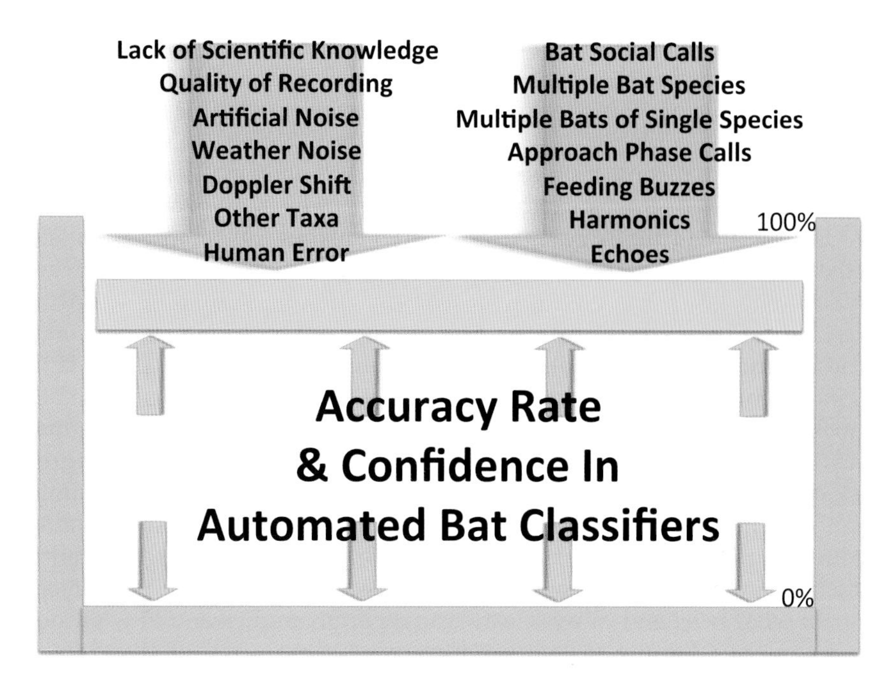

Figure 1.15 Factors impacting upon (i.e. forcing downwards) the potential accuracy rate and confidence in automated call classifiers.

In the right hands and being used for the right purposes, with appropriately qualified 'health warnings' placed against any outputs, these classifiers can be very powerful, especially considering the amount of data that can be mined extremely quickly. So when someone asks me, 'Are you for or against classifiers?', despite what I might say in an off-the-cuff remark, my considered response would be that I recognise there are flaws, and I recognise there are benefits. But, irrespective of the negatives and the positives, what I do really care about is how they are built, and then, as importantly, how they are applied and the experience levels of the user. In the right hands they are extremely useful and beneficial, while in the wrong hands they can be extremely dangerous with regards to the outputs presented and how these outputs may be interpreted by an uninformed audience, be that at a scientific conference or in a local authority planning department. It is up to us, as experienced professionals, to always apply quality checks (i.e. audits), to be transparent about our processes, and to qualify our results and thoughts in such circumstances.

For classifier outputs to be relied upon by inexperienced interpreters without appropriate auditing by experienced technicians is potentially very risky. Adopting such an approach is 'buying in' to an unrealistic expectation that it is possible to identify every bat call to species level, and safely discard everything else that is labelled as not being a bat.

It is all a question of balance and trade-offs. Do you batch process everything using a classifier, and accept (and tell your audience) that there will be errors, or do you manually process everything, taking considerably longer to achieve a supposedly more accurate result? 'Supposedly', because who is to say that the human processing the data manually won't make mistakes, and who is to say that when the bat classifier picks up an unusual species for a site, that it is wrong?

Is it feasible that some bats get misidentified as other species? Is it feasible that bats get missed? Is it feasible that things that aren't bats get misidentified as bats? Is the author actually going to answer any of these questions? The answer to all of these is yes, most definitely, and at times potentially to a high degree. As you go through this book you will see some things that definitely look and sound like bats to your eyes and/or ears, but they aren't. Unfortunately, many of these same things also look like bats to automated classifiers. Now if that last statement has surprised you slightly then I am afraid you are in for a bit of a shock later on. But, having said all of that, these systems also identify many things correctly, if not to an individual species level, at least to a species group level. Also, in some cases the systems find and identify bats more confidently than the technician (depending on the technician's level of experience). And of course humans can also get tired, distracted, have a bad day etc., something that a software program shouldn't do, provided it doesn't crash, of course.

So, at what level and in what way do you apply audits, and at what point do you accept what the system is telling you? What is the solution? In any filtering and automated classification process, at the very least there definitely needs to be an audit for levels of accuracy, and corrective action taken if the audit establishes that there has been an issue, either with the analysis process or with the skill levels of those involved. And while we are at it, going back to a point made earlier referring to the basic principles of bird identification, it isn't safe to accept that a classifier is always accurate for a species that the technician would not be able to confidently identify, unaided. You can only carry out an audit if you know, independently, the answers to the items you are checking. Otherwise all you are doing is accepting the initial answer as being correct, without the potential for correction. In such a case it's pointless doing an audit at all, because that approach isn't actually an audit.

There are different methods that could be followed in carrying out audits. For example, all calls where the confidence level is below a certain threshold could be audited, or, taking a bottom-up approach, all calls could be audited until the classifier is demonstrating a high level of accuracy. These approaches are all definitely along the right lines, in my opinion. There would of course be some debate as to where the thresholds are set, and my suggestion would be to set them as high as possible. Something, however, that I haven't heard many people discuss in this respect would be applying different thresholds within a project, driven at a 'species-by-species' level. In other words, the threshold set for soprano pipistrelle (*Pipistrellus pygmaeus*) may differ from the one set for common pipistrelle (*P. pipistrellus*), for example, where the potential for other species occurring with similar call parameters could be higher. Other aspects can also be built in on a case-by-case basis. For example, it would be useful to remove any bat species from the classifier options that is unlikely to occur at the survey location. Another consideration is what to do about results that show similar ranges of probability across more than one species. Depending upon the species involved, it could be risky to go with a classifier's interpretation, however high the confidence level, where there is a range of options which are close to each other. For example, a result saying 69% whiskered bat (*Myotis mystacinus*), 66% Brandt's bat (*M. brandtii*) and 62% Daubenton's bat (*M. daubentonii*) should be viewed with caution, in that it would be wrong to think that the probability of the bat being a whiskered bat was high. One final example to consider would be where the classifier is able to provide a degree of confidence on a pulse-by-pulse basis. In this situation it can prove useful to see if there is a pattern to how it is determining the allocation of species. For example it may be regularly saying common pipistrelle, but as the bat call progresses from search phase through to approach phase and then a feeding buzz (see Appendix 1, Figure A1.4), the calls are being allocated, somewhat understandably perhaps, to a *Myotis* species.

Figure 1.16 shows a flow chart that considers the process at a high level, within which different levels of human interface, manual analysis and audit are proposed, depending on the dataset being considered. In adopting such a process, it would be hoped, broadly speaking, that the best of both approaches (automated classification and manual analysis) is taken advantage of, while simultaneously allowing for the worst of both approaches. Is it perfect? Most definitely not. What would make it perfect? Having a classifier that is able to diagnostically identify every species of bat without missing one, and simultaneously being able, on every occasion, to recognise and categorise all other conceivable noise (natural or otherwise) in order to safely remove these from the process. And while we are at it, everyone trained to a level whereby everything that is known is fully understood and there is nothing new left to find out. Is all that possible? I don't think so, if for no other reason than that there appear to be too many areas of overlap in echolocation between certain bat species, and in some cases between bats and other taxa. But if you had asked me 25 years ago if we would ever have the technology we have today, I would probably have said the same thing. I am now hearing you say something like, '*I think the author is now balancing precariously on a barbed-wire fence. I am sure he said in the preface that he wouldn't do that.*' Point accepted. Strike one against me, and thank you for having read the preface.

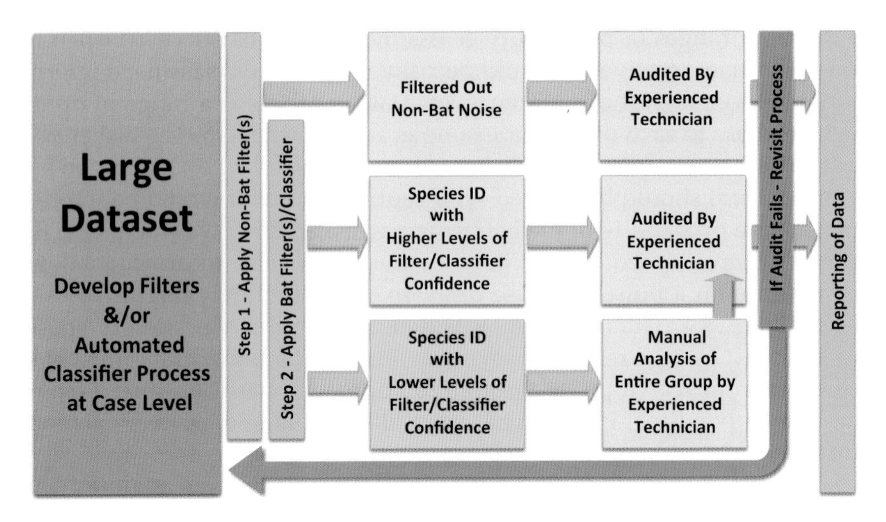

Figure 1.16 Bat call analysis – analysing large datasets. An example of a good-practice process.

When it all comes together, only to fall apart

We have now discussed quite a few different aspects of bat-related work that are current in today's world. Next, I will take a fairly commonly encountered scenario and use this to demonstrate a number of potential issues that can occur. To achieve this we will look at two regularly occurring and widespread genera, *Pipistrellus* and *Nyctalus*. High levels of activity of a *Nyctalus* species on a development site, for example, could influence the concluded level of impact resulting from a proposed development. So, if *Nyctalus* calls were misidentified as *Pipistrellus* social calls, or vice-versa, then the potential for misinformed conclusions or inappropriate mitigation definitely exists.

The emission of pipistrelle type D advertisement social calls is a regular feature during many bat surveys, especially in spring and autumn (Middleton 2006), and we have already discussed the potential to confuse these in the field with *Nyctalus* echolocation when using a bat detector (Table 1.3). Now let's consider what can occur during the desk-based sound analysis phase of the work. As before, it would be beneficial from a learning perspective if you worked through this process as follows.

First of all, look at Figures 1.17 and 1.18, which show the same *Nyctalus* species echolocation bat pass. In this instance the bat, a noctule (*Nyctalus noctula*), was flying close to woodland canopy. Figure 1.17 shows the pass in true time, with the time between calls (i.e. empty space) still visible, while Figure 1.18 shows it in compressed mode, with the empty space between the calls removed. Compressed mode is an option that is becoming more commonly available on bat analysis software (it has always been an option on AnaLookW). Having put this pass through an automated classifier, the result came back as a *Nyctalus* species with 88% confidence, and my audit agreed with this as being accurate, as I can confirm, independently, that the bat was a noctule, which I had also observed and recorded flying in the open, at less than 20 kHz peak frequency.

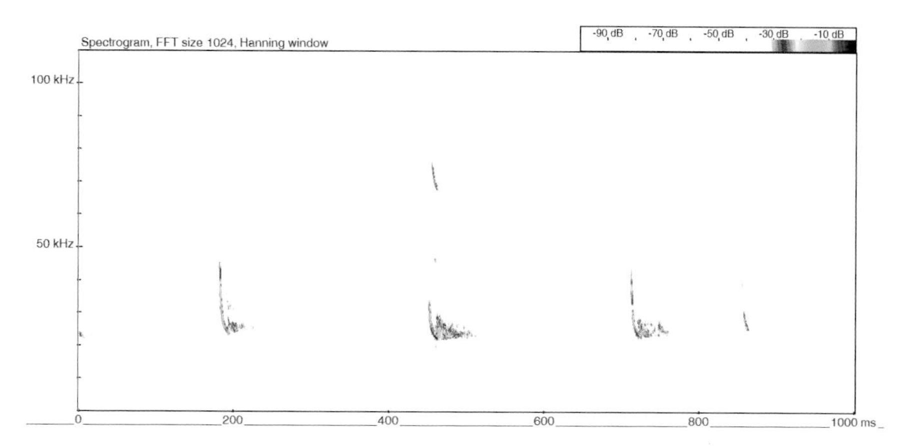

Figure 1.17 *Nyctalus* – echolocation, uncompressed (frame width 1000 ms).

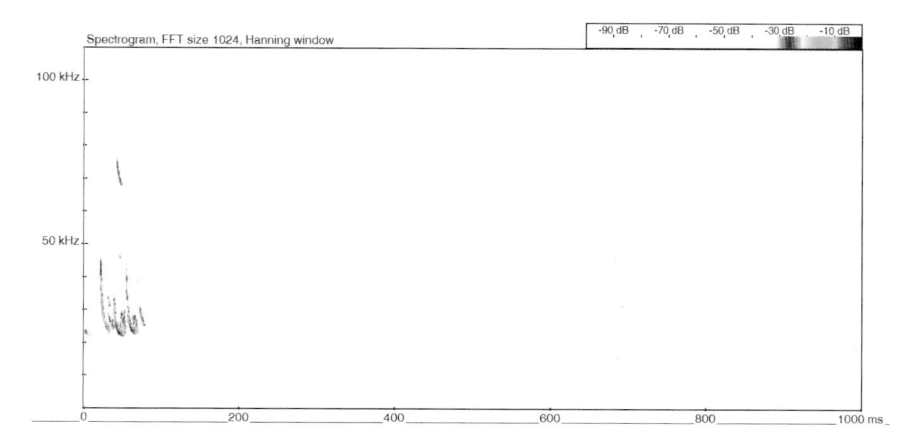

Figure 1.18 *Nyctalus* – echolocation, compressed (frame width 1000 ms).

Now look at Figure 1.19, which shows a *Pipistrellus* species (in this instance a soprano pipistrelle) emitting two type D advertisement calls in true time. The individual call sequences in Figure 1.19 look very similar to those in Figure 1.18. However, it's a totally different species of bat, performing a totally different sound, for a completely different purpose. Having put the soprano pipistrelle call through the same automated classifier as before, the result came back as a *Nyctalus* species with 68% confidence, with my audit claiming 0% accuracy, as I can confirm, independently, that the bat was a soprano pipistrelle.

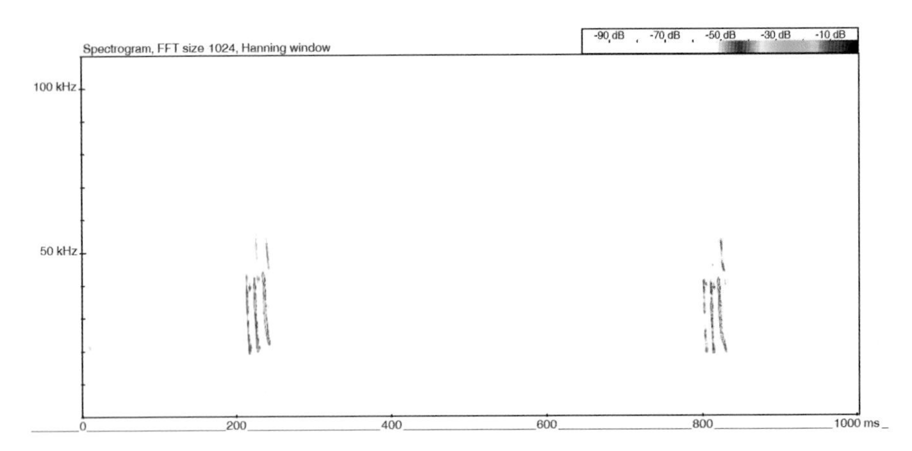

Figure 1.19 Soprano pipistrelle – type D social call, uncompressed (frame width 1000 ms).

So, in conclusion, there is the potential for *Pipistrellus* social calls to be misidentified as *Nyctalus* echolocation, for example, by inexperienced or 'under pressure of time' experienced technicians, especially if they are using a compressed-mode view, not listening to calls, and relying on the automated classifier as being accurate. Whether using a manual process (in compressed mode) or an automated approach, there is the potential to misidentify these en masse. The importance of an audit by an experienced technician has now, hopefully, been demonstrated (see Figure 1.16).

The example discussed here is one of many that could be used, as many types of bat social calls, from a range of species, have the potential to be misidentified as echolocation of another species altogether, or classed as noise files through filtering processes. Accordingly, a knowledge about social-call structures would be extremely beneficial to those carrying out sound analysis at any level (Middleton *et al.* 2014). In addition, as you will discover later on, there is also the potential for sound coming from things other than bats to be identified confidently as a bat, either manually or by an automated approach.

That's weird – other stuff that happens while recording sound

During this chapter we have considered, in a broad manner, sounds that don't belong to bats, and then we went on to discuss sound that did come from bats. To complete the picture, it will be useful to consider sound that occurs while bats or other target species are being recorded, but for various reasons, what has been recorded is not exactly what the animal directly produced.

What follows is a range of scenarios that someone might be presented with while doing sound analysis. Although I have related them all to bat encounters, much of what follows is equally relevant, no matter what is being recorded. These situations tend to pop up now and again, and when they do, for those not in the know they can be quite perplexing – and they certainly have the potential to waste time or give an inexperienced technician the wrong steer. Even for more experienced people, some of what we are about to discuss can, occasionally, prove challenging.

Comb filtering

Comb filtering gets its name from the comb-teeth-like shape created, as seen on a power spectrum (see Appendix 1, Figure A1.1), when this effect is evident (Figure 1.20). It occurs when an original sound is recorded simultaneously with the echoes of that sound (i.e. overlapping occurs). 'It is most dramatic when the two signals are of nearly equal amplitude' (Chris Corben, personal correspondence, 2019). This results in what can be referred to as constructive and deconstructive interference. In bat-related scenarios it can arise because a bat, or a bat detector, is close to a smooth surface (e.g. a window) from which echoes have bounced.

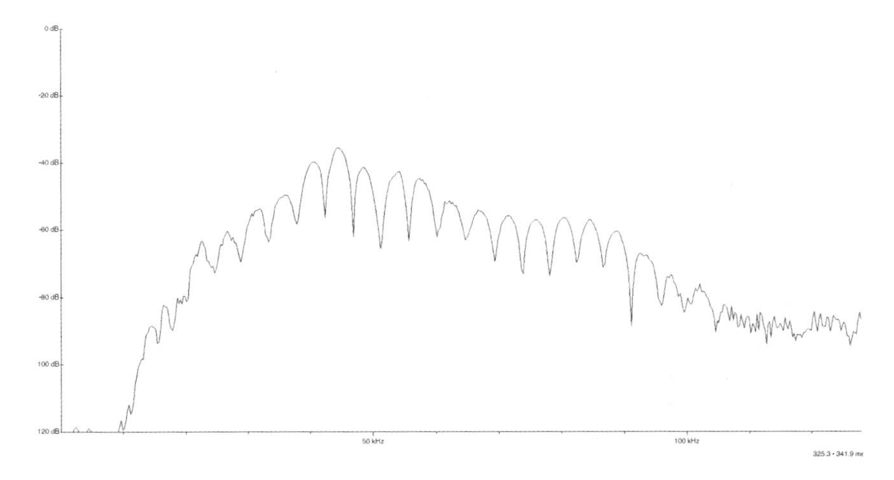

Figure 1.20 Comb-filtering effect of a single Daubenton's echolocation pulse as presented on a power spectrum.

To reduce the risk of comb filtering, it is best to have the detector positioned well away from obvious reflective surfaces (Brabant *et al.* 2018). Of course, as a surveyor you don't have the same control over where the bat has positioned itself. However, if your detector is pointing towards a horizontal reflective surface (e.g. calm water, solid ground surface) then having it angled upwards (e.g. at 45 degrees) would help in reducing such occurrences, as would (in any setting) placing the detector above a non-reflective surface (e.g. vegetated ground) and at least 2 metres away from reflective surfaces such as calm water, glass, or smooth-surfaced structures.

The comb-filtering effect could occur, albeit relatively rarely, with any species of bat, provided it is flying close enough to a smooth surface. It is, however, a more regular feature when recording bats that typically fly low over calm water surfaces (e.g. Daubenton's bat). This may appear initially to be quite frustrating to the surveyor, but it can assist with species identification if, for example, a *Myotis* has been recorded on site when the bat detector was positioned facing over the water. The presence of comb filtering gives a strong steer, in such circumstances, that it was a Daubenton's bat, a species which likes to forage close above the surface of calm water. That is not to say that it couldn't be another species behaving in a similar manner. Also, be aware that other European bat species (e.g. pond bat, *M. dasycneme*, and long-fingered bat, *M. capaccinii*) can show similar habitat interaction, thus producing a similar result. When seen on a spectrogram, irrespective of the species involved, the effect gives a bubbled impression, as shown in Figure 1.21.

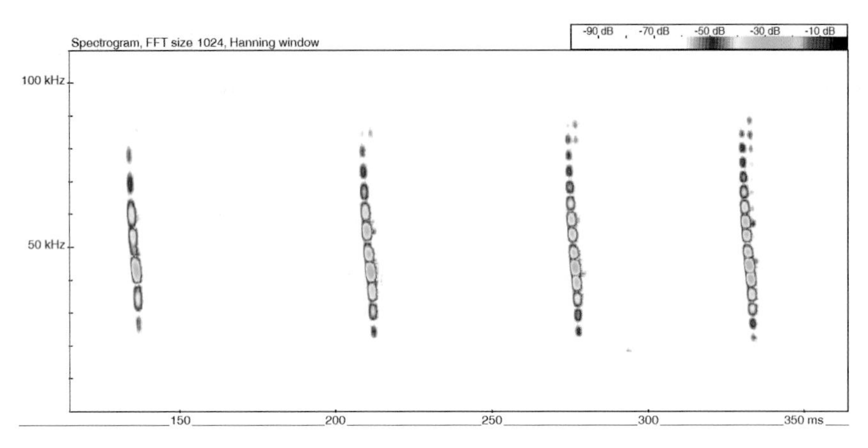

Figure 1.21 Bubbled impression as presented on a spectrogram of four Daubenton's bat echolocation pulses, from a bat flying over calm water (frame width 250 ms).

Comb filtering in the sense discussed above is fundamentally different from the 'comb filter' function that appears, for example, in Anabat products (Titley Scientific, Australia), where the same words are used in a different context to describe a different process. The reason for mentioning it here is because I don't want to give anyone the impression that the detector's 'comb filter' functionality is anything to do with what has been described above. It isn't. I am grateful to Chris Corben for the following explanation of the 'comb filter' option in Anabat products (e.g. Walkabout detector and Anabat Insight). In these products, think of the effect you are listening to as being similar to a heterodyne bat detector, but with the 'comb filter' being another way of producing an audio signal from an ultrasonic signal, allowing you to hear it. The output is like combining the results from a number of heterodyne detectors, each set to different frequencies. Compared to a typical heterodyne detector, you don't have to tune it, but the sensitivity is reduced by the overall broad bandwidth. And, of course, Anabats are not the only products that create this type of output, in that other manufacturer's machines can have similar functions, creating the same or similar outputs (e.g. auto-heterodyne).

Echoes

It is interesting to think that when we are recording bats we are focusing all of our attention and expensive bat detectors on sounds that the bats themselves don't listen to. Each time a bat produces a loud echolocation pulse (usually occurring many times per second), which is what we hear on our detectors, it contracts muscles within its middle ear in order to prevent it from being deafened, and the muscles are then relaxed as it listens for the returning echoes (Kick & Simmons 1984). It is these echoes, giving the information about its surroundings, that the bat is really interested in. Bats and people, both out listening for bat-related sound in the dark, but each focusing on a different part of the process, and, of course, for entirely different reasons.

Our equipment does manage, occasionally, to capture the weak echoes that the bat is listening out for. When this happens, unlike the effect created during comb filtering (where the original sound and the echo overlap), there is clear separation in time between the original pulse and the returning echo. As a result, constructive and deconstructive interference does not occur, and we see the returning echo as a separate sound registration, appearing quickly after the original pulse emitted by the bat.

Picking up echoes doesn't happen that often, but when it does you need to be confident that it is in fact an echo you are looking at and not another bat echolocating slightly further away. Figures 1.22 and 1.23 demonstrate the difference between these two situations.

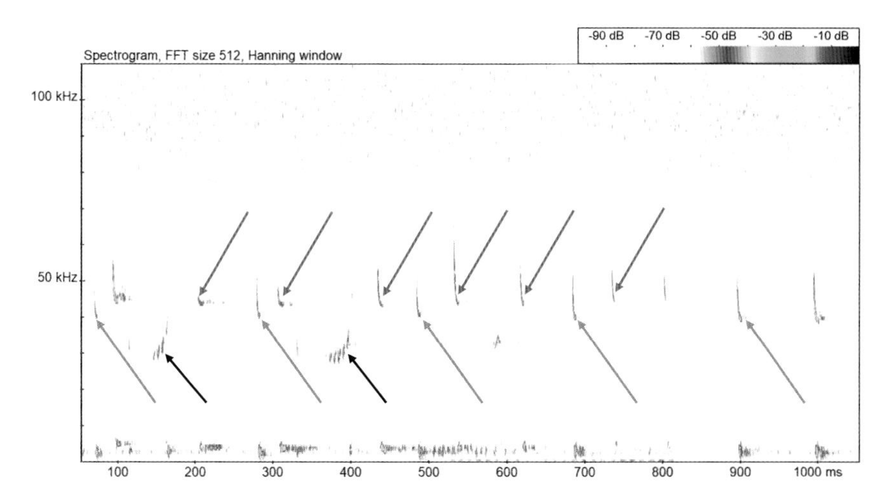

Figure 1.22 Multiple bat echolocation sequences (blue and red arrows), with social calls (black arrows).

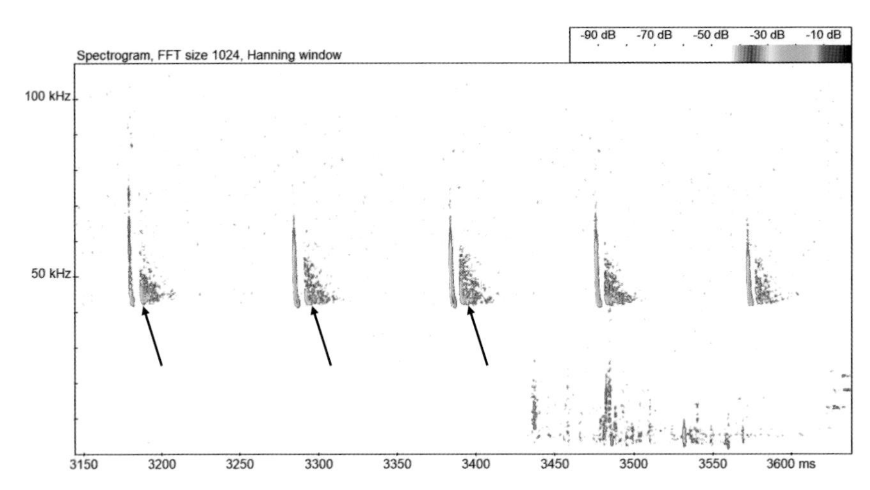

Figure 1.23 Single bat with returning echoes (black arrows) being picked up.

In the first of these figures (Figure 1.22), we have at least two separate bats. There are clearly two echolocation sequences, as shown with the blue and red arrows, as well as social calls (black arrows), probably from one of the two echolocating individuals. Focusing on the echolocation sequences, what are the clues? Well, first of all there are two typical frequencies, signal strengths and progressing call structures occurring as we look at the spectrogram, from left to right. Secondly, the inter-pulse interval within each of the two sequences is fairly evenly spread. To interpret this as a series of echolocation pulses from a single bat, you would have to accept considerable variation in the inter-pulse interval. Conversely, when it is considered as two bats it can be seen that each is working to a slightly different repetition rate. So, all in all, Figure 1.22 is a classic example of multiple bats. If you had more than two bats you would follow a similar approach in order to establish how many you actually have.

Turning now to Figure 1.23 – here you can see a clear echolocation sequence, with each pulse being followed by a weaker pulse (black arrows). Are these weaker pulses being made by another bat? Well, if they were, it would be an extreme coincidence that this second bat was producing a series of pulses in exact synchronisation to bat number one. If you measure the distance between the pulses in each pair, you will find that it is similar for each occurrence. Also, the weaker call always follows the stronger one. The most logical conclusion is that there is only one bat, with echoes of its pulses being picked up by the detector.

As with comb filtering, when echoing occurs as a result of where the bat is positioned within its habitat (e.g. flying close to a building) there isn't very much that can be done to avoid it. However, if your detector is pointing towards a horizontal reflective surface (e.g. calm water, solid ground surface) then having it angled upwards (e.g. at 45 degrees) would help in reducing such occurrences, as would, in any setting, placing the detector above a non-reflective surface (e.g. vegetated ground) and at least 2 metres away from reflective surfaces such as calm water, glass, or smooth-surfaced structures.

Lower-frequency sound in tandem with echolocation or social calls

The bulk of occasions when lower-frequency sound occurs in tandem with bat-related noise appear to be tracked back to situations when those involved have not been using headphones (see Table 1.4). John Harrison-Bryant (HB Bat Surveys) very kindly sent me good examples of this occurring (Figures 1.24 and 1.25). If you listen in real time to the call sequence through Windows Media Player (or similar) you can clearly hear the audible sound that a surveyor would be listening to, coming directly from the detector's speaker.

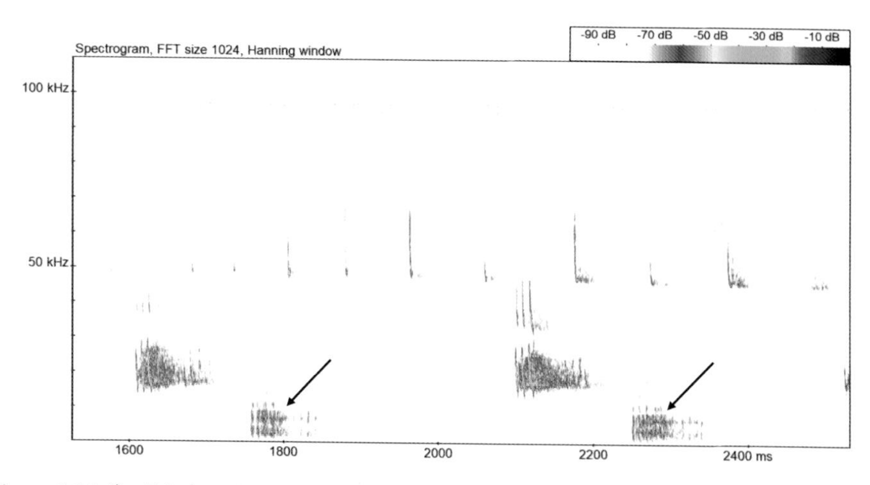

Figure 1.24 🔊 Echolocation sequence with social calls, the latter being picked up audibly from the bat detector's speaker (J Harrison-Bryant, HB Bat Surveys; frame width 1000 ms).

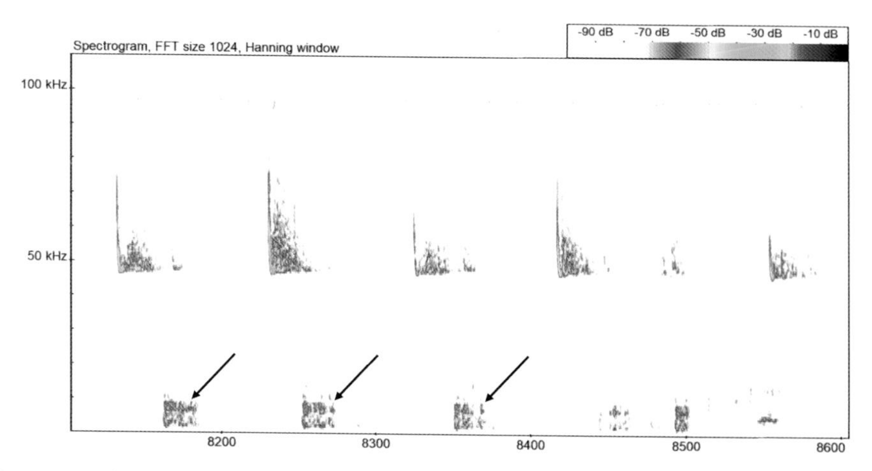

Figure 1.25 🔊 Echolocation sequence being picked up audibly from the bat detector's speaker (J Harrison-Bryant, HB Bat Surveys).

In both Figures 1.24 and 1.25, it can be seen that the audible noise (black arrows) is uncannily synchronised with the higher-frequency sounds. It looks similar to what happens with returning echoes, but the frequency is totally wrong. Although there are potentially other causes of scenarios like this, the most likely explanation usually is that the detector microphone is also picking up the audible sound coming from the detector's speaker, while it is recording the bat in real-time ultrasound (Chris Corben, personal correspondence). Of course, this audible sound would differ according to the detector model, or the audio settings of a detector chosen by a surveyor (e.g. heterodyne, time expansion, frequency division).

Therefore, with the speaker turned on, the bat detector microphone may also record speaker noise, and this additional noise may occur, for example, ever so slightly out of sync, but in tandem with the actual bat call recording the machine is storing internally. The effect can be exacerbated if the detector is also close to something that reflects (echoes) the noise from the speaker. This could, in theory, cause a double registration of the audible sound.

In order to prevent this happening, we should always use headphones or not have the speaker volume on, for example if the detector is being left unattended somewhere. And there is no point in trying to prevent such occurrences happening yourself, if the person standing next to you is recording the same bat with a speaker full blast without headphones. Or to make matters even more complicated, if that person is recording another bat calling from a different direction, and your detector picks up the audible output.

Using headphones greatly increases your awareness and the clarity of the sound that you are picking up, and irrespective of what has just been described (as if that alone is not reason enough) it is something to be recommended. When listening purely to your own machine, and not being audibly confused by a mix of various machines that are all blasting out their sounds simultaneously, it becomes easier to understand what you are listening to. The added benefit of having the external speaker off is that you save power on your device, which means longer uninterrupted use for machines holding a charge, or less expenditure on batteries. Having said all that, you do also need to appreciate that you become less aware of other things happening around you, with potential health and safety considerations and/or ineffective communication with your colleagues. It is all about taking a balanced approach. There is always, of course, a halfway house, where you have one ear with the headphone over it, and the other ear headphone-free.

Clipping

Clipping occurs when the sound being recorded is too loud for the recording device to cope with. This results in a recording input volume which overloads the system. It can be prevented by lowering the recording input volume to a level where clipping no longer occurs, and with this in mind, some good-quality digital recorders allow you to control the recording input volume manually, while others may do this automatically.

As a consequence of clipping, some of the original sound quality can be lost, as well as distortion and artefacts occurring. In particular, clipping can cause even-numbered harmonics to feature less strongly than odd-numbered ones, as well as creating spurious higher-frequency harmonics beyond the range in which a bat, or anything else for that matter, would be expected to naturally produce sound.

Figure 1.26 shows an example of a call sequence which starts off relatively strong, but still within (just!) acceptable recording parameters, but by the fourth pulse, clipping has began to occur (arrow). This can be clearly seen within the oscillogram, where the peaks in amplitude have exceeded the 100% reading on the scale, resulting in a straight-edged appearance against the top and the bottom of the oscillogram window. In addition, greater levels of distortion can be seen corresponding with these echolocation pulses, as shown within the spectrogram window. In Figure 1.27 the clipping is so bad that not only are the original pulses clipped, but so also are the returning echoes (arrows).

Ideally you should seek to control the recording input volume to a level where clipping is avoided. When clipping occurs everything is happening at a high amplitude and, as a consequence, any higher-frequency spurious harmonics will get picked up more strongly, and reflected back down within the spectrogram parameters you are viewing. This effect is known as aliasing – and (after you read the next section on that subject) if you look closely at Figure 1.27, you will see an element of it at about 90 kHz.

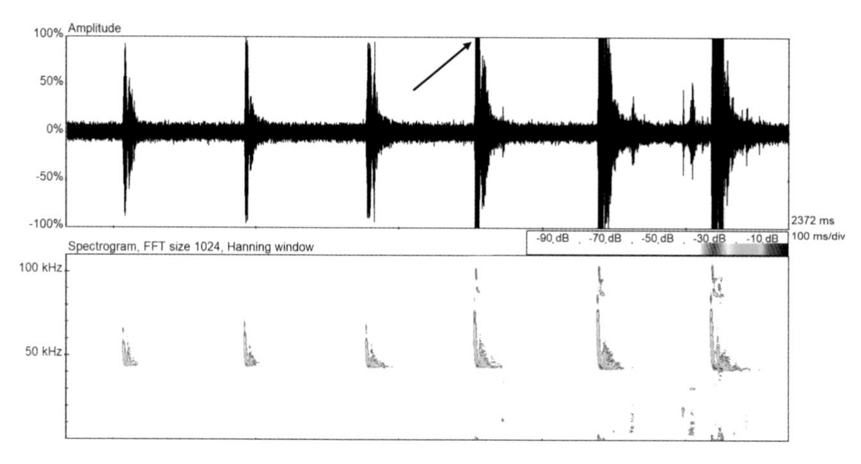

Figure 1.26 Example of clipping, as demonstrated by oscillogram (above) and spectrogram (below).

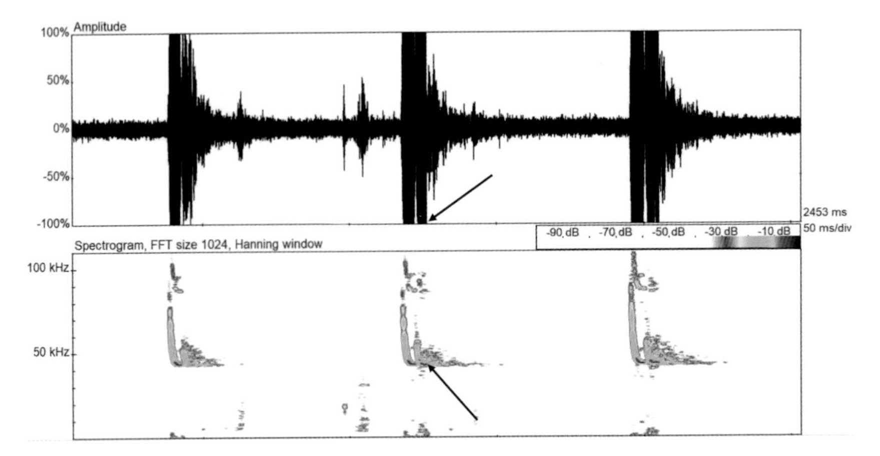

Figure 1.27 Clipping example, showing echo (arrows) clipped as well as the original pulse.

Aliasing

Aliasing occurs when the sample rate used while recording a sound is too low to cope with what's being recorded. In effect this causes an artefact that can make the spectrogram appear odd, or quite different to what you would normally expect to see. Now, provided you don't panic and are able to focus on the fundamental part of the call (i.e. ignoring any additional visible harmonics, and mirrored effects – see *Glossary*) you can still, quite often, probably establish what you would need to know for identification purposes. But having said that, it can be very useful to know how aliasing impacts upon recordings and how to recognise it when it occurs. This is helpful in order to satisfy yourself as to whether or not you are looking at something that a bat actually did, as opposed to an artefact.

So how do we know? Well, first of all we need to understand that in order to successfully record and accurately reproduce sound, the frequency components of that sound need to be, ideally, less than 50% of the sampling rate. Or to put it another way, 'the signal must be sampled at twice the highest frequency contained with the signal' (Olshausen 2000). If the source sound exceeds half of the sampling rate then aliasing occurs. If you then add in a very loud encounter with a bat, causing clipping to occur (see previous section), it can complicate matters even further in that the results of aliasing can appear to be as strong as the actual bat call, and in some instances make you feel that what you are seeing must have been produced by the bat naturally.

When aliasing does occur, it happens at exactly the same point in time on a spectrogram as where the actual sound was recorded (unlike echoes, it isn't delayed), and gives a visual, at least in part, of an upside-down version of the real call. In Figure 1.28 we have an example that shows the original call produced by a pipistrelle bat (blue arrow, minimum frequency, *c.* 43 kHz), immediately followed by an echo (red arrow). Above the original fundamental (or first harmonic) call (see Appendix 1, Figure A1.5), at double the frequency (*c.* 86 kHz), we can see the second harmonic. All of this, so far, was actually produced by the bat … '*OK Neil, but what's the other stuff?*'

On the right-hand side of the spectrogram I have positioned black arrows against three points where the bat appears to have produced a call that is curved at the top, after having first of all swept up through a range of frequencies. What you are actually seeing at these three points is aliasing, whereby higher-frequency spurious harmonics (caused by clipping), occurring beyond 50% (156 kHz) of the sampling rate (312,000 samples per second, or 312 kHz) have been mirrored back down into the visible range of the spectrogram. Spectrogram settings are normally set within software to 50% of the level of the sampling rate. This level is also referred to as the Nyquist frequency (Grenander 1959). The results produced beneath the Nyquist rate are mirrored representations of what has been picked up beyond 50% of the sampling rate. These artefacts happen at a distance equal to how far above the Nyquist rate they actually occurred. And while we are at it, I have deliberately missed out another point where this has also occurred. If you look at the very bottom of the spectrogram you can see a fourth aliasing impression at a maximum frequency of *c.* 10 kHz.

In Figure 1.28 we thus have aliasing occurring due to spurious harmonics, created as a result of clipping, occurring beyond 50% (156 kHz) of the sampling rate. These 'harmonics' have then been mirrored back down within the visible spectrogram. The pivot point of the mirroring occurring at a Nyquist frequency of 156 kHz.

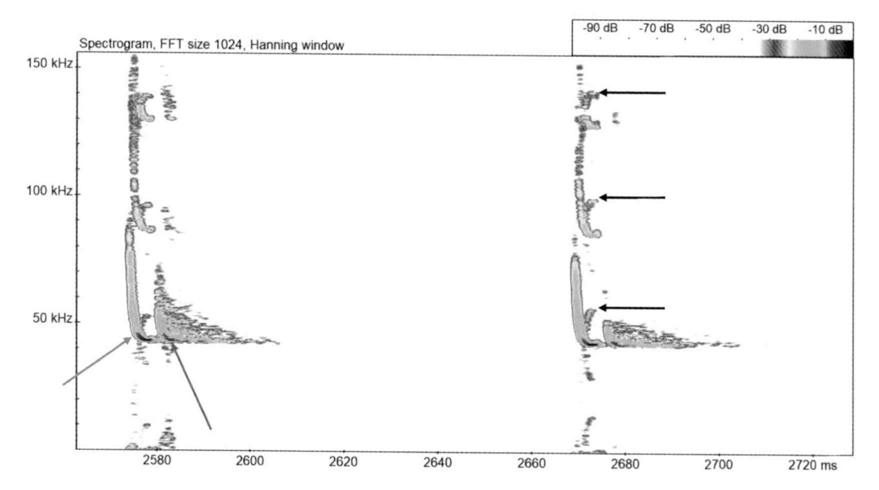

Figure 1.28 Pipistrelle – echolocation demonstrating the visual effect of aliasing (black arrows) (frame width 150 ms).

If you haven't quite grasped all of this just yet, don't panic, as I am about to use three artificial examples (Figures 1.29 to 1.31) to demonstrate further how it works in different circumstances.

For the artificial examples I have chosen a sampling rate of 192 kHz, meaning that only frequencies up to half this rate (i.e. 96 kHz) are measured accurately, beyond which aliasing would occur. In practice, this means that any signal recorded above the Nyquist frequency of 96 kHz is subtracted from the Nyquist frequency, folded over (i.e. mirrored) and presented beneath the Nyquist frequency at the same distance below, as it would have appeared above. In each of the figures I have provided more specific detail as to what is occurring.

Figure 1.29 (soprano pipistrelle) is a very similar situation to that which is shown in Figure 1.28, albeit I have deliberately chosen a different frequency for the call, as well as a different sample rate. But hopefully you can see a similar shape presenting itself, but more clearly in Figure 1.29, as only one alias has been included.

Figure 1.30 is even clearer, as we are dealing with a relatively long, and predominantly constant-frequency call, meaning that there is no collision between the aliased artefact and the true call components. This example was in fact a very real scenario in recent years for those of us with one particular version of the Song Meter SM2 bat detector (Wildlife Acoustics, USA). In that model the sampling rate was exactly as in the examples shown (192 kHz), which meant that it wouldn't pick up the true representation of lesser horseshoe bat (*Rhinolophus hipposideros*) calls. At that time we all learned how to recognise them in their aliased form (i.e. upside down and at *c.* 80/85 kHz). Bear in mind it is usually only the second harmonic of these bats that gets recorded by bat detectors, so unlike the artificial example, in which I have shown the first harmonic, normally all you have to go on is the second harmonic. Anyway, in this situation we always had to double-check that what was recorded wasn't actually a true representation of greater horseshoe bat (*R. ferrumequinum*), which produce their calls at similar frequencies to aliased lesser horseshoe bats.

Finally, Figure 1.31 demonstrates where things can become far more complicated, potentially steering the technician into thinking that he or she is actually looking at something that the bat did. In this example, aliasing is more likely to look completely natural when the signal produced travels, uninterrupted, through the Nyquist frequency.

In conclusion, therefore, we need to be aware of how aliasing works and how it can potentially throw us in the wrong direction if we are not considering it. Apart from thinking about it in a visual way, it is also beneficial to know the sampling rate of your recording equipment, so that you can calculate unusual occurrences yourself in order to help convince you, one way or another, as to whether or not the part of a call you are looking at is more likely to have truly occurred or is merely an aliasing artefact.

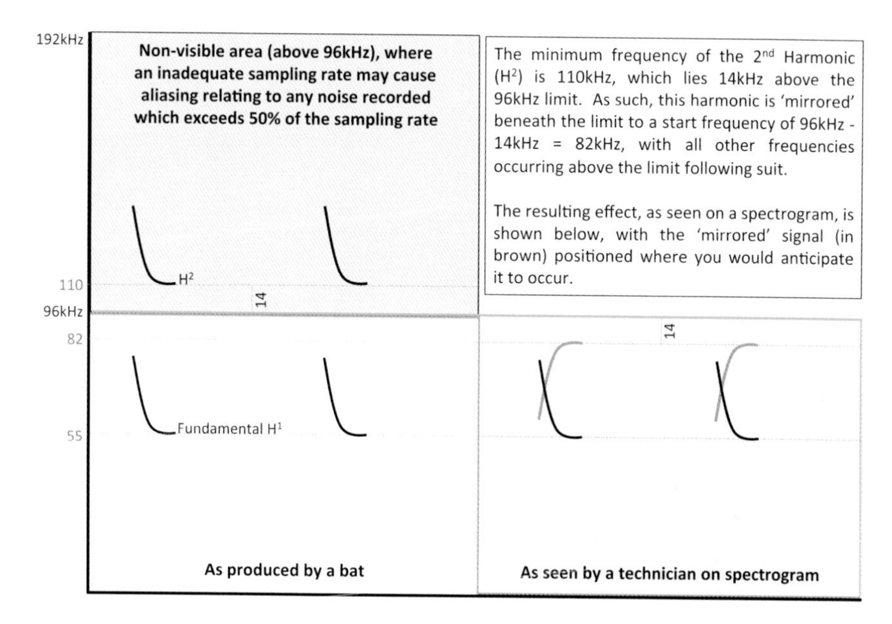

Figure 1.29 Aliasing impacting upon soprano pipistrelle echolocation recording.

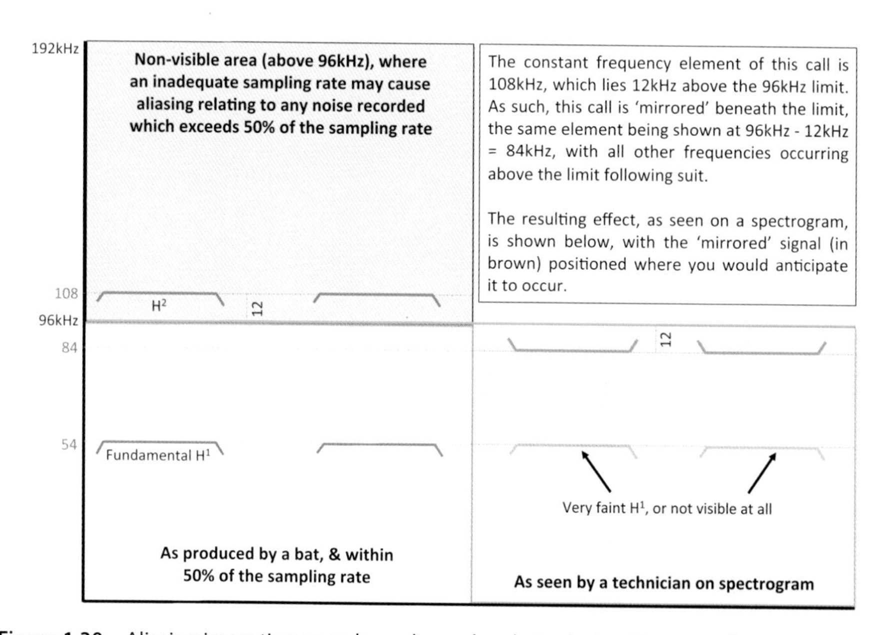

Figure 1.30 Aliasing impacting upon lesser horseshoe bat echolocation recording.

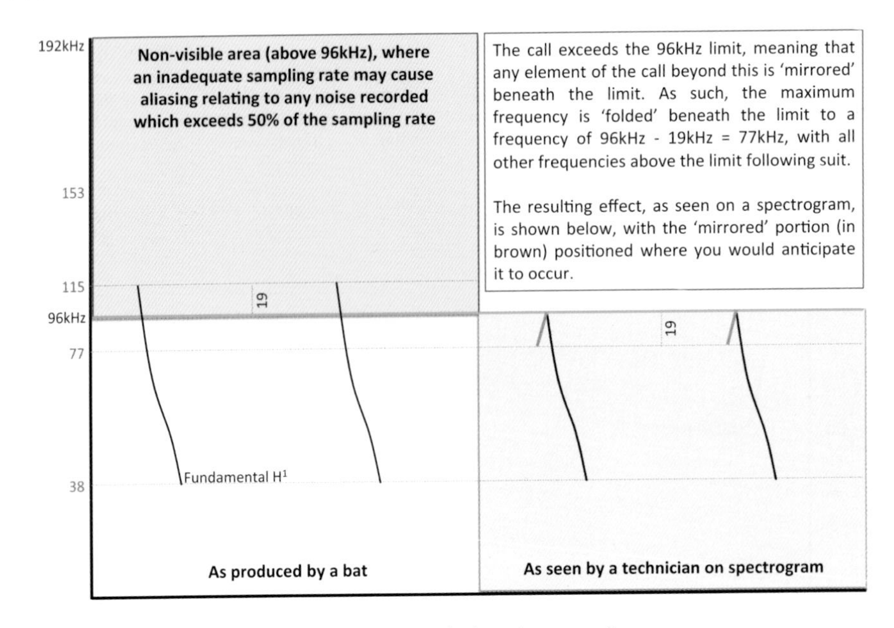

Figure 1.31 Aliasing impacting upon *Myotis* echolocation recording.

In an ideal world

Bearing in mind everything that we have discussed earlier in this chapter, is there anything a bat surveyor can do in order to try and reduce the risks of echoes and artefacts occurring, and as a result increase the likelihood of accuracy during the analysis process? Figure 1.32 pulls much of this together, and hopefully also enables you to see how a number of the points raised are interrelated. Table 1.5 outlines many of the scenarios from a problem-solving perspective.

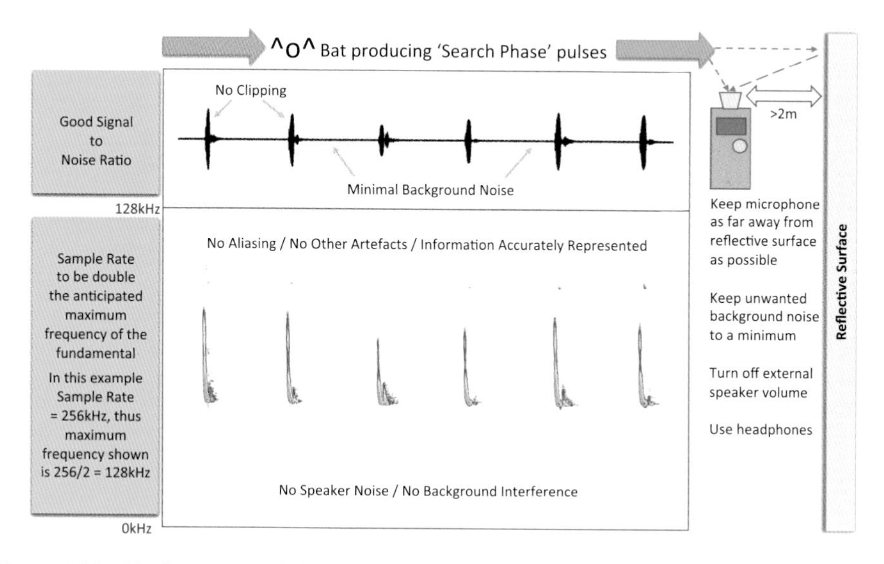

Figure 1.32 Ideal scenarios while recording bats and carrying out sound analysis.

Table 1.5 Potential solutions to reducing the recording of echoes and artefacts.

Subject	Problem	Potential solutions
Comb filtering	Echoes overlapping with original pulse causing constructive and deconstructive interference Potentially creating problems when measuring call parameters	Position microphone > 2 m away from reflective surfaces Position microphone above/next to non-reflective surfaces Position microphone at a 45-degree angle to horizontal reflective surfaces Avoid using such calls when measuring call parameters for identification purposes
Echoes	Echoes picked up by detector and potentially misinterpreted as a separate bat, or adjoining the original pulse and giving the impression that it is a longer call Potential for double-counting number of bats present, or for misidentification of calls	Position microphone > 2 m away from reflective surfaces Position microphone above/next to non-reflective surfaces Position microphone at a 45-degree angle to horizontal reflective surfaces Avoid using echoes for identification When carrying out audit of calls identified, be extra-vigilant when echoes are present
Speaker noise	Creating confusion during analysis	Turn off external speaker volume Use headphones Ensure those around you do not create their own speaker noise that can be picked up by your recording device
Clipping	Distortion Even-numbered harmonics less obvious or missing Spurious harmonics created at higher frequencies Potential to increase the noticeability of aliasing	Choose a recording input volume that reduces the risk of clipping When taking call measurements for identification/research purposes, do not rely on clipped calls
Aliasing	Higher frequencies of original fundamental call are not accurate if sample rate isn't high enough Confusing representations on spectrogram Potential for errors or falsely describing call structure	Ensure sample rate represents species likely to be recorded, with particular emphasis on the highest frequencies likely to be recorded Avoid clipping When in doubt always check sampling rate and Nyquist frequency, as well as for the presence of clipping

It's alright to say 'I just don't know'

So, here is a thing that I would like us all to be clear on, and this may come as a surprise to some less experienced folk, or those who have been in the company of people who always manage to get every call allocated to a species.

It is alright not to know. In my opinion, a real expert in the field of bat-related sound analysis understands that there will be many times when there just is not enough information to safely identify a call to species, genus or group level – e.g. noctule (*Nyctalus noctule*)/Leisler's (*N. leisleri*)/serotine (*Eptesicus serotinus*). In fact to aim do so is unrealistic, and you certainly shouldn't be beating yourself up when you can't. Also, bear in mind that many extremely experienced people carrying out sound analysis for non-consultancy purposes regularly dump a large percentage of their recordings because they are unable to work with the quality of what has been recorded, or because the call is just not safely separable.

When you truly don't know, you have options. First of all you can refer it to a more experienced person. But if you do this, then ask them to explain why they have arrived at their conclusion, and how confident they are in their assessment. But what about if you have explored these avenues and you are not that much further down the road? If you know it's a *Myotis* or a *Nyctalus*, for example, then say what you know, and no more (e.g. say 'it's a *Myotis*'). If you suspect, with good reason, that it is more likely to be a particular species, based on call parameters being a better fit for 'A', or a visual observation giving you more confidence that it is 'B', then in that case it is alright to give your opinion using a statement such as:

> The call belongs to a *Myotis* species, and based on call structure and measured parameters, it is more likely to be from a whiskered bat

or:

> The call resembles examples that are known to be produced by whiskered bat, and as such this is possibly the species occurring in this instance, although other species should not be ruled out.

These statements don't box you into a corner, and are far more defensible if an error has been made. Apart from that, they are also far more credible, unlike the thing that most definitely should never be done, given the same information. That is to say 'It's definitely a whiskered bat.' If you do this based on echolocation alone, then you run the risk of being wrong at least some of the time.

How often are you prepared to risk your professional reputation? If for example you say you are 80% confident it's a whiskered, so therefore let's say it is definitely a whiskered, that means that once you have made 100 similar judgement calls you were possibly 100% wrong 20% of the time, and you can't say, for sure, which judgements fell into that 20% wrong bucket. Would you employ a solicitor or an architect or an airline pilot on that basis? The most professional approach, and the most accurate approach, is to say what you know to be fact, at whatever level that is, and no more. Not what you want it to be, or what you hope it is. And if someone tells you that you are an inadequate bat worker because you couldn't make the call, then perhaps you actually know more than they do, or at the very least you are credible in your approach. In such circumstances, they may be the one who is inadequate, not you. They may be refusing to recognise that there may be lots of things that they, and the rest of us, just don't know as yet, and indeed may never be able to fully understand. In fact, sometimes, the information you have in front of you is just so limited that the best you can say, in order to maintain credibility and accuracy, is no more than '*that is a bat*'.

Author's diary note

In September 2018 I had the opportunity to deliver a workshop on social calls to a small audience at the National Bat Conference in Nottingham. This was one of the first sessions on the subject I had delivered since I had begun writing this book. For that session I prepared an overarching statement that I wanted to leave in people's minds that day. I can't recall the exact words I used when it was delivered, but it was along the following lines:

Our bat detectors and associated software should be considered as 'educated idiots'. Very intelligent, but, on occasion, totally lacking any common sense. In some respects the technology isn't good enough as yet (e.g. automated classifiers). Either that, or it is too good (e.g. too fast, requiring the surveyor to think in real bat time, juggling with vast amounts of information and options). As well as all of this, so many of us don't use our ears often enough, when most of the software today allows us to do this easily.

Taking all of this into account, there is one part of the process that is in our ultimate control. That is the person involved, at the sharp end, making decisions about what is actually occurring. You have the most important role to play in the whole process. Your finger hovers above the 'Common Sense Button' (you won't find it on your keyboard, unfortunately!). You need to keep pressing that button, before jumping in with the wrong conclusion and subsequent decisions.

If you don't have a Common Sense Button in your office, then drop me an email and I'll send you one.

CHAPTER 2

Terrestrial mammals

These mice are just like bats, but without the wings

By and large bat workers don't often hear that many mammals (other than bats) when they are out surveying, and for that reason in this chapter I am concentrating mostly on the animals that are more likely to be picked up by bat detectors (especially those placed remotely, for longer periods of time) due to the frequency of the sounds being emitted. And it's not only in outdoor scenarios we need to think about this. It is not uncommon for bat workers to be placing static bat detectors in more confined areas (e.g. lofts or underground) where small mammals may also be present. These small mammals are usually quite quiet and scurry away when humans are about, only later returning to their normal behaviour, as well as to investigate the new piece of furniture (i.e. the bat detector) that's been left behind.

Of course, these issues, if or when they occur, are not going to be transparent to an alert and well-informed bat worker until afterwards, during the analysis stage of the process. Even during this phase many would not necessarily be thinking that anything recorded that looks like a bat call could potentially come from something else altogether. Now I don't want to appear to be excluding myself from this group, as I know I have made some mistakes of interpretation in the past (and no doubt will unwittingly continue to do so in the future). We only 'know what we know', after all, and therefore we are all blissfully unaware of what we don't know.

Before considering the higher-frequency squeaks etc. and getting into the details of sound analysis, however, we should cover at least a small number of the more likely audible mammal noises that we can expect to hear when out at night.

Audible noise made by larger mammal species

Given the right circumstances, pretty much any mammal out there (wild or domestic) has the potential to make noise during the period from dusk to dawn. And probably the noisiest of all is one that is widespread, common, and always present when we are around, *Homo sapiens*. If we are doing a structure survey, for example, we could have a group of people centred around a specific location, all in hi-vis, using head torches, communicating with walkie-talkies, and munching on our favourite survey confectionery (not something that I would ever confess to!). This is hardly the stealth required to have meaningful encounters with most of our terrestrial mammals, and as such, any mammal worth its highly developed senses would usually be long gone before we have even taken our headphones out of our rucksack. Alternatively, we might be doing a transect, and again there would be torchlight, hi-vis, more chocolate, lots of tramping through vegetation etc. Some of this anthropogenic noise, along with other sources of sound not related to non-human animals, is covered in Chapters 6 and 7.

So, apart from ourselves, what else are we most likely to hear? Well, the possibilities are wide-ranging, but in most instances, sadly, not a regular feature of many bat surveys. There are, however, a few species that do pop up more often if you are in the right place and you know what you are listening to. With this in mind, Table 2.1 contains a brief description of a small selection of sounds that are potentially more likely to be heard audibly. Each of these species occurs widely across the British Isles, with the exception of roe deer, which is absent from Ireland. Knowing what they sound like may either be beneficial for developmental impact purposes (e.g. badger and otter), or just from a 'satisfaction of curiosity' point of view. In each case, audible sound files are available in the Sound Library, so that you can acquaint yourself with these sounds and be more suitably informed, should you ever encounter them when in the field.

Table 2.1 Brief notes on night-time audible noise made by relatively common and widely distributed mammal species occurring in the British Isles.

Species	Distribution and typical habitat	Notes on communication
Fox *Vulpes vulpes*	Widely distributed throughout Britain and Ireland[1][2] Generalist[1]	A wide range of adult vocalisations have been documented. Peak period for vocalisation is normally November through to March, when both peak dispersal, resulting in territorial disputes, and mating occur.[3] Two of the more recognisable sounds heard during this period are triple barking and scream-like noises.[4]
Badger *Meles meles*	Widely distributed throughout Britain and Ireland[1][5] Unimproved grassland Broadleaved woodland Urban and gardens Arable and horticulture[1]	Badgers generally are not regarded as being particularly noisy, but they do have a wide range of sounds, described as a 'mixed vocal repertoire', which have been documented.[6] Adult calls have been described as 'bark', 'chitter', 'churr', 'growl', 'grunt', 'hiss', 'kecker', 'purr', 'snarl', 'snort' and 'yelp', with additional call types relating purely to cub behaviour. They tend to be more vocal from February to June,[5] when territorial activity peaks, coinciding with mating activity. It is also during this time of year when the cubs start emerging above ground, which also results in the potential for the occurrence of associated vocalisations.
Otter *Lutra lutra*	Widely distributed throughout Britain and Ireland, although less so in England[1][7] Rivers and wetlands Coastal[1]	Can be relatively more vocal during mating activity, which can take place at any time during the year.[8] Calls described in adults are a 'whistle' contact call, a 'hah' alarm call, and a 'chittering' threat call.[7][9] Typically, vocalisations occur at frequencies < 10 kHz.[9] 'The contact call is the one that is usually heard. It's a repeated short whistle, often sounding like "an insistent dunnock" and carries for some distance. The other common noises are very quiet. One is a very quiet "murmur/chitter" between mum and cub when they are very close to each other, and the other is the "huff". These are unlikely to be heard in isolation, without the contact noise. There are various "mewing" noises too, but again rare in isolation. Young will repeatedly call for mother if separated, with the call described as a bit like someone air kissing.' (Melanie Findlay, Findlay Ecology Services, personal correspondence, 2018).

(cont).

Table 2.1 (cont).

Species	Distribution and typical habitat	Notes on communication
Roe deer *Capreolus capreolus*	Widely distributed throughout Britain; absent from Ireland[1][10] Coniferous woodland Broadleaved woodland[1]	Rut occurs from July through to September. Males make grunts and rasping noises especially during breeding season when pursuing a female. Roe deer 🔊 can emit short repeated barks when disturbed (especially at dawn or dusk), either occasionally and irregularly, or averaging at 14 per minute.[10] These calls could be confused with fox barks 🔊 or Reeve's muntjac (*Muntiacus reevesi*), the latter of which is known to make single dog-like barks that can be repeated a few times or persistently (hundreds of times).[11] Other widely distributed deer species do not commence rutting until September (Sika, *Cervus nippon*,🔊 and Red deer, *C. elaphus*), or October (Fallow deer, *Dama dama*).[12][13][14]

References

[1] Mathews *et al.* 2018; [2] Baker & Harris 2008; [3] Newton-Fisher *et al.* 1993; [4] Harris & Lloyd 1991; [5] Delahay *et al.* 2008; [6] Wong *et al.* 1999; [7] Jefferies & Woodroffe 2008; [8] Green *et al.* 1984; [9] Gnoli & Prigioni 1995; [10] Hewison & Staines 2008; [11] Chapman 2008; [12] Putman 2008; [13] Staines *et al.* 2008; [14] Langbein *et al.* 2008.

Emission of sound by small terrestrial mammals

As bat workers we are often challenged with the statement '*Bats are just like mice, but with wings.*' Upon hearing such a statement, we all quite rightly explain why this, most definitely, is not the case. There have, however, been many moments while working on this chapter when I have muttered to myself, in frustration, that '*these mice are just like bats, but without the wings.*' The small mammals that we are about to consider truly do have the potential to show many of us bat workers that we are maybe just not as smart as we thought we were.

Ultrasonic vocalisation by small terrestrial mammals was little understood until fairly recently (Anderson 1954). In fact the development of bat detectors helped to establish that many of these other mammals make noises that are not audible to humans (Brudzynski 2009). There are reasons why ultrasonic noise developed in these mammals, primarily to reduce the risk of attracting predators (at least those that can't hear at higher frequencies), as higher frequencies may not only be less likely to be heard, but will attenuate (see Appendix 1, Figure A1.2) more quickly over a shorter distance. In fact, it has been demonstrated that many of the calls emitted would not be picked up much further than 5–6 metres away at best (Ancillotto *et al.* 2014). Attenuation would be even more dramatic (i.e. effective in being less likely to being heard at a distance) when the animal is within dense vegetation or underground within a burrow.

A number of studies have been carried out in order to demonstrate that small mammals can emit both audible and ultrasonic sound while communicating and interacting with each other both in conspecific (i.e. between animals of the same species) and heterospecific (i.e. between animals of different species) situations (Sales & Pye 1974, Stoddart & Sales 1985, Kapusta *et al.* 2007, Ancillotto *et al.* 2014). Such circumstances for sound to be emitted would include, but not be restricted to, the following: aggression, mating, distress, contact and territorial (Sales 2010). Naturally there is no point in being able to emit a sound if your conspecifics are unable to hear it. In many species of small mammal it has been shown that they can hear at both audible and ultrasonic frequencies (e.g. brown rat and harvest mouse can hear up to *c.* 60 kHz, and house mouse up to *c.* 100 kHz). In some species there have been shown to be two regions of hearing sensitivity, one within the audible (sonic) range and another within the ultrasonic range (Thomas & Jalili 2004).

As already discussed in Chapter 1, one thing that bat workers need to remember is that the production of ultrasound does not necessarily mean an animal is using that sound for echolocation purposes. We know, for example, that bats emit ultrasonic noise in a social context (Middleton *et al.* 2014), as well as for orientation and/or foraging purposes within their environment (echolocation), the latter being by far the majority of what we record the bulk of the time. Likewise, small terrestrial mammals produce ultrasonic noise, but this does not mean they are echolocating. In fact, the bulk of these ultrasonic sounds are considered to be what we would call, in its widest sense, social calls. There are exceptions, for example in shrews, where echolocation is known to occur (von Merten 2011), as well as social calls.

Although small terrestrial mammals can be quite vocal at times, it should be borne in mind that they do not need sound to travel too far (comparative to bats), and because they are often at ground level within vegetation, as previously discussed, any sound emitted is quickly absorbed (attenuated) by its immediate surroundings, probably travelling only a few metres at best. As such, picking up these sounds on a bat detector is not that easy, unless of course it has been left *in situ* at ground level as part of a long-term study. In this situation any inquisitive small mammals that are present may very well investigate, at least initially, new additions (such as a bat detector) to their habitat. Bearing all of this in mind, it is worth remembering that many small terrestrial mammals are considerably more abundant than many of our bat species (Table 2.2). So as 'unusual' as it may be considered to record these animals with a bat detector, it should be factored in that there are almost always going to be many more of them within a survey site boundary than there will be bats, and certainly more than meets the eye. Having looked at Table 2.2, take a moment to consider how often you see a bat, and then how often you see a small terrestrial mammal. With this in mind, it should be fairly apparent that recording a non-bat mammal is certainly feasible, especially on a static detector.

Table 2.2 Examples of the relative abundance, within Britain, of small terrestrial mammals compared to bat species (Mathews *et al.* 2018).

Group	Species	Estimated population	
Small terrestrial mammals	Brown rat	7,070,000	
	Wood mouse	39,600,000	
	Field vole	59,900,000	
	Common shrew	21,100,000	
	All species within group (16)	**165,249,000**	**(94.1%)**
Bats	Common pipistrelle	3,040,000	
	Soprano pipistrelle	4,670,000	
	Daubenton's bat	1,030,000	
	Brown long-eared bat	934,000	
	All species within group (18)	**10,312,100**	**(5.9%)**
	Total both groups (34)	**175,561,100**	**(100%)**

'All species within group' refers to the total for all species present within Britain, with the total number of species considered being shown in brackets. The individual species shown are given as examples that the reader can easily relate to.

Species-specific examples

We are now going to consider, species by species, some calls known to have been made by small terrestrial mammals, incorporating notes and spectrograms as appropriate. Compared to many of the other sources of sound covered elsewhere in this book I am going to go into relatively more detail here, as the sounds made by these small mammals have the potential to directly impact upon the sound analysis phase of a bat project, and accordingly I have tried to be appropriately detailed when it comes to reference material etc. Please bear in mind, though, that what follows is only a limited range of examples and certainly should not be considered to be the entire vocal repertoire for any one particular species.

Spectrograms in Chapter 2

Spectrograms in this chapter use the following scales, unless otherwise indicated in the figure legend:

Time (x-axis): 1 second (1000 ms)
Frequency (y-axis): 0–110 kHz

When a figure legend includes the 🔊 symbol, this means that the figure has been created from a file in the Sound Library. The figure number matches the file number there. For more information about how to access and download files from the Sound Library please see pages xv–xvi.

Rat species

Ultrasonic sound made by rats has been researched and reported upon extensively, with many studies available for reference. The occurrence of emitted sound is dependent upon the animal's age and sex, and also the presence or absence of conspecifics, and their age and sex (Sales & Pye 1974, Quy & Macdonald 2008, Wöhr & Schwarting 2013). In the British Isles, two species of rat occur (Table 2.3).

Table 2.3 Rat species occurring in the British Isles.

Species	Distribution and typical habitat	Notes on communication/recordings
Brown (common) rat *Rattus norvegicus*	Common and widely distributed throughout Britain and Ireland[1][2] Urban and buildings[2]	🔊 Can produce a variety of ultrasonic calls (> 20 kHz) with duration being from 3 to 300 ms.[3] Very vocal during aggressive encounters with conspecifics, as well as predators, with various squeaks, grunts and whistles noted. Variety of vocal calls noted during mating activities.[1][4] Two main call types regularly encountered (22 kHz alarm call, and 50 kHz positive interaction call), both of which are discussed in more detail in the text. During my own observations of a long-term captive colony (> 50 individuals), these mammals were very vocal, producing a wide range of calls.
Black (ship) rat *Rattus rattus*	Very rare, with a fragmented distribution. Only occurs in a low number of small and isolated populations, and on some offshore islands[2][5] Generalist[2]	🔊 Known to produce a variety of ultrasonic calls (> 20 kHz) during mating activities.[4] During my own observations of a long-term captive colony (*c.* 50 individuals), these mammals were very vocal. A range of call structures were recorded, many of which were not obviously dissimilar to those of brown rat.

References

[1] Quy & Macdonald 2008; [2] Mathews *et al.* 2018; [3] Halls 1981; [4] Sales 2010; [5] Twigg *et al.* 2008.

Of all of the small mammal species that I have personally received correspondence about, by far the most regular is the brown (common) rat. This species is more likely than many others to be picked up on a bat detector, for a number of reasons. Firstly, they are widely distributed and abundant (see Table 2.2). They are also regarded as being more vocal. Finally, the fact that they are a bit bigger means they are louder and therefore can be picked up by a bat detector from further away.

We are now going to look at some spectrograms for brown rat, as well as consider any immediately apparent potential confusion points when considering bat call analysis. Given the restricted range of the black rat, we are not going to consider this species any further, but recordings are available in the Sound Library.🔊

Brown rat

Relative to most of the other small mammals discussed in this chapter, there has been much academic research on the vocalisations of the brown rat (*Rattus norvegicus*) (e.g. Halls 1981, Burgdorf *et al.* 2008, Takahashi *et al.* 2010, Coffey *et al.* 2019). Rats are highly social, forming large colonies, meaning that vocalisation is an important aspect of their interactions with each other. The research mainly focuses on two groups of more common call types that they make, described as alarm (22 kHz) calls and appetitive or 'friendly' (50 kHz) calls (Brudzynski 2009).

The 22 kHz call – which is a long-duration, near-constant-frequency (CF) call – serves as an alarm call emitted by a member of the colony when it senses danger or experiences anxiety (Kim *et al.* 2010). Such calls may also occur during aggressive interactions. Brudzynski (2009) describes these calls and the 50 kHz call types in more detail. The 22 kHz calls last from 100 ms to 3000 ms, with a frequency of maximum energy (FmaxE) usually within the range 20–23 kHz (bandwidth 1–4 kHz), although frequencies beyond this can occur. Normally a series of two to seven calls is emitted, and immediately afterwards this may be followed by shorter calls (*c.* 100 ms) at the same frequency. An interesting point of note, for bat workers, is that when using a heterodyne bat detector (especially automated heterodyne as available on many modern detectors) these near-CF calls 🔊 sound similar to what you would experience when listening to a horseshoe species. Double-checking the peak frequency should alert you to the fact that something odd is occurring, but less experienced surveyors might immediately be drawn into concluding that they had encountered a bat.

The 50 kHz calls appear to be emitted in more positive circumstances (e.g. sexual contact, feeding, non-aggressive interaction with conspecifics). The average duration of these calls is much shorter than that of the 22 kHz calls, with these higher-frequency emissions being 30–40 ms in duration. Although they are described conveniently as 50 kHz calls, there is a variation in the frequency where they occur, with typical FmaxE range being 45–55 kHz (easy to remember for bat people!), but higher frequencies are also known to occur (up to *c.* 70 kHz). The bandwidth is wider than that of the 22 kHz calls, at 5–7 kHz (i.e. these calls are more FM in appearance).

One study concluded that the calls traditionally referred to as 50 kHz calls ranged in peak frequency (FmaxE) from 35 to 75 kHz (mean at *c.* 55 kHz) (Burgdorf *et al.* 2008). This study went on to separate the calls at these levels into two sub-categories, named (1) flat (i.e. near-CF) 50 kHz, and (2) FM 50 kHz. In differentiating these sub-categories, the authors described the flat calls as having only a flat (CF) 50 kHz (or similar) component, while the FM calls could also contain the flat component but always a trill or stepped component. In analysing the behaviour of rats in conjunction with these call types they concluded that the FM 50 kHz calls were related to '*positively valanced appetitive behaviour*' during mating, play or aggressive encounters, and the reward value of such encounters (also discussed in Wöhr 2018). The flat 50 kHz calls were not related to the same behaviours, but appeared to be more evident during aggressive interactions.

The examples in Figures 2.1 to 2.6 are typical of ultrasonic recordings made from known brown rat encounters. In addition, an audible squeak is shown (Figure 2.7) for comparative purposes, where the call is distorted and displays numerous harmonics. As well as what is shown within the text, additional sound files are provided in the Sound Library.🔊 Also, variations of the call types shown are known to exist, and this species can in addition produce other call types (i.e. what is provided here is not intended to be the full repertoire).

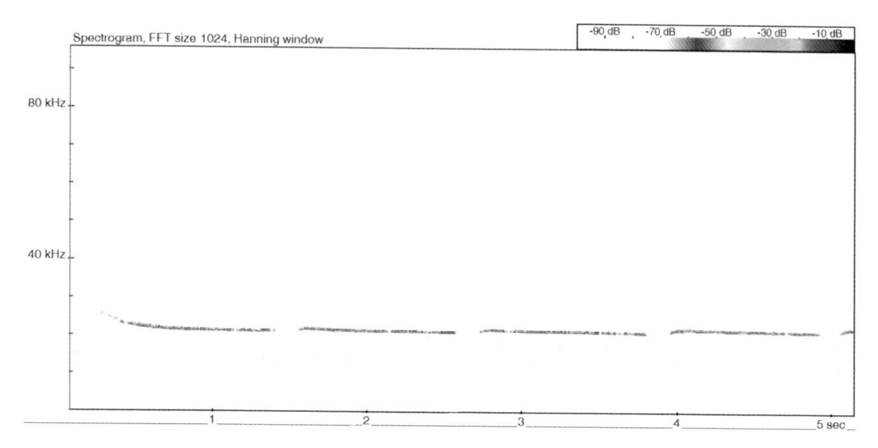

Figure 2.1 🔊 Brown rat – distress call (A Dexter, 2018; frame width 5000 ms, frequency scale 0–96 kHz).

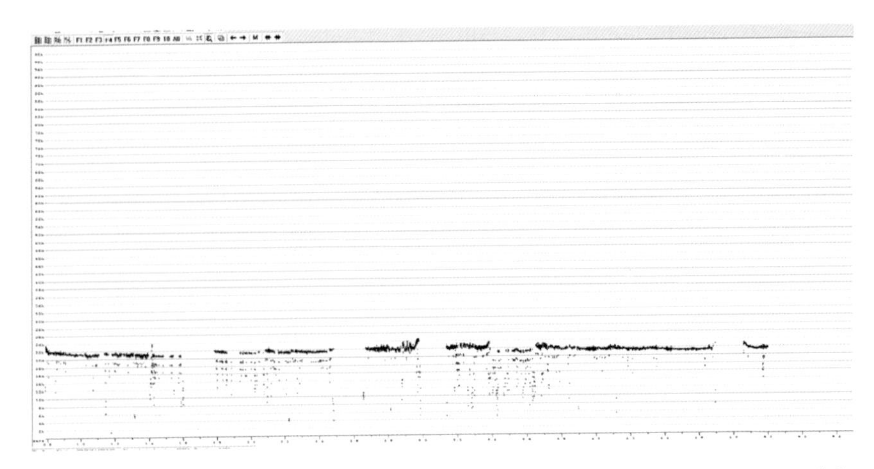

Figure 2.2 🔊 Brown rat – distress call (ZCA file, viewed in AnaLookW, F4, uncompressed; frame width 5000 ms, frequency scale 0–100 kHz).

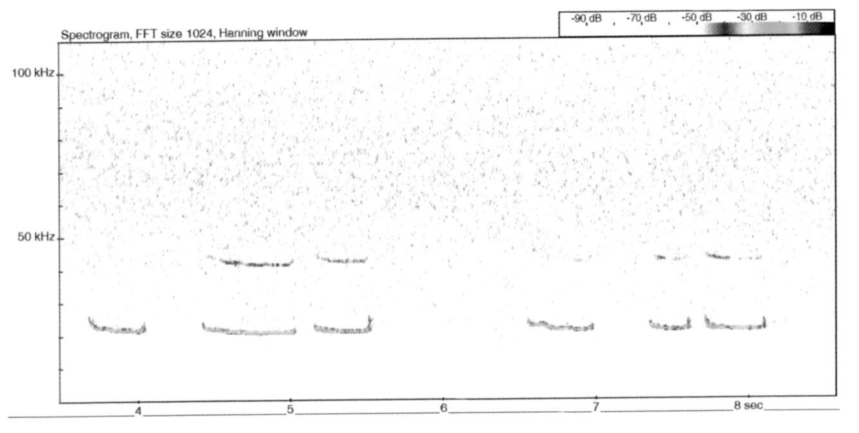

Figure 2.3 🔊 Brown rat – variation of 22 kHz call type (frame width 5000 ms).

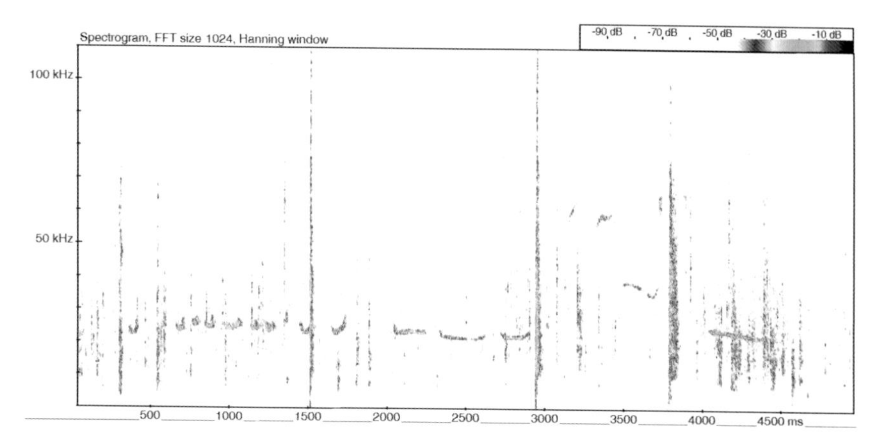

Figure 2.4 🔊 Brown rat – shorter 22 kHz call type, with other associated noise (frame width 5000 ms).

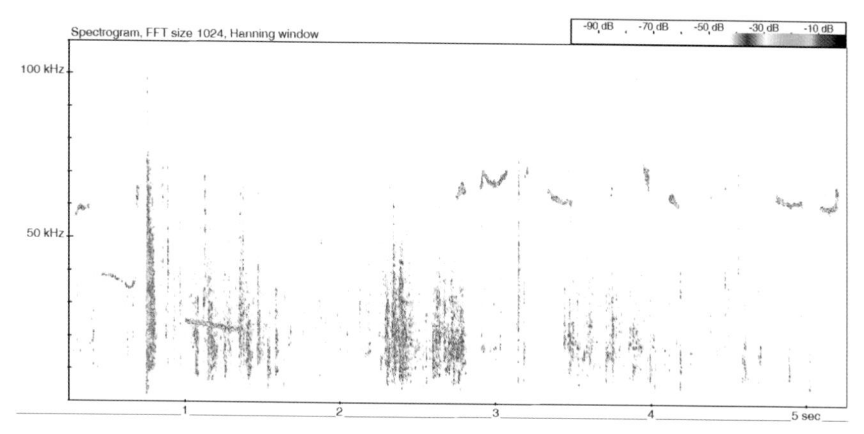

Figure 2.5 🔊 Brown rat – including 50 kHz call types, with other associated noise (frame width 5000 ms).

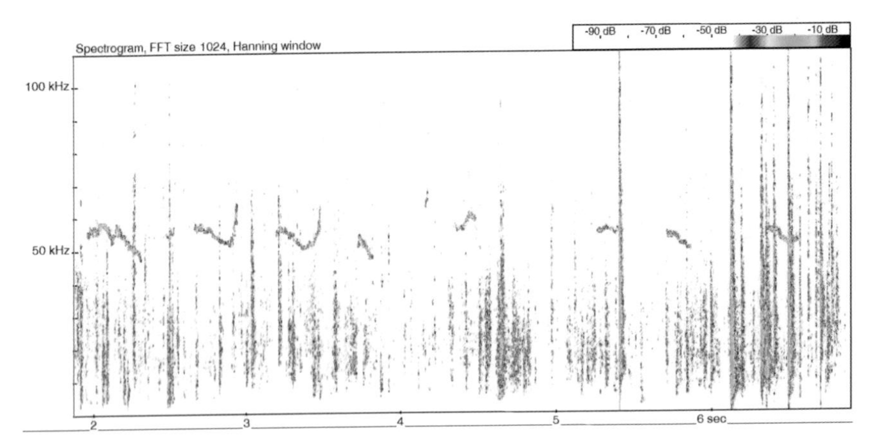

Figure 2.6 Brown rat – examples of 50 kHz call types, with other associated noise (frame width 5000 ms).

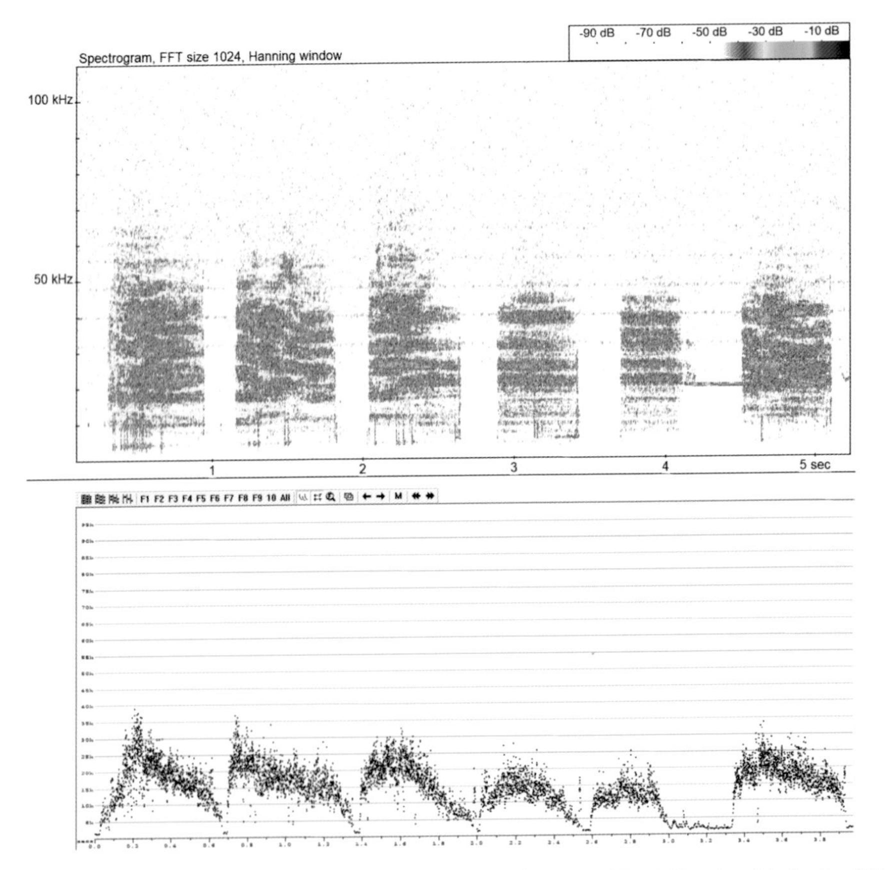

Figure 2.7 Brown rat – example of an audible call (frame width 5000 ms), with AnaLookW example below (F3, uncompressed, frequency scale 0–100 kHz).

Differentiating between rat 22 kHz type calls and Nyctalus species

Considering the calls shown in Figures 2.1 to 2.4, it is not beyond the imagination that someone, or more likely something (i.e. an automated bat classifier – see Table 2.10), could look at these 22 kHz type calls (especially shorter-duration examples) and quickly label them as a *Nyctalus* species of bat (noctule or Leisler's) (Stuart Newson, personal correspondence, 2018). This could have quite a dramatic effect on a survey result, following through to any associated impact assessment. Imagine having automated bat detectors deployed as part of a wind farm development, and the site is full of brown rats. Your filtering process has identified them all as *Nyctalus* sp. You didn't know what rats looked like on a spectrogram, and there was no training or a proper audit carried out in this respect. You have hundreds of these bat (oops, rat) passes. The potential for inaccurate reporting, and inappropriate measures going forward, is definitely there.

OK, hopefully that's got your attention. The frequency is right for a *Nyctalus*. If you don't look too closely they are typical *Nyctalus*-like 'open-habitat', CF pulses. And if you are using a software package that allows you to compress (or clear empty space), it's even more convincing, depending upon how far zoomed in/zoomed out you are. You quickly scroll through, under the pressure of time that is a regular feature of your working environment. Yes, it's another *Nyctalus*, and that's another point against this wind turbine's location. So we have to be careful.

Let us now compare these options (rat vs. *Nyctalus*) side by side. I am going to use the examples shown in Figures 2.8, 2.9 and 2.10. The latter two belong to a noctule, showing a typical open-habitat bat pass. Figures 2.8 and 2.9 look fairly similar, don't they? But notice the time frame along the bottom. The rat (Figure 2.8) is a 5000 ms (i.e. 5-second) sequence, whereas the noctule (Figure 2.9) is 500 ms, and is in compressed mode. Now look at how they compare to each other over the same period of time, and uncompressed (Figures 2.8 and 2.10). The difference is far more obvious. Just to be certain, now listen to the calls in time expansion (× 10).🔊 They should sound very different to your ears. The noctule produces a shorter whistle compared to the long, distant jet-engine-like noise produced by the rat. If there was any doubt, then all of these tests taken together should help determine whether it is, or is not, a bat/rat.

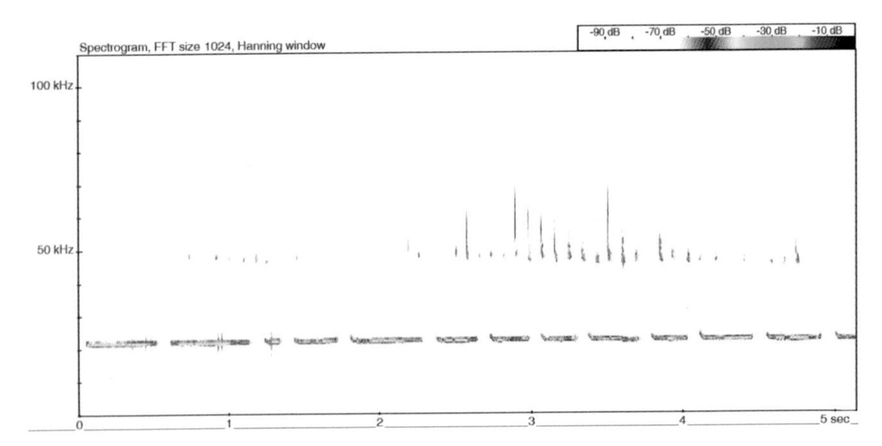

Figure 2.8 🔊 Brown rat – distress call (*c.* 22 kHz), with common pipistrelle echolocation above (S Newson, 2018; frame width 5000 ms).

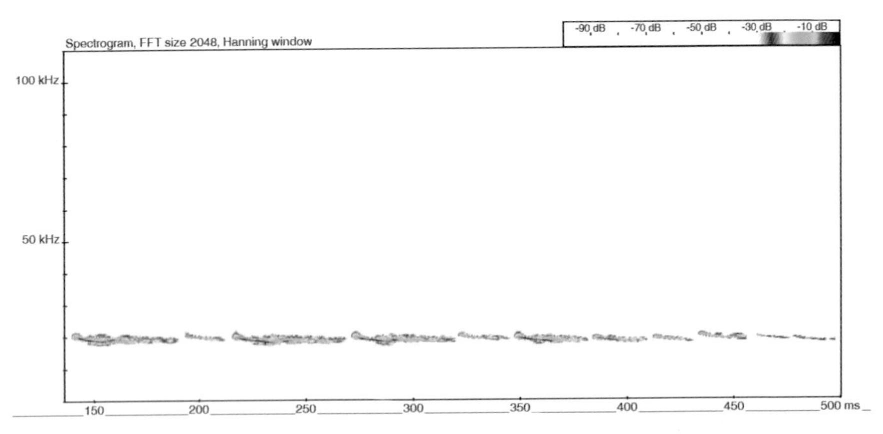

Figure 2.9 Noctule – open-habitat echolocation (compressed; frame width 500 ms).

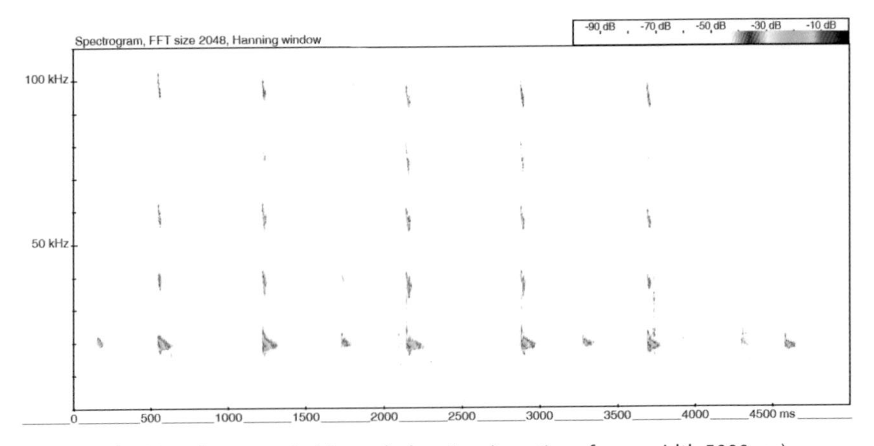

Figure 2.10 🔊 Noctule – open-habitat echolocation (true time; frame width 5000 ms).

Mouse species

Ultrasonic sounds made by mice, like those of rats, have been researched and reported upon many times (Coffey *et al.* 2019). There are numerous studies for researchers to refer to, in particular relating to house mouse (e.g. von Merten *et al.* 2014, Lupanova & Egorova 2015, to name only two of so many). The other mouse species that occur within our area have not had anywhere close to the same amount of attention regarding the study of their vocalisations.

As with rats, the occurrence of emitted sound is dependent on the animal's age and sex, and also the presence or absence of conspecifics, and their age and sex (Wöhr & Schwarting 2013). As well as ultrasonic sound, all of the mouse species discussed here make audible noise.

Not much research of any value relevant to this book exists for harvest mouse, although I was fortunate enough to obtain some interesting calls of these myself. Regarding wood mouse and yellow-necked mouse, some earlier research looked at the ultrasonic behaviour of these species (Hoffmeyer & Sales 1977, Gyger & Schenk 1983, 1984). More recently a study of their distress calls (carried out in Italy) demonstrated a very high rate (98%) of accurate identification, differentiating these two species from each other 'based on a combination of acoustic variables and body mass' (Ancillotto *et al.* 2017). The study established that the approach taken by its authors was more reliable than morphology alone. They used a Pettersson D500X bat detector, carrying out analysis using BatSound software. It is not uncommon for researchers studying bioacoustics in small terrestrial mammals to use bat detectors and related software. Call parameters being measured were similar to those we would use when measuring bat calls. Table 2.4 provides more information about calls recorded in the mouse species occurring in our part of the world.

Table 2.4 Mouse species occurring in the British Isles.

Species	Distribution and typical habitat	Notes on communication/recordings
Harvest mouse *Micromys minutus*	Occurs throughout much of England and Wales, and also southern Scotland. Absent from Ireland[1][2] Arable and horticulture Unimproved grassland Hedgerow[2]	🔊 Compared to other species discussed, fewer studies have been documented regarding bioacoustics in this species. Audible calls have been noted during courtship and mating behaviour, as well as between conspecifics.[1] Ultrasonic sound up to 114 kHz has also been recorded.[3]
Wood (long-tailed field) mouse *Apodemus sylvaticus*	Occurs throughout Britain and Ireland[2][4] Broadleaved woodland Coniferous woodland Unimproved grassland Urban and gardens Hedgerow[2]	🔊 Ultrasonic sound known to be emitted during exploration, mating, chase sequences, aggressive behaviour and contact with conspecifics.[5][6] Exploratory calls are higher in frequency than in yellow-necked mouse, usually in the region of 70 kHz, and occurring relatively more often.[7] Related to mating behaviour, known to make arched calls (upside-down 'V' shaped) in a series, with maximum frequency of *c.* 90–100 kHz.[7] Audible calls also known from distressed animals.[4][8]

(cont).

Table 2.4 (cont).

Species	Distribution and typical habitat	Notes on communication/recordings
Yellow-necked mouse *Apodemus flavicollis*	Occurs in southern England and Wales. Absent from Scotland and Ireland[2][9] Broadleaved woodland Urban and gardens Coniferous woodland Hedgerow[2]	🔊 Known to make audible and ultrasonic noise (adults, 40–50 kHz).[5][9] Exploratory calls are lower in frequency than wood mouse, usually in the region of 40–50 kHz, and occurring relatively less often.[7] Audible calls also known from distressed animals.[8]
House mouse *Mus musculus*	Occurs throughout Britain and Ireland[2][10] Buildings[2]	🔊 Known to make audible noise (e.g. distress) and ultrasonic noise across many social contexts, including during mating and contact with conspecifics.[10][11][12][13] Call structure can be complex and variable.[12][13][14] Calls used during courtship may also serve a territorial function.[14]

References

[1] Trout & Harris 2008; [2] Mathews *et al.* 2018; [3] Trout 1978; [4] Flowerdew & Tattersall 2008; [5] Stoddart & Sales 1985; [6] Sales 2010; [7] Gyger & Schenk 1983; [8] Ancillotto *et al.* 2017; [9] Marsh & Montgomery 2008; [10] Berry *et al.* 2008; [11] Lupanova & Egorova 2015; [12] von Merten *et al.* 2014; [13] Lahvis *et al.* 2011; [14] Hammerschmidt *et al.* 2012.

We will now look at some spectrograms for each of the species discussed in Table 2.4, as well as consider any potential confusion points when considering bat call analysis.

Harvest mouse

Figures 2.11 to 2.14 show some typical examples of ultrasonic sound of harvest mouse (*Micromys minutus*) that I regularly recorded during *c.* 15 hours of studying a captive and established colony. Figure 2.15 gives an example of an audible sound. Additional examples are provided in the Sound Library.🔊

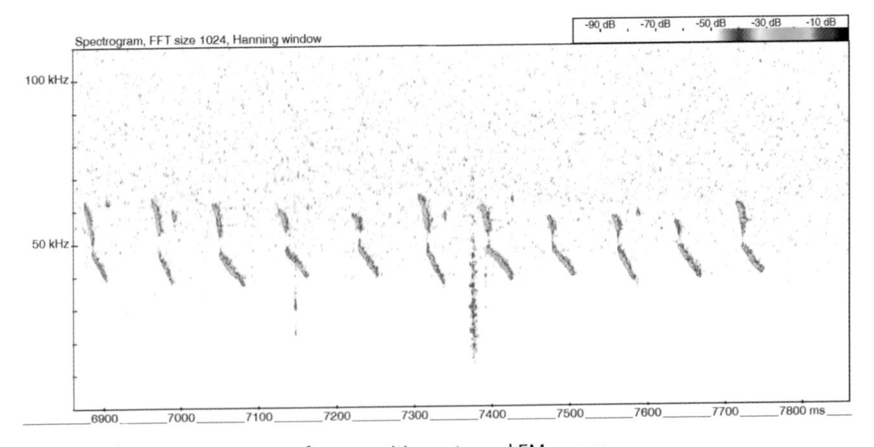

Figure 2.11 🔊 Harvest mouse – fast repetition rate and FM sweep.

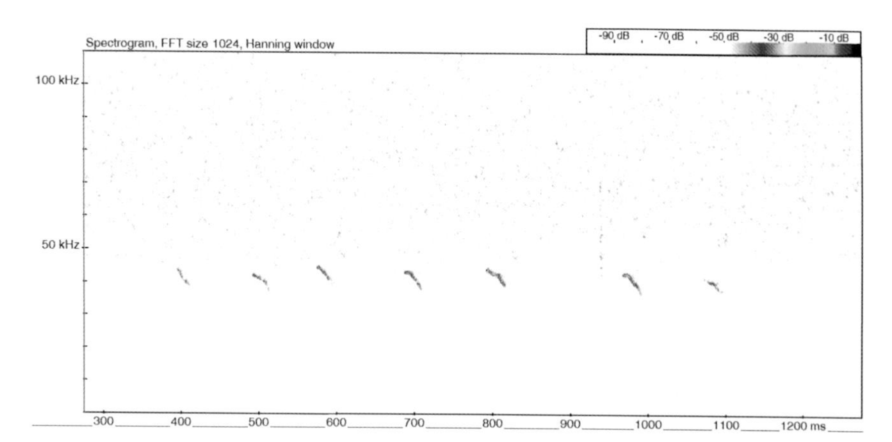

Figure 2.12 Harvest mouse – FM downward, slightly humped in appearance.

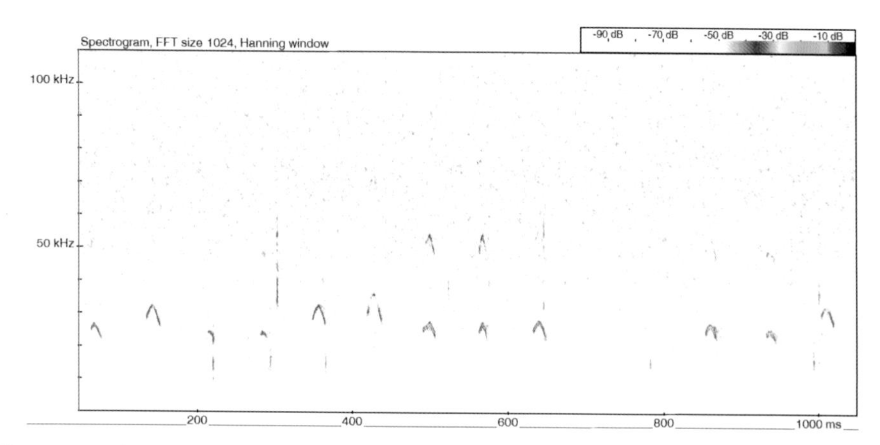

Figure 2.13 Harvest mouse – arched FM.

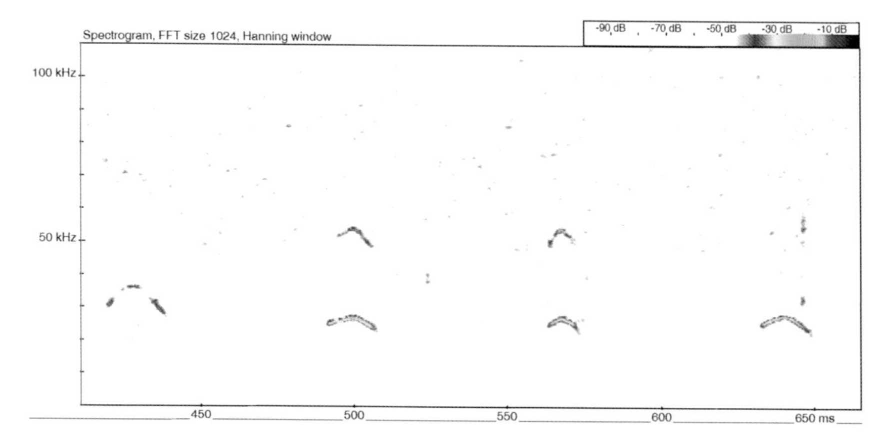

Figure 2.14 Harvest mouse – arched FM (frame width 250 ms).

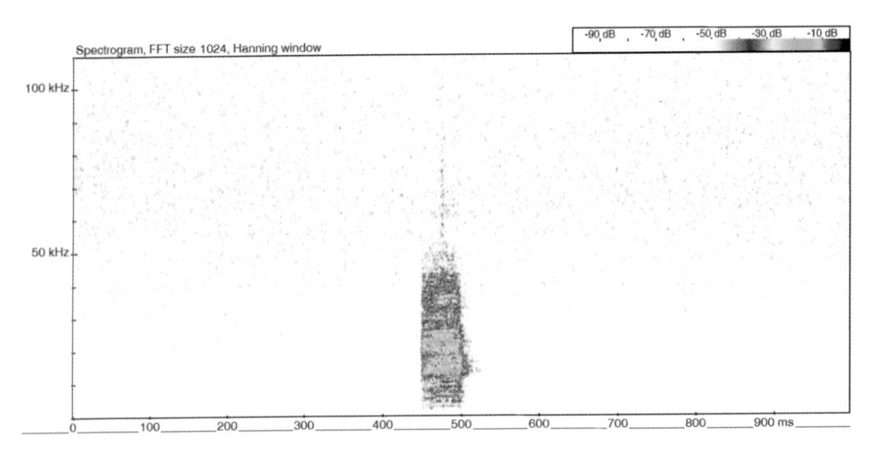

Figure 2.15 🔊 Harvest mouse – audible call.

Differentiating harvest mouse from bat echolocation and Daubenton's social calls

Of all the rodent species, this one in particular surprised me as to its potential to be confused with bats generally, and specifically in relation to a couple of scenarios. Bear in mind that here we only have one species of mouse, with recordings from a single colony over a short period of time. It would be foolish of me to suggest that these are going to be the only types of call produced, by any species of mouse, that could be confused with bats at some level.

The call type shown in Figure 2.11 was encountered on numerous occasions, and I think most of us would agree that to an untrained eye, glancing at it quickly, it looks a little like bat echolocation. A more experienced eye would immediately think that it doesn't look quite right in terms of structure, albeit the repetition rate (*c.* 12 per second) is echolocation-like. Ignoring the human factor for a moment or two, what would an automated classifier say? I chose three of the more regularly used ones, all of which told me, with a high degree of confidence, that it was a bat (see Table 2.10), the most likely contenders being common pipistrelle (Figure 1.4) and soprano pipistrelle (Figure 2.16). So, potentially these calls could get lumped into a pipistrelle dataset without being noticed, if the person involved relied on a filter or a classifier without carrying out an audit. Who would question someone for not auditing pipistrelle calls? We are, after all, often given the impression that call classifiers are pretty accurate when it comes to this species group. The pulse rate is right, the frequencies are right, the call structure is sort of right-ish. Please recall, however, that classifiers can only allocate calls based on comparisons within their call library (see Chapter 1), and as such a call library bat example is going to be a far better fit than the noise file option.

Next, we have Figure 2.12. This type of call was noted a few times during recording sessions. It is a little barbastelle-like, if it was being considered that it was a sequence showing the higher of the two alternate calls (type A and type B) these bats can produce when echolocating (Russ 2012, Barataud 2015).

Compare it against Figure 2.17, which shows a classic barbastelle echolocation sequence. So, what about an automated classifier (see Table 2.10)? What would that process tell us in this instance? Well, not a barbastelle, but a common pipistrelle (Figure 1.4), or perhaps Nathusius' pipistrelle (Figure 2.23).

Finally, just to really kick us when we are down, Figures 2.13 and 2.14. It looks a bit like a Daubenton's bat social call (Figure 2.18). As bat workers we all know confidently what they look like. Nothing else looks like a Daubenton's hooked social call. Well perhaps not another bat species, but I would suggest an extra degree of caution from now on. In fact, within weeks of recording these calls I was sent a Daubenton's bat 'hook' by an enquirer, and I had to tell them that I didn't feel on that occasion it was. And to make matters worse, now go and check Figures 2.19, 2.56, 2.60 and 2.61. As I said earlier, these mice are just like bats, but without wings. Also, as I have said previously, the research that unearthed these calls was very limited, and I would not be confident that what is shown in this chapter is anywhere close to the full picture in any respect. I am just trying to open eyes, highlighting potential scenarios that could throw people off in the wrong direction, in the same way that my eyes were opened when I first started encountering examples such as these.

With this in mind, how would we go about trying to resolve whether a call being looked at is a Daubenton's bat or indeed a similarly structured call but from something else? In Appendix 2 I have provided a process that could be followed when faced with such circumstances.

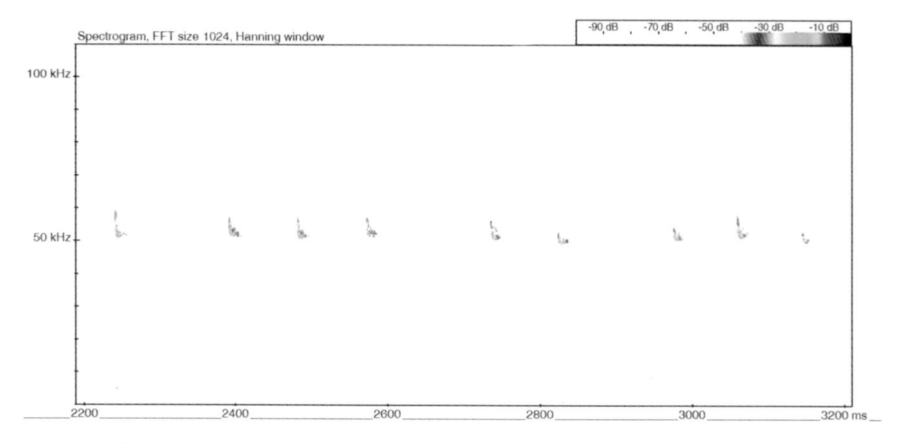

Figure 2.16 🔊 Soprano pipistrelle – echolocation.

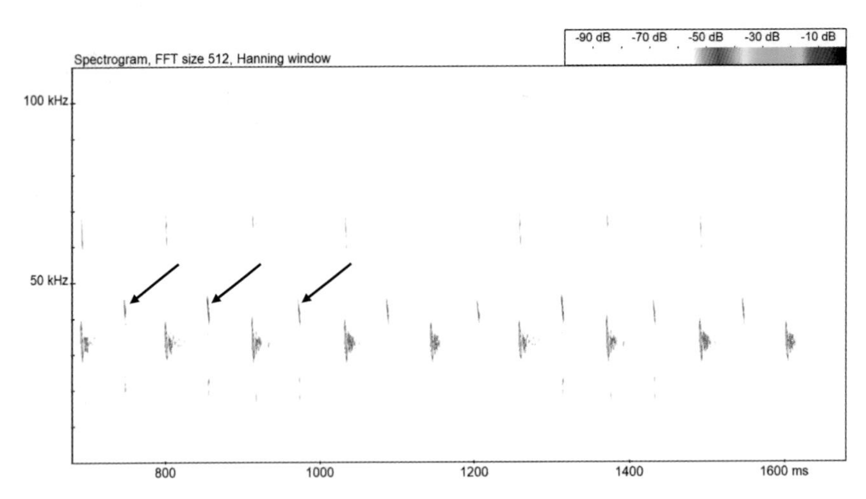

Figure 2.17 Barbastelle – echolocation, higher alternate call marked with arrows (K French, 2013).

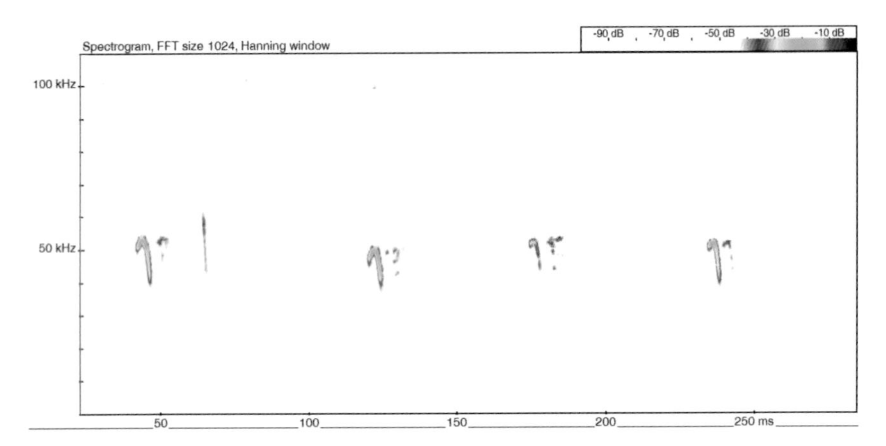

Figure 2.18 Daubenton's – hooked social call (frame width 250 ms).

Wood mouse

Figures 2.19 to 2.22 show some typical examples of ultrasonic sound of wood mouse (*Apodemus sylvaticus*) that I regularly recorded during *c.* 15 hours of studying a captive and established colony. Additional examples are provided within the Sound Library.🔊

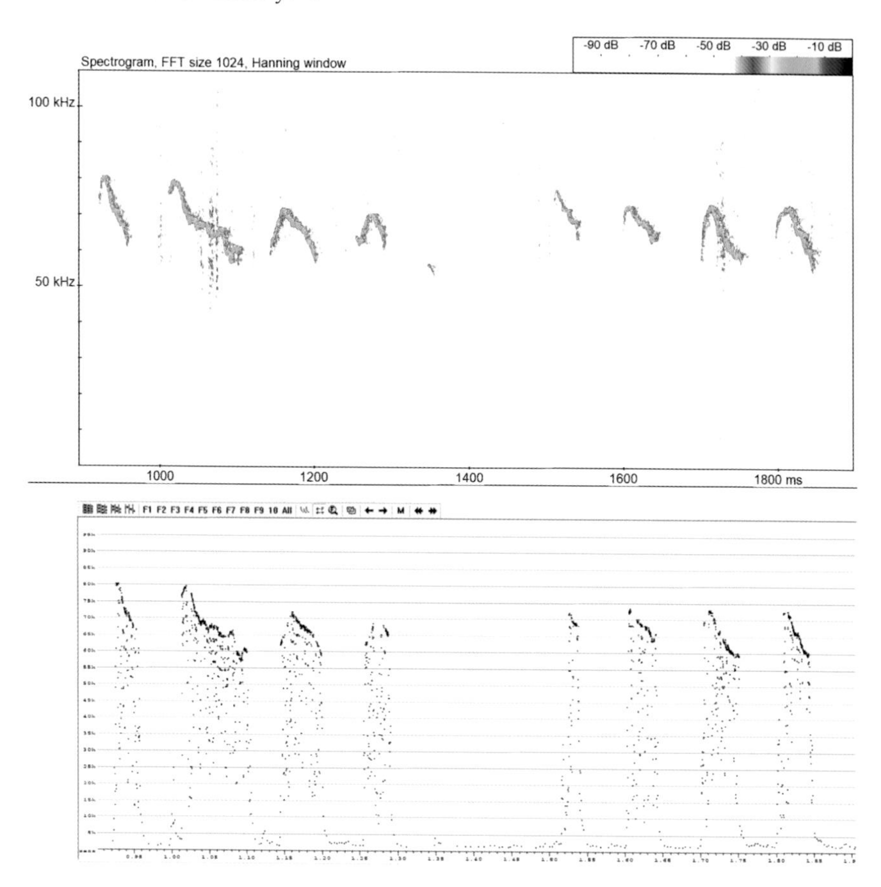

Figure 2.19 🔊 Wood mouse, with AnaLookW ZCA example below.

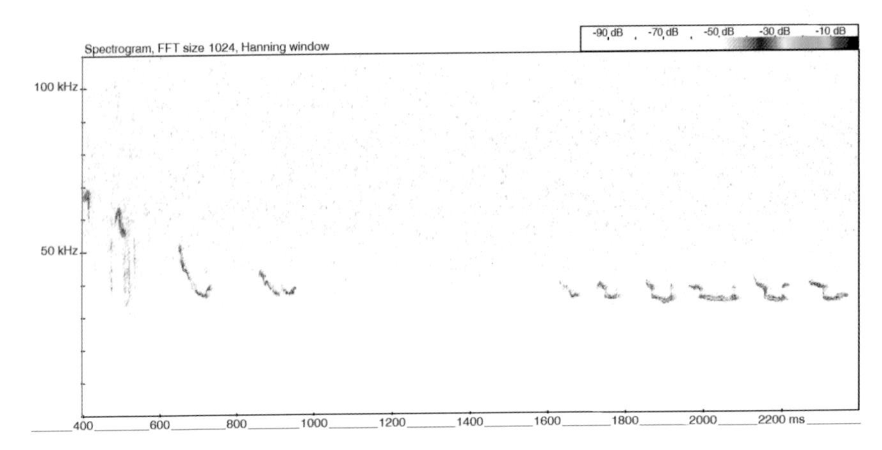

Figure 2.20 Wood mouse (frame width 2000 ms).

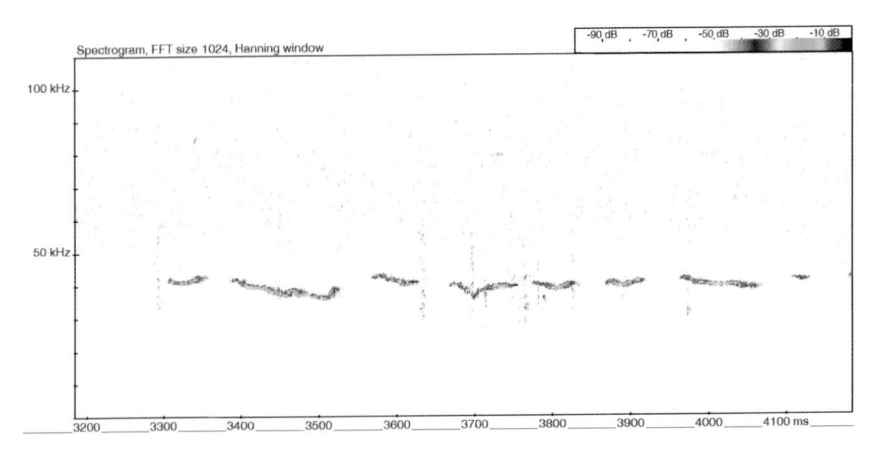

Figure 2.21 Wood mouse.

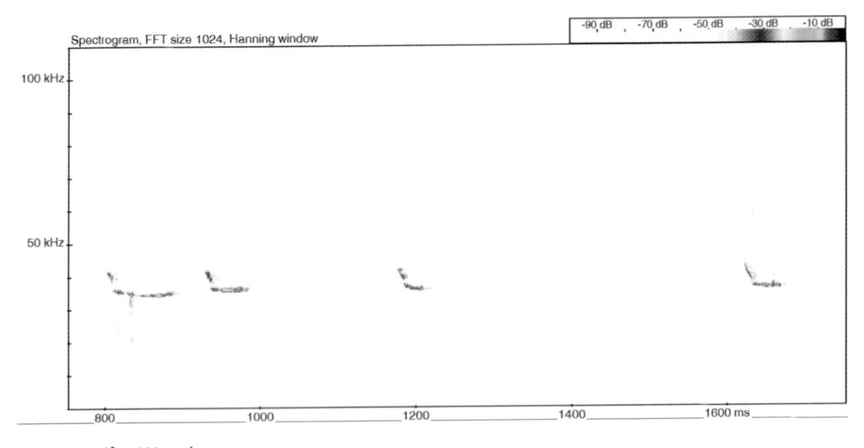

Figure 2.22 Wood mouse.

Differentiating wood mouse from Nathusius' pipistrelle echolocation and brown long-eared type D1 social call

Figures 2.21 and 2.22 have minimum frequencies in an area you would perhaps expect Nathusius' pipistrelle to occur (Figures 2.23 and 2.24). In Figure 2.21 these calls are relatively more CF, as you would normally expect from a bat in open habitat, but, amongst other things, the pulse rate is too fast (> 8 pulses per second), and the call duration is too long. Apart from that, the structure is variable, and in many respects not what you would expect from an echolocating bat. Then in Figure 2.22 the call structure is almost pipistrelle-like, in that there is a downward FM sweep that then flattens to CF at the end of each pulse. Again, however, it just doesn't look quite right to an experienced person.

Let us now consider a brown long-eared bat type D1 social call (Middleton *et al.* 2014). Compare Figure 2.19 (wood mouse) with Figure 2.25 (brown long-eared bat). The bat call is quite distinctive, and there are some classic brown long-eared pulses included within the sequence. Also, bear in mind when comparing these two examples that the scales on the y-axis differ from each other.

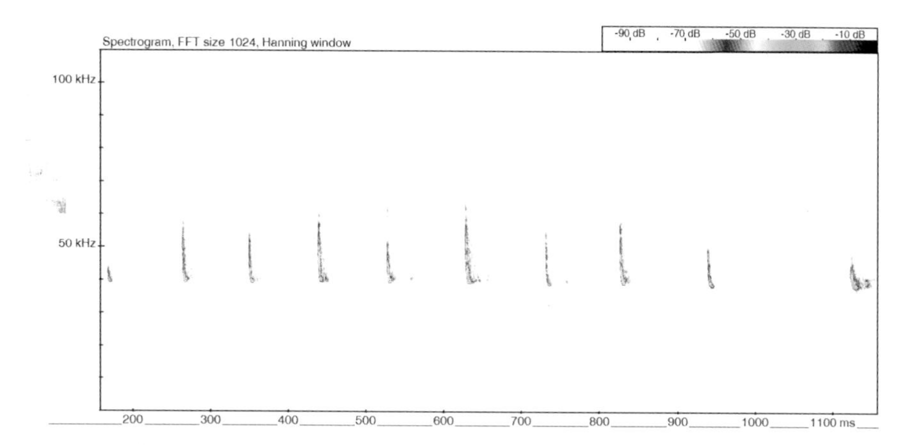

Figure 2.23 🔊 Nathusius' pipistrelle – edge-habitat echolocation.

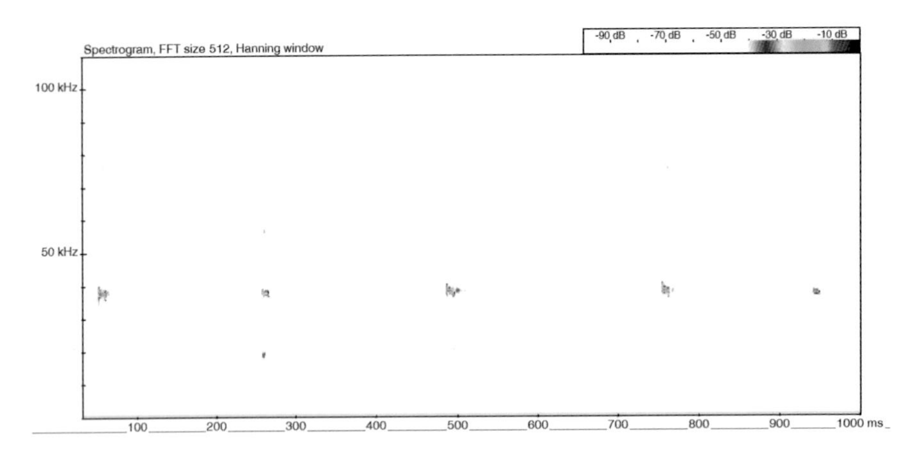

Figure 2.24 Nathusius' pipistrelle – open-habitat echolocation (A Froud, 2008).

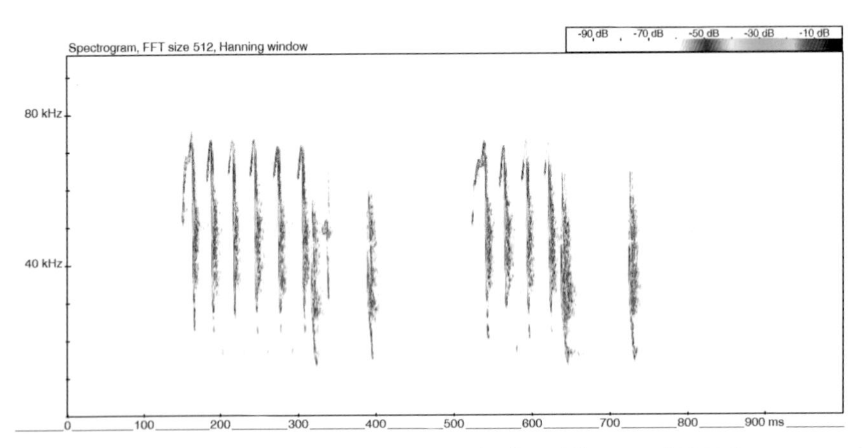

Figure 2.25 🔊 Brown long-eared bat – type D1 social call (Middleton *et al.* 2014; frequency scale 0–96 kHz).

Yellow-necked mouse

Figures 2.26 to 2.30 show some typical examples of ultrasonic sound of yellow-necked mouse (*Apodemus flavicollis*) that I regularly recorded during *c.* 15 hours of studying a captive and established colony. Additional examples are provided in the Sound Library.🔊

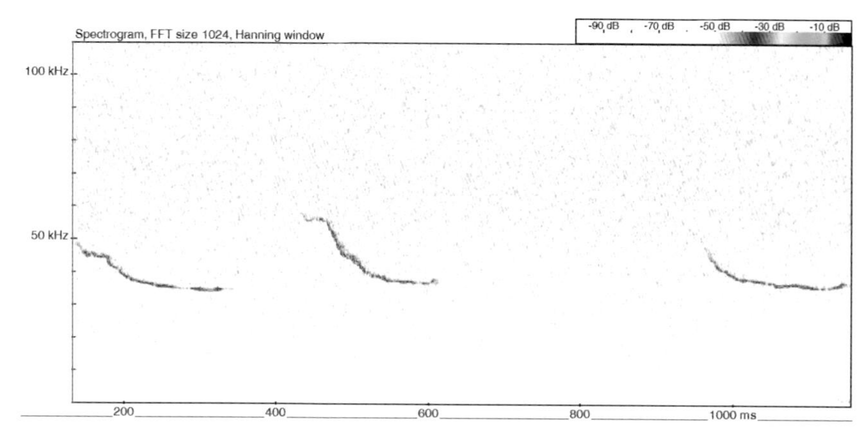

Figure 2.26 🔊 Yellow-necked mouse.

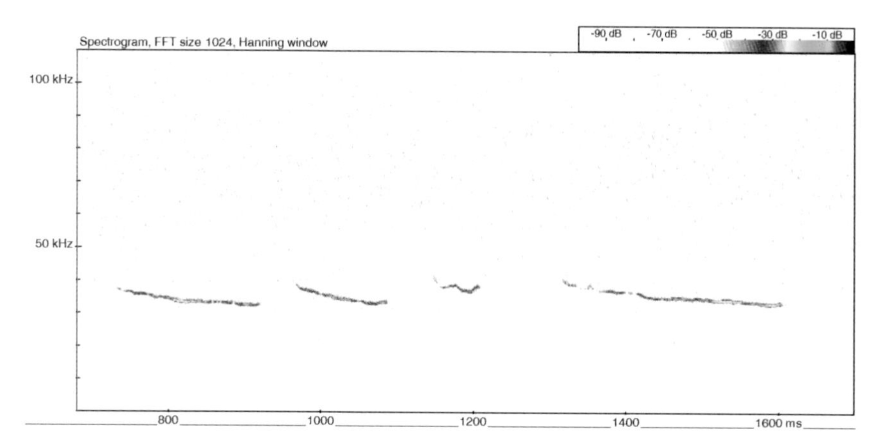

Figure 2.27 🔊 Yellow-necked mouse.

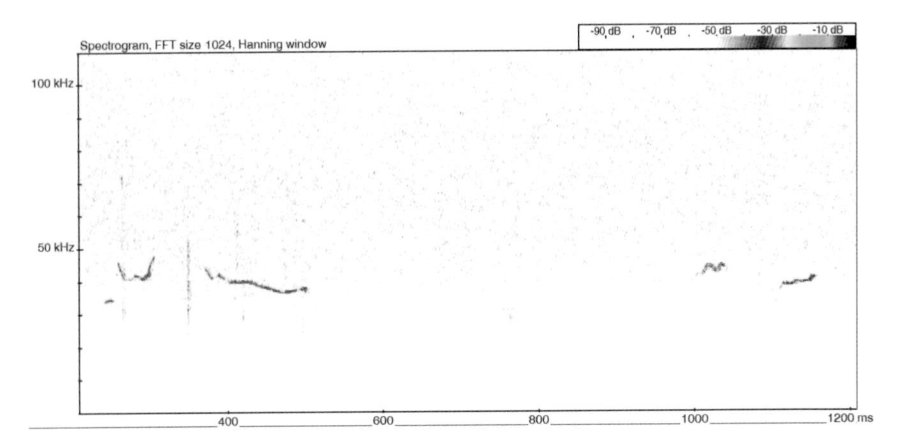

Figure 2.28 🔊 Yellow-necked mouse.

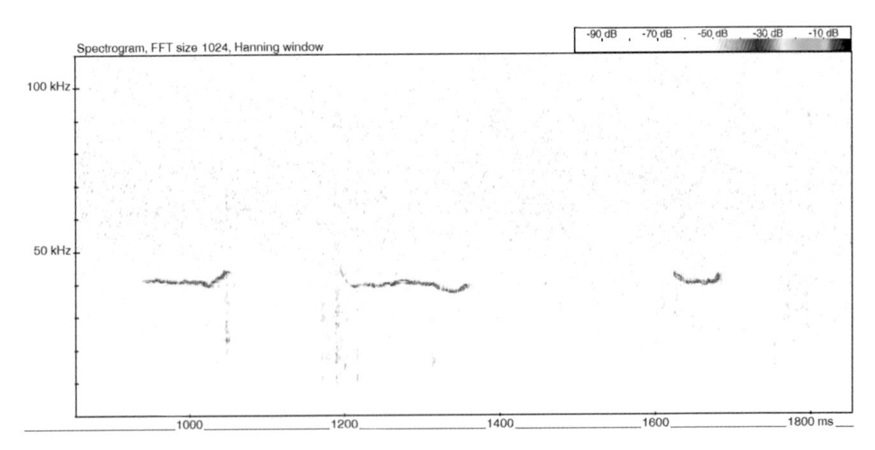

Figure 2.29 🔊 Yellow-necked mouse.

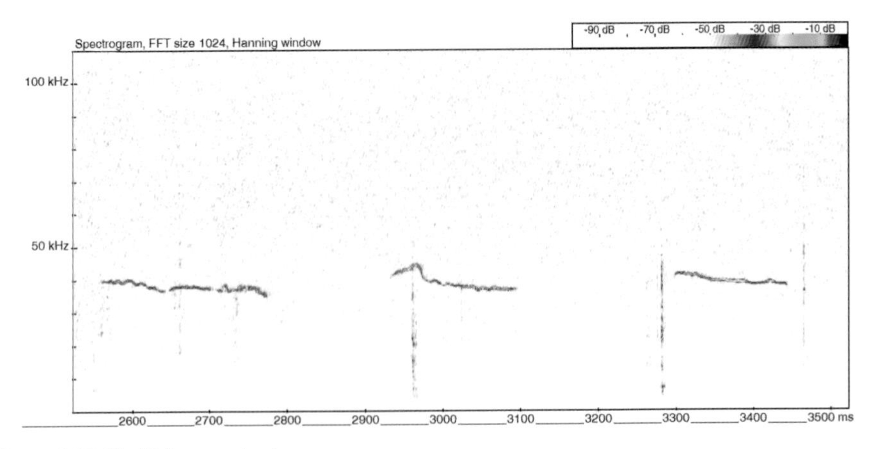

Figure 2.30 🔊 Yellow-necked mouse.

House mouse

Over time, the house mouse (*Mus musculus*) has become a regularly researched rodent when it comes to studying ultrasonic vocalisation. Figure 2.31 shows a typical example of relatively high ultrasonic sound (compared to other mouse species) that I regularly recorded. Based on my recordings, these calls occurred at higher frequencies than the other mouse species recorded during this project. Figure 2.32 gives an example of an audible sound. Additional examples of each of these call types are provided in the Sound Library.🔊 In addition to these call types, it should be expected that other call structures will occur.

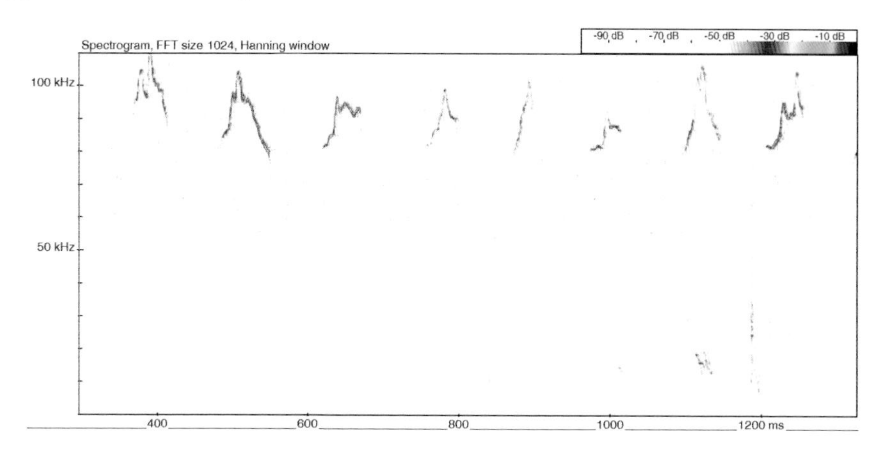

Figure 2.31 🔊 House mouse – high frequency, FM arched.

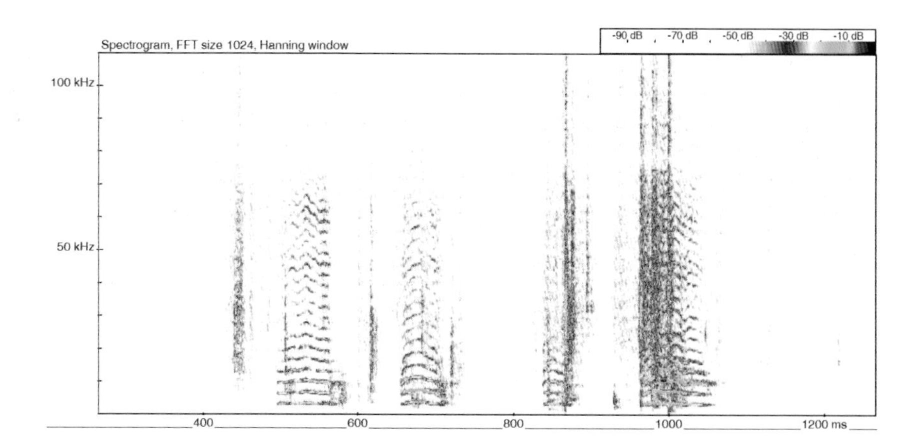

Figure 2.32 🔊 House mouse – audible call.

Shrew species

Three species of shrew occur within mainland Britain and one in Ireland, as shown in Table 2.5. As well as using sound for communication purposes (Simeonovska-Nikolova 2004), shrews have been shown to use acoustics for echolocation. An interesting piece of research that demonstrated this established, in laboratory conditions, that common shrews 'used broadband echolocation pulses to locate protective cover', as well as to have the potential to detect obstructions within tunnel systems (Forsman & Malmquist 1988). These calls were described as being effective up to a distance of 200 mm. The research carried out for this study reinforced similar research carried out on four North American shrew species (Gould *et al.* 1964). Another piece of research reported upon in 2009 also used common shrew (as well as white-toothed shrew, *Crocidura russula*) in its experiments, and demonstrated echolocation functioning in the behaviour of these species, describing shrew echolocation as not being as sophisticated as in bats, but using 'call reverberations for simple, close-range spatial orientation' (Siemers *et al.* 2009). In their discussion, the authors suggested that this mechanism would be important for route planning, barrier avoidance and establishing escape routes. They also discussed that it was not likely that the echolocation served any purpose in hunting for prey. As such, the term 'echo-orientation' has been used to describe how shrews use sound to navigate within their environment (von Merten 2011).

A further study which explored the differences in call structure of six different shrew species in Europe (Zsebok *et al.* 2015) went on to suggest that these twittering calls may have a dual function (echo-orientation and conflict avoidance), but more studies in this respect were needed. This same piece of research demonstrated that although, taking numerous measurements, it was possible, to a degree, to separate calls to species level (with *c.* 66% accuracy), there was a wide range of overlap between the species studied, including the three species discussed in Table 2.5. It is interesting that Siemers *et al.*, von Merten and Zsebok *et al.* all failed to find any evidence of the click calls described by earlier researchers, the suggestion being that more modern technology, now being used in research, is finding calls and measuring their parameters more accurately than earlier researchers would have ever been able to achieve.

Table 2.5 Shrew species occurring in the British Isles.

Species	Distribution and typical habitat	Notes on communication
Common shrew *Sorex araneus*	Occurs throughout Britain; absent from Ireland[1][2] Bog Unimproved grassland Urban and gardens Broadleaved woodland Hedgerow[2]	Known to produce high-pitched, barely audible calls, when in contact (i.e. communication) with conspecifics and heterospecifics, or when in distress. More often heard during the breeding season, April to August.[1][3] Echolocation initially described as ultrasonic clicks in earlier studies (perhaps attributable to constraints in technology[3]). More recently described as audible (sonic) twittering calls, variable in structure.[3][4][5]
Pygmy shrew *Sorex minutus*	Occurs throughout Britain and Ireland[2][6] Unimproved grassland Broadleaved woodland Bog Urban and gardens Hedgerow[2]	Less vocal than common shrew, but known to emit a short audible 'chit' when alarmed or in contact with conspecifics.[6] Echolocation described as audible twittering calls, variable in structure.[3][5]
Water shrew *Neomys fodiens*	Occurs throughout Britain; absent from Ireland[2][7] Rivers and wetland[2]	Heard more frequently during summer months, when it is noted for making audible squeaks and a rolling 'churr-churr' during threat or alarm behaviour.[7] Echolocation described as audible twittering calls, variable in structure.[5]
References [1] Churchfield & Searle 2008a; [2] Mathews *et al.* 2018; [3] von Merten 2011; [4] Siemers *et al.* 2009; [5] Zsebok *et al.* 2015; [6] Churchfield & Searle 2008b; [7] Churchfield 2008.		

The naming of the shrew

Referring to Table 2.5, it may very well be possible to increase your level of confidence in species identification based on where you are geographically. For example, only pygmy shrew is present in Ireland. Please note, however, that an 'invasive' species, greater white-toothed shrew, has been recorded and is expanding its range within Ireland. Other than this, actually putting the precise species name confidently against a recording with absolute certainty, based on calls alone, is problematic. The previously referenced Zsebok *et al.* (2015), which achieved high degrees of accuracy, suggested that there were wide degrees of overlap adopting the methods they tested.

We are now going to look at some spectrograms produced by this species group as a whole, as well as one particularly frustrating confusion point with a bat (noctule fast trill social call) that certainly had me and a few others perplexed in the lead-up to preparing this book. Figures 2.33 to 2.36 show some typical examples of the shrew calls most regularly picked up by bat detectors or heard audibly. In my experience, calls such as these are audible easily from *c.* 3 m away, and at this distance a bat detector picks them up clearly, as was the case regarding the call shown in Figure 2.33. Additional examples are provided in the Sound Library.

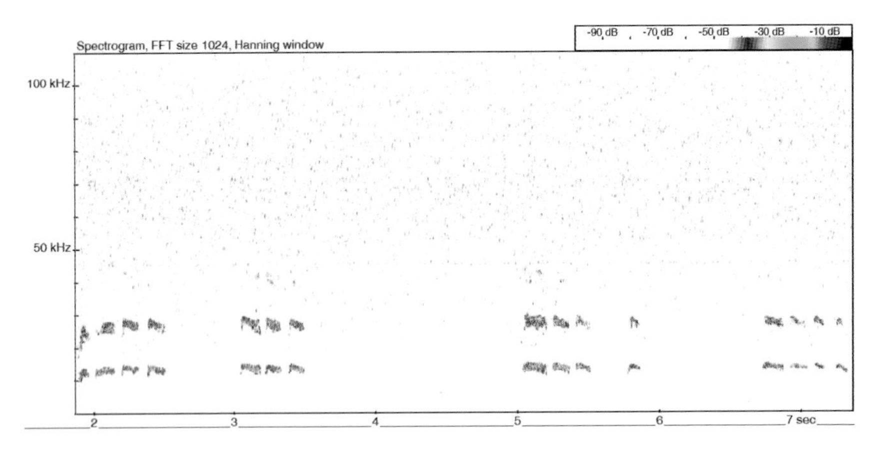

Figure 2.33 Shrew – trill call (frame width *c.* 5000 ms).

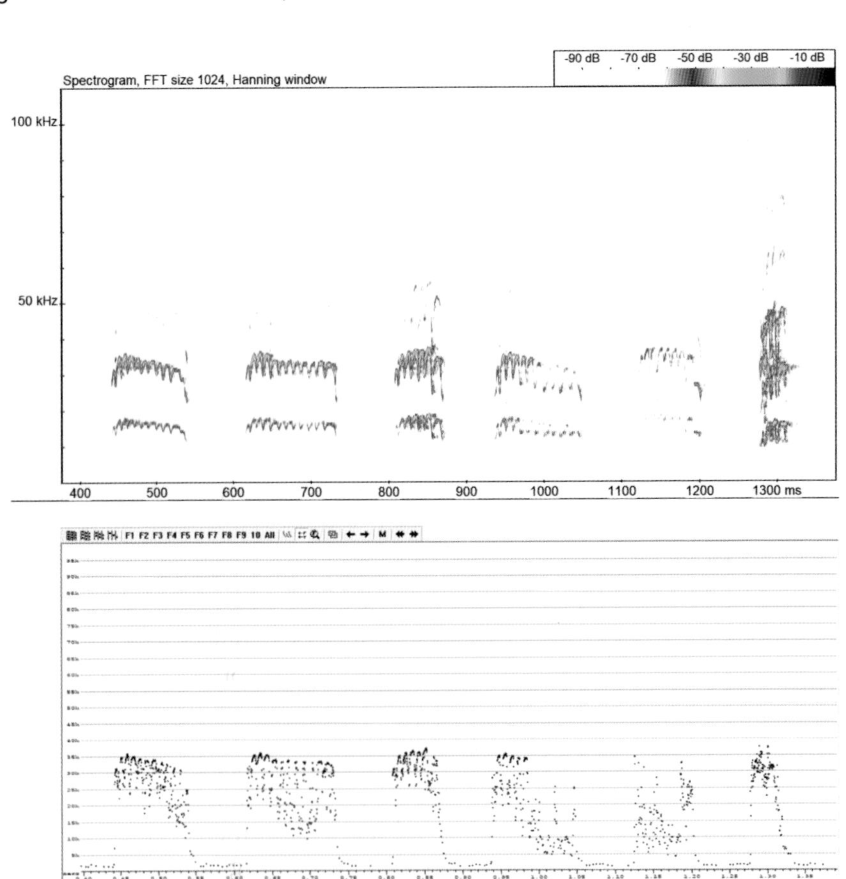

Figure 2.34 Shrew – trill, with AnaLookW ZCA example below (G Slack and D Wildsmith, 2017).

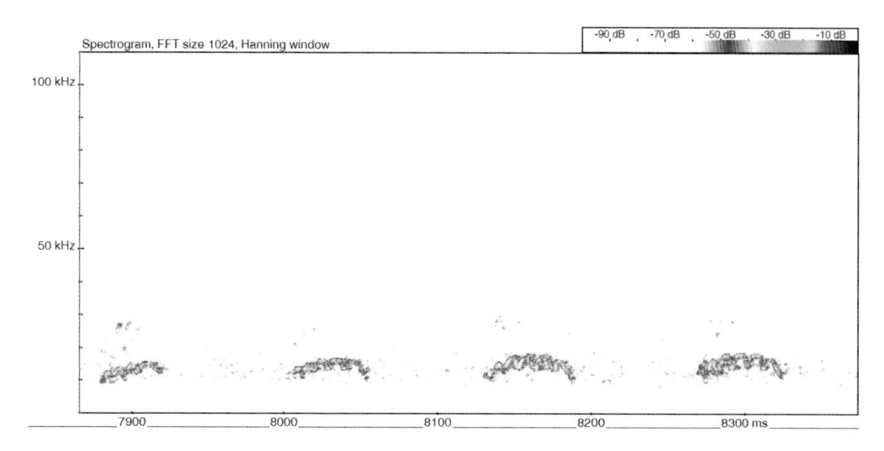

Figure 2.35 Shrew – trill call 1 (A Ellis, WSP, 2018; frame width 500 ms).

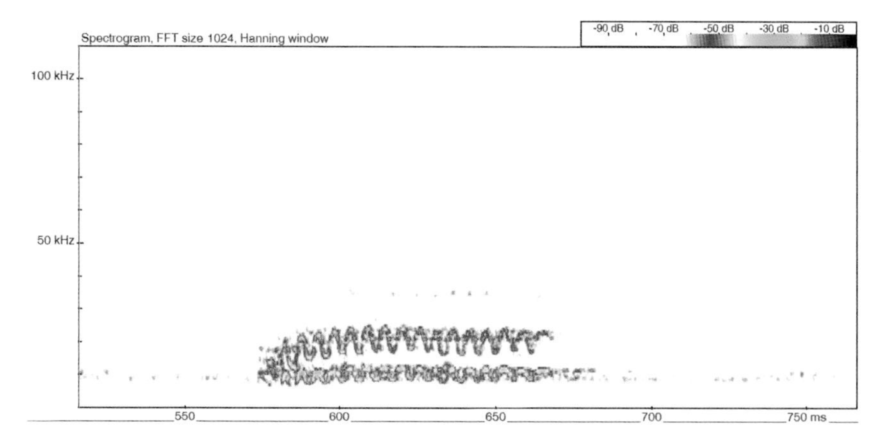

Figure 2.36 Shrew – trill call, thought to be water shrew (H Lehto, Finland, 2015; frame width 250 ms).

Differentiating shrew calls from noctule fast trill

The issue in this case is that these trill-type calls produced by shrews (Figures 2.33 to 2.36), and so often picked up by bat detectors, look and sound (in time expansion) extremely like noctule fast trill calls (Middleton *et al.* 2014), as shown in the example in Figure 2.37. Considering what we have here, there are a few things that may help in differentiating shrews from noctule fast trills. First of all, quite often the shrew trills are repeated, and occasionally grouped. Secondly, if it was a shrew call being looked at, then there is a fair chance that there would not be any associated 'noctule-type' echolocation in the vicinity of the call. Unless, that is, unluckily, a noctule happened to be flying past at the same time. Thirdly, with noctule fast trills, there would usually be something else going on in conjunction with the fast trill (e.g. preceded by a slow trill, as in Figure 2.37), and it would be unusual to see noctule fast trills repeated at regular intervals. I would like to think that it will be possible to separate them more convincingly using call parameters etc., but at this stage I have not seen enough examples of these, in either camp, to develop something convincing. This isn't helped by what appears to be a wide range of variability between individual animals among both noctules and shrews.

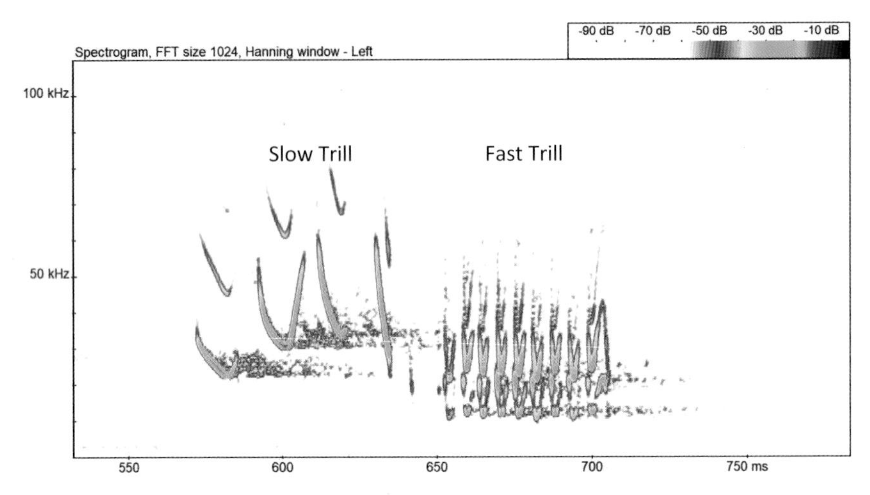

Figure 2.37 🔊 Noctule – 'slow trill' followed by 'fast trill' call (frame width 250 ms).

Shrew versus bat – which is the smallest?

Shrews quite often come up in bat-related talks, when we discuss Kitti's hog-nosed bat (*Craseonycteris thonglongyai*) as being the smallest bat on the planet, and arguably the smallest mammal. The latter is a bone of contention, owing to competition from the smallest species of shrew (Jürgens 2002).

The Etruscan shrew (*Suncus etruscus*), which occurs as close to us as southern Europe, is regarded as the smallest mammal by mass, weighing in at *c.* 1.8 grams. This small shrew's distribution range goes far to the east, including Thailand, where it meets its competitor for the Light Flyweight title, Kitti's hog-nosed bat, also known as the bumblebee bat, which typically weighs 2.0 grams. This bat is regarded as the smallest mammal when considering measurements of skull size and body length. All of which is not really relevant to the subject matter of this book, but more of an interesting diversion that may be of interest to bat workers.

Vole species

Three species of vole occur within mainland Britain. Voles are absent from Northern Ireland. Within the Republic of Ireland, only bank vole occurs (in the southwest). Table 2.6 describes these in more detail, and thereafter spectrograms (Figures 2.38 to 2.51) describe some typical calls recorded by myself, as well as a couple of water vole contributions from Craig Macdonald (CSM Ecology & Environmental Assessment Ltd). Additional call examples are provided in the Sound Library.🔊

Some of the calls described here have been known to be labelled by automated classifiers as Leisler's bat. The water vole calls shown in Figures 2.46 and 2.48 are good examples of false Leisler's bat, where the classifier was very confident it was a bat.

Table 2.6 Vole species occurring in the British Isles.

Species	Distribution and typical habitat	Notes on communication
Bank vole *Myodes glareolus*	Occurs throughout Britain, including many surrounding islands. Present in southwest of Ireland[1][2] Broadleaved woodland Urban and gardens Coniferous woodland Dwarf shrub heath Hedgerow[2]	🔊 Adults are known to produce high-pitched sounds (e.g. between males during aggressive encounters),[3] as well as a sound described as teeth chattering.[4] High-frequency calls have also been described during mating.[5] In other studies, adults have been described as producing both ultrasonic and audible sounds.[6][7][8] Ultrasonic calls occur at a minimum frequency of 24/30 kHz, lasting approximately 61/70 ms. These calls are lower than those of field vole. As well as this, bank vole are generally less vocal.[5] Audible sounds have a longer duration (*c.* 50–113 ms), occurring at a minimum frequency of 1.5–2.7 kHz. Infants are known to make noise while in their nest, these sounds being both audible (< 8 kHz) and ultrasonic (*c.* 20–35 kHz).[4][9]
Field vole *Microtus agrestis*	Occurs throughout Britain, including many surrounding islands. Absent from Ireland[2][10] Unimproved grassland Urban and gardens Broadleaved woodland Coniferous woodland Hedgerow[2]	🔊 Adults have been described as producing both ultrasonic and audible sounds.[4][7] Ultrasonic calls occur at a minimum frequency of 42/46 kHz, lasting approximately 61/73 ms. High-frequency calls have also been described during mating,[5] and are considerably higher than those recorded from bank voles. Field vole is generally far more vocal than bank vole.[5] Audible sounds have a longer duration (82–138 ms), occurring at a minimum frequency of 2.2–3.2 kHz.
Water vole *Arvicola amphibius*	Occurs throughout Britain, including many surrounding islands. Absent from Ireland[2][11] Rivers and wetlands[2]	🔊 During agonistic encounters this species will make irregular calls at frequency range from 2.5 kHz to 4.4 kHz, with duration of *c.* 100 ms.[11] The audible plop noise heard when this species dives is thought to act as a warning to conspecifics.[12]

References

[1] Shore & Hare 2008; [2] Mathews *et al.* 2018; [3] Miska-Schramm *et al.* 2018; [4] Stoddart & Sales 1985; [5] Kapusta & Sales 2009; [6] Kapusta *et al.* 2007; [7] Sales 2010; [8] Osipova & Rutovskaya 2000; [9] Marchlewska-Koj 2000; [10] Lambin 2008; [11] Woodroffe *et al.* 2008; [12] Strachan 1999.

Bank vole

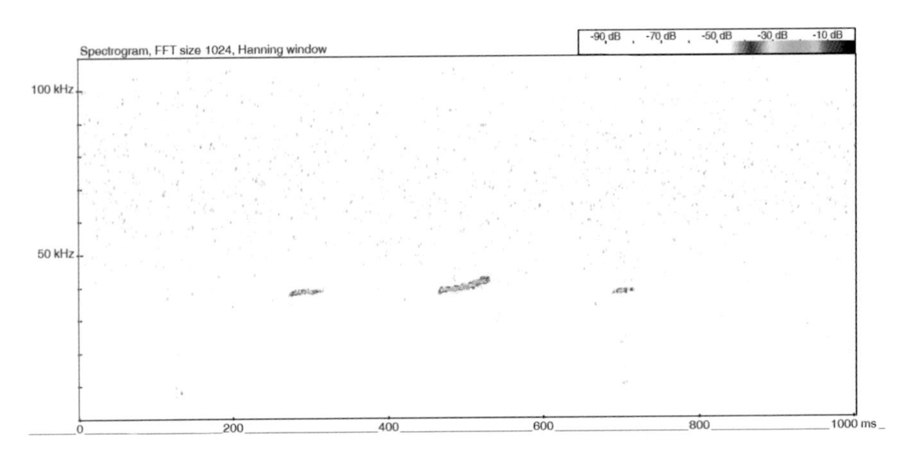

Figure 2.38 🔊 Bank vole – near-CF calls occurring at *c.* 40 kHz.

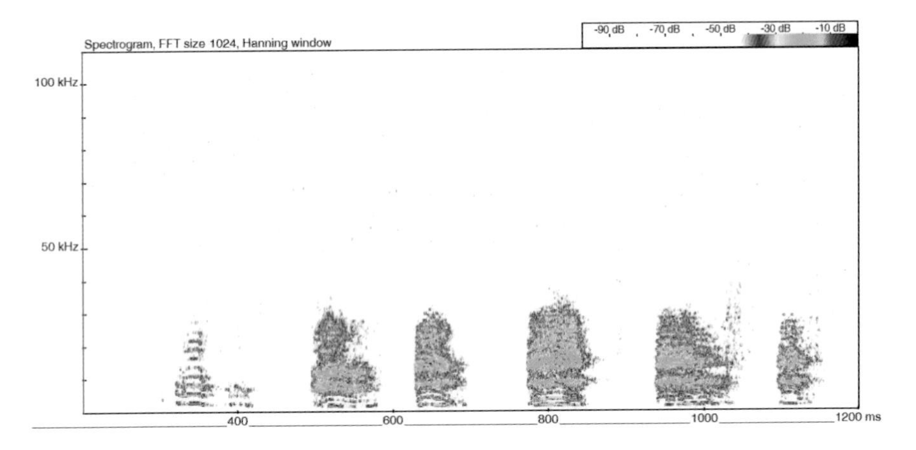

Figure 2.39 🔊 Bank vole – audible calls.

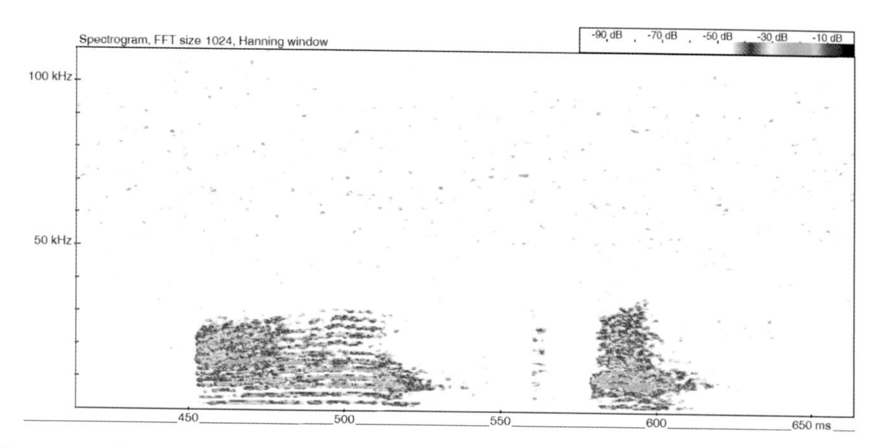

Figure 2.40 Bank vole – audible calls (frame width 250 ms).

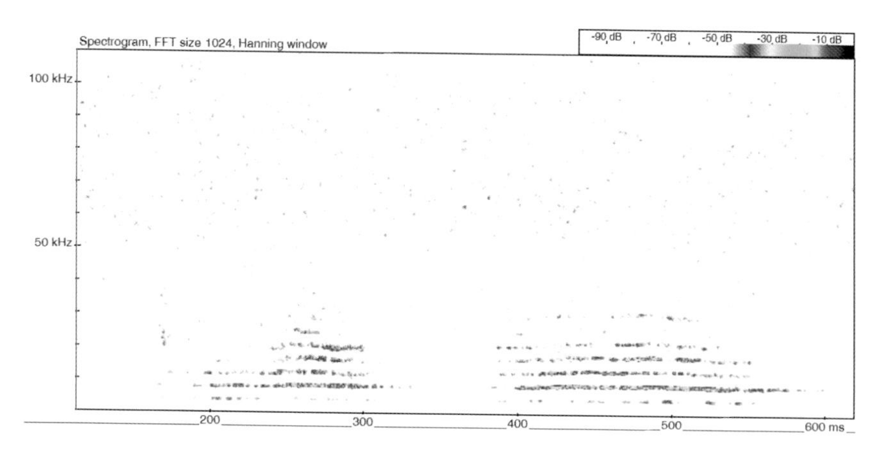

Figure 2.41 Bank vole – audible whistle (frame width 500 ms).

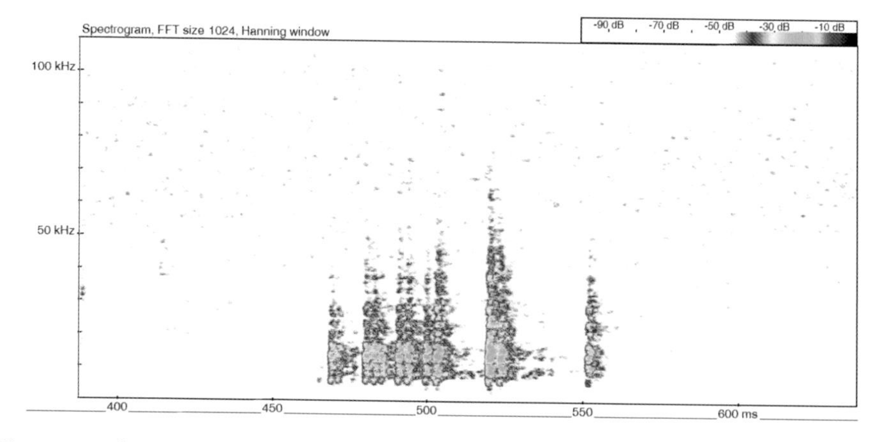

Figure 2.42 Bank vole – audible grinding noise (frame width 250 ms).

Field vole

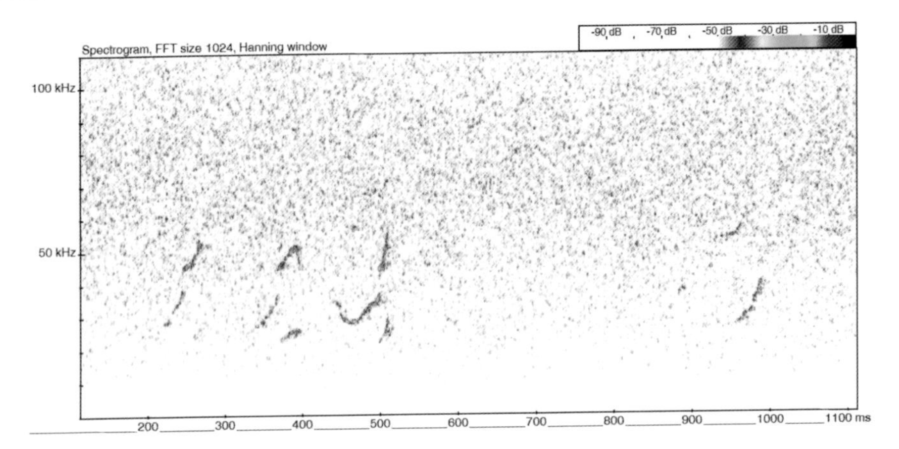

Figure 2.43 🔊 Field vole – complex call sequence.

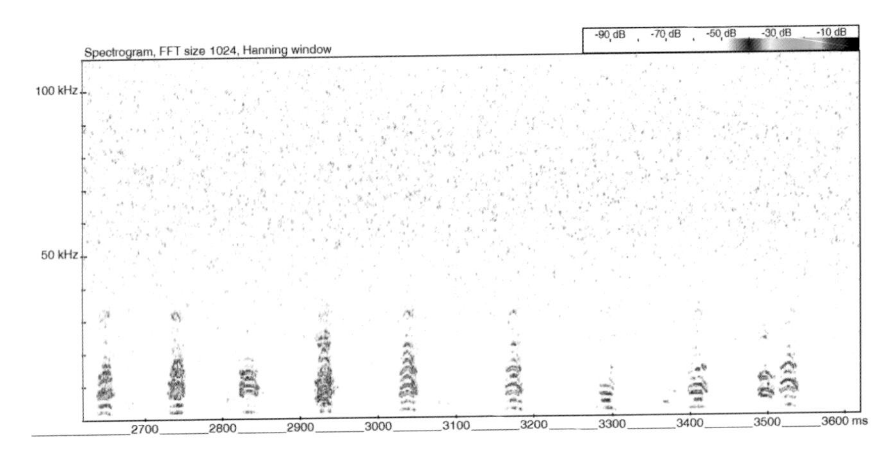

Figure 2.44 🔊 Field vole – audible calls.

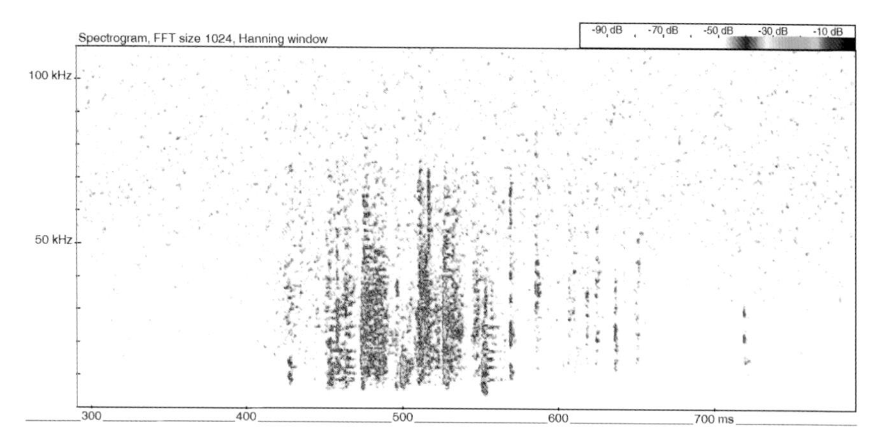

Figure 2.45 Field vole – grinding noise (frame width 500 ms).

Water vole

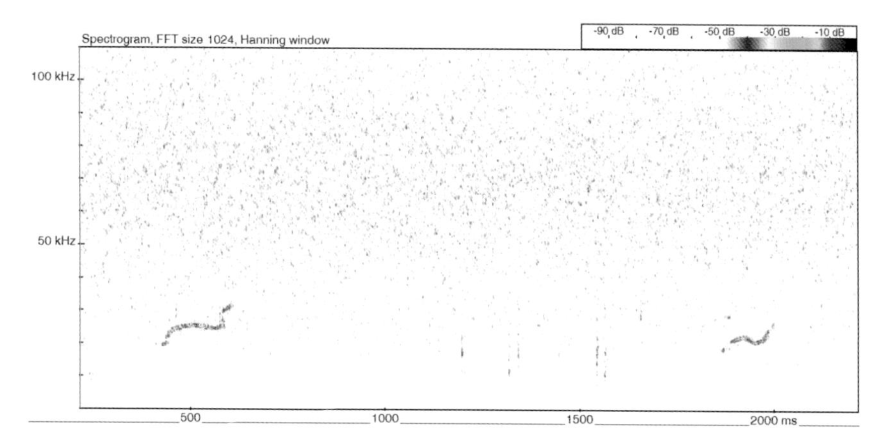

Figure 2.46 Water vole – rising FM stepped calls (frame width 2000 ms).

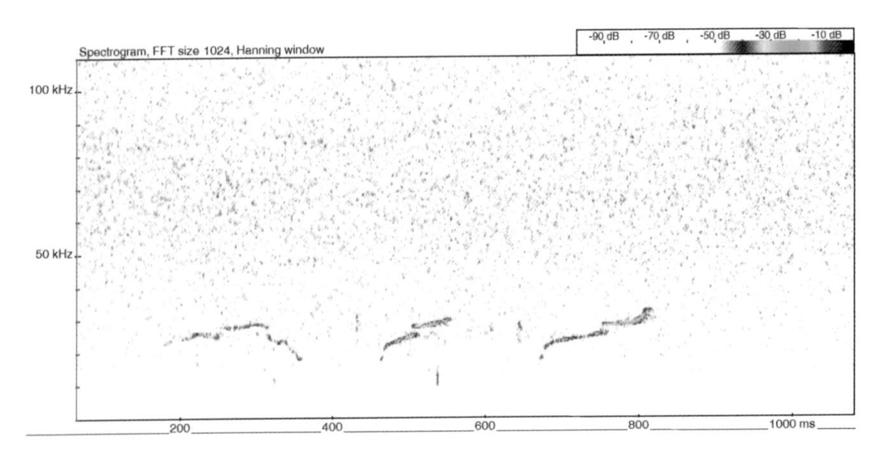

Figure 2.47 Water vole – rising and falling stepped calls.

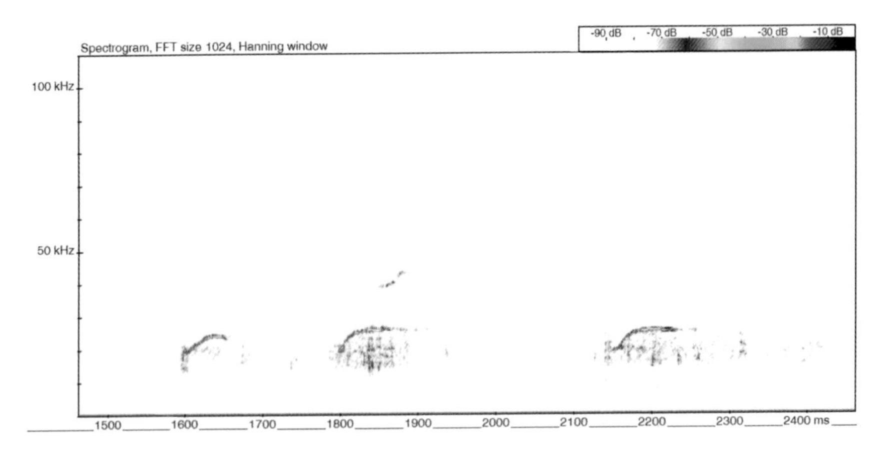

Figure 2.48 Water vole – rising FM call (C Macdonald, 2019).

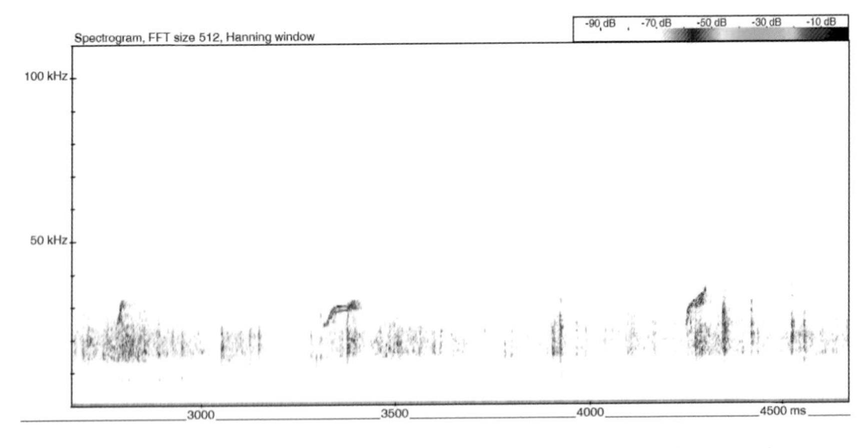

Figure 2.49 Water vole – rising FM call (C Macdonald, 2019; frame width 2000 ms).

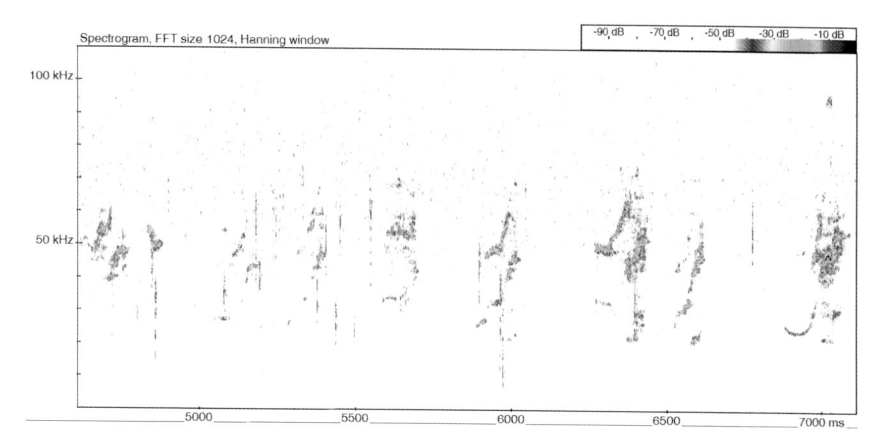

Figure 2.50 Water vole – complex series (frame width 2500 ms).

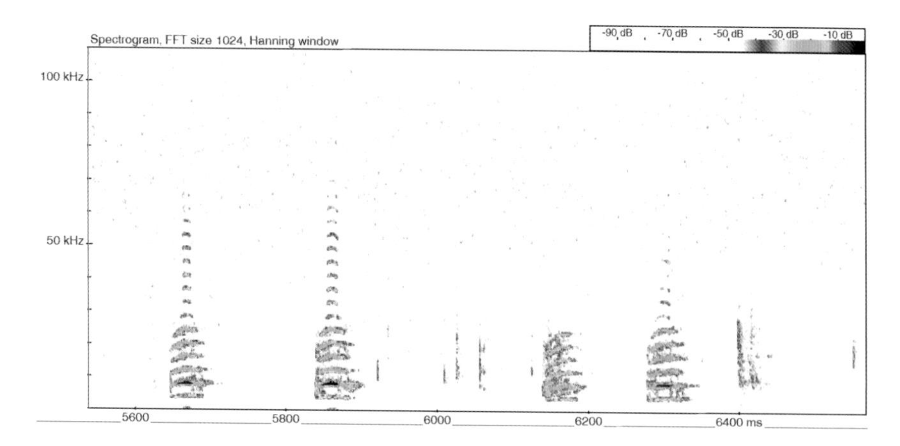

Figure 2.51 Water vole – audible calls.

Dormouse species

Both of our species of dormouse (Table 2.7) are normally nocturnal and arboreal when active. During colder months, they are inactive and enter into a long period of hibernation, normally from October through to May.

Table 2.7 Dormouse species occurring in Britain.

Species	Distribution and typical habitat	Notes on communication/recordings
Hazel (common) dormouse *Muscardinus avellanarius*	Patchily distributed across southern England and Wales. Isolated populations in northern England. Absent from Scotland and Ireland[1][2] Broadleaved woodland Coniferous woodland Hedgerow[2]	🔊 Hazel (common) dormouse is not regarded as being particularly vocal, and is quieter than edible dormouse (see below).[1] Hazel dormice have been known to make calls during aggressive/chase encounters. Young in the nest can emit ultrasonic sounds within the range 20–30 kHz.[3] As well as audible sounds, ultrasonic sounds have been recorded, occurring in the frequency range of 25–100 kHz including harmonics.[3] A study carried out with captive specimens described six types of vocalisation, five of which typically occurred within the frequency range 18–48 kHz.[4] The sixth call type related to mother/infant 'clucking' communication, occurring within the audible range of 6–8 kHz. These calls are discussed in more detail in Table 2.8.
Edible dormouse *Glis glis*	Considerably more restricted in its range than hazel dormouse, occurring primarily in an area centred around Tring (Chilterns)[2][5] Broadleaved woodland[2]	🔊 There appears to be far more vocal activity in this species, compared to hazel dormouse. Edible dormouse will call from trees, and can be particularly vocal during their mating season (June to August). A variety of sounds are known to be emitted.[6][7] Of note, a threat call described as an explosive churring, and a distinctive raucous squealing ('chirp' call[7]) emitted from a high vantage point, which may be territorial.[5] Audible calls are useful in order to survey for this species, helping to locate their presence at night.[7][8][9]

References

[1] Bright & Morris 2008; [2] Mathews *et al.* 2018; [3] Juškaitis & Büchner 2013; [4] Ancillotto *et al.* 2014; [5] Morris 2008; [6] Hutterer & Peters 2001; [7] Iesari *et al.* 2017; [8] Ciechanowski & Sachanowicz 2014; [9] Rodolfi 1994.

We are now going to look more closely at each of these species, paying more attention to hazel (common) dormouse, as edible dormouse has a restricted range in Britain (Table 2.7).

Hazel (common) dormouse

I know that the hazel dormouse (*Muscardinus avellanarius*) is a species that attracts quite a lot of attention in Britain. I am therefore going to provide a bit more detail in this instance compared to many of the other species tackled in this chapter. The most useful study I was able to source for reference purposes was carried out by Ancillotto *et al.* (2014). That study describes examples of the six different types of ultrasonic call for hazel dormouse. The research was carried out on a small group ($n = 5$ adults + 3 pups) of captive animals, using a Pettersson D500X bat detector to record sound, and study the associated relationships and interactions. To a bat worker, not knowing what other things may produce sound at similar frequencies, some of the calls could easily be thought to have been emitted by a bat, especially when considering bat social calls. In Table 2.8 I have provided artificial examples of the five adult-generated ultrasonic calls described by Ancillotto *et al.* (2014).

It should be borne in mind that the call parameters and examples shown in Table 2.8 are approximate, and also taken from a study of a small group of animals, over a short period of time (10 days). It should be anticipated that variation to what is shown here could occur, and the authors of the study themselves say that 'further analysis relying on a larger sample size would be beneficial in order to create a more complete assessment.'

Table 2.8 Typical call structure of hazel dormouse (adapted from Ancillotto *et al.* 2014).

Call type and context	Description of structure	Artificial representation
'A' produced by solitary males and females	Described as an 'arising stepped call', or shallow FM stepped rising, commencing at *c.* 18 kHz and rising to *c.* 34 kHz	18 kHz ← *c.* 600 ms →
'B' produced by solitary males and females	Described as an 'arising, smooth call', or shallow FM smooth, commencing at *c.* 19 kHz and rising to *c.* 25 kHz	20 kHz ← *c.* 600 ms →
'C' produced by male with female present	'Arising sloped call', giving a slightly humped impression, commencing at *c.* 35 kHz and rising to *c.* 43 kHz	35 kHz ← *c.* 400 ms →
'D' courtship song. Male with female present	'Down-sweep' call, comprising up to 62 individual notes, split over two phases. Phase 1: numerous steep FM sweep calls of short duration, commencing at *c.* 40 kHz, dropping steeply to *c.* 18 kHz, followed by phase 2: more complex variety of call structures, occurring within a similar frequency range to phase 1	40 kHz 20 kHz Phase 1 → Phase 2 ← up to *c.* 3200 ms →
'E' produced by solitary males	'Flat wiggled' call, occurring within the frequency range *c.* 44 kHz to *c.* 53 kHz	50 kHz ← 500 ms →

Figures 2.52 to 2.62 show some typical examples of ultrasonic sound recorded by contributors Lorna Griffiths (Nottinghamshire Dormouse Group) and Kari Bettoney (Mid Devon Bat Rescue Centre), as acknowledged. Numerous additional examples are provided in the Sound Library.🔊

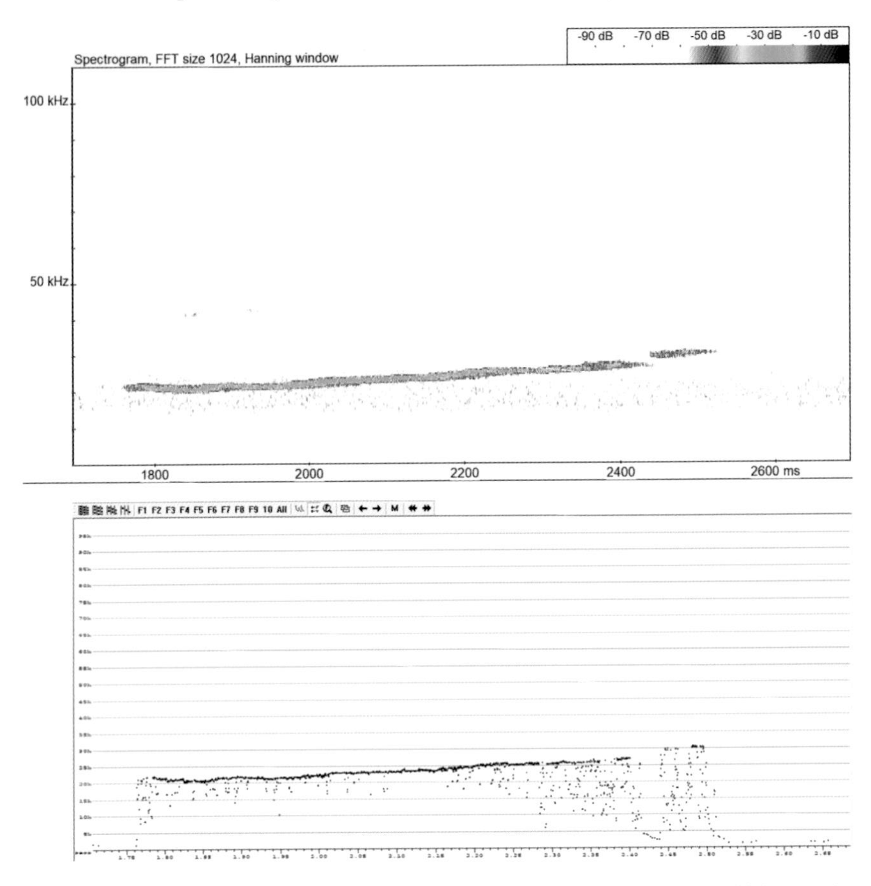

Figure 2.52 🔊 Hazel dormouse – rising FM with AnaLookW version below (L Griffiths, 2018).

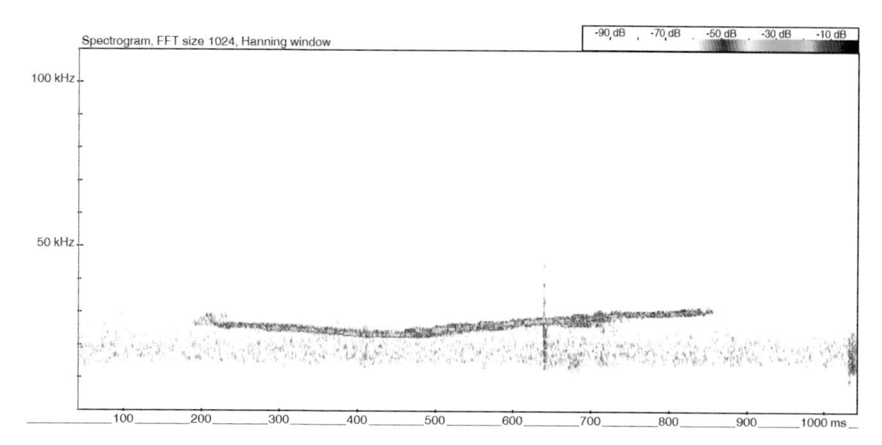

Figure 2.53 🔊 Hazel dormouse – rising FM (L Griffiths, 2018).

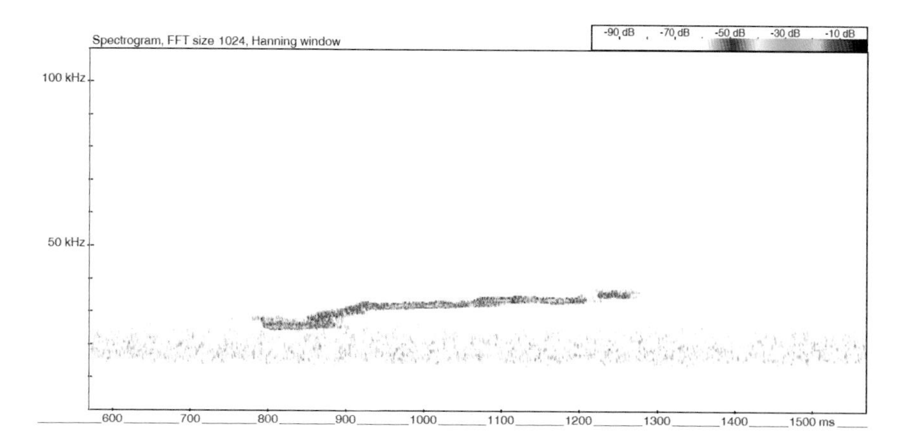

Figure 2.54 🔊 Hazel dormouse – rising stepped FM (L Griffiths, 2018).

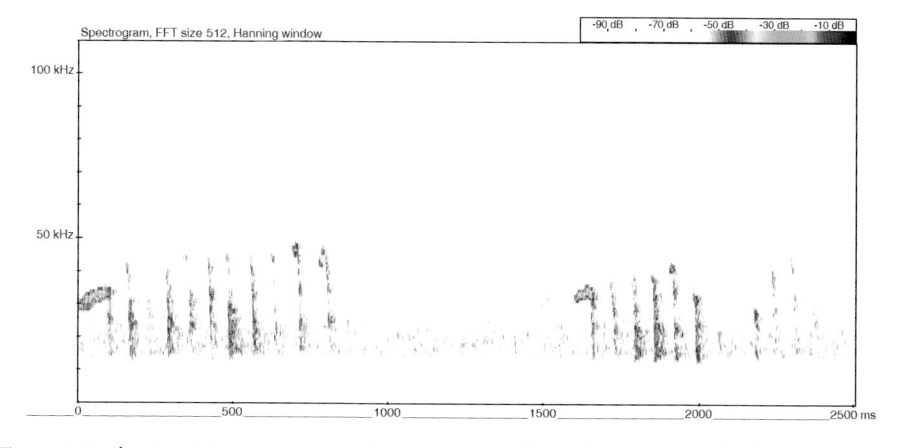

Figure 2.55 🔊 Hazel dormouse – type D sequence (L Griffiths, 2018; frame width 2500 ms).

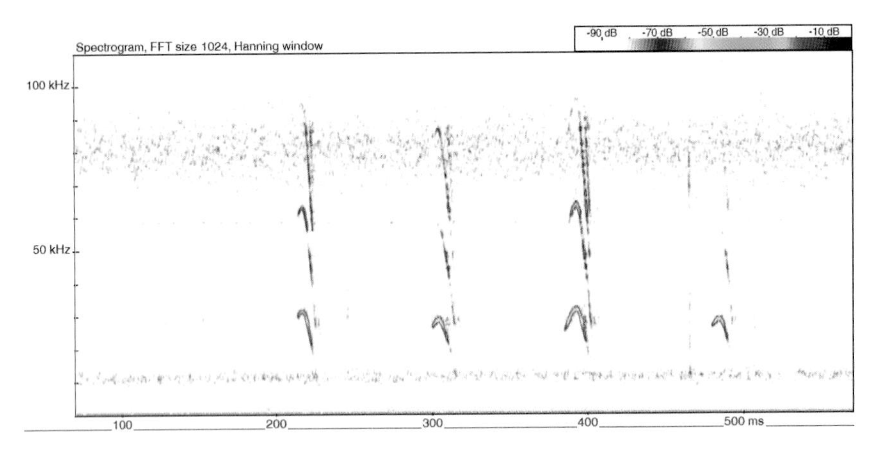

Figure 2.56 🔊 Hazel dormouse (juvenile) – rising then falling (K Bettoney, 2018; frame width 500 ms).

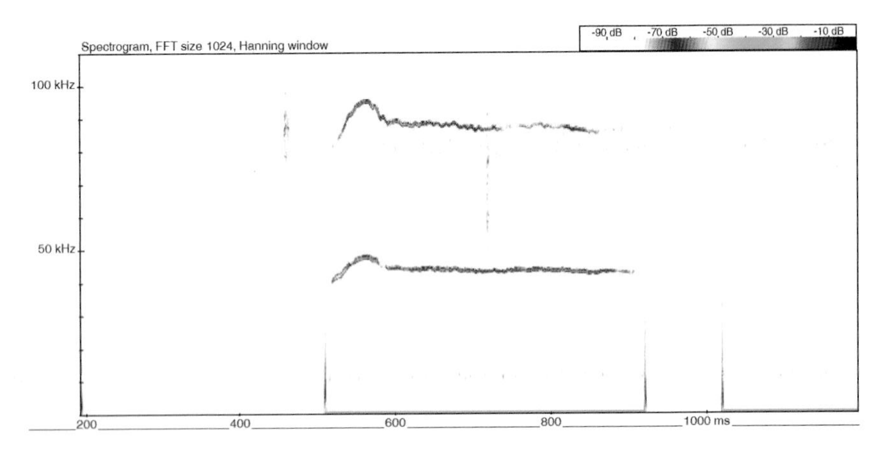

Figure 2.57 🔊 Hazel dormouse (juvenile) – variation (K Bettoney, 2018).

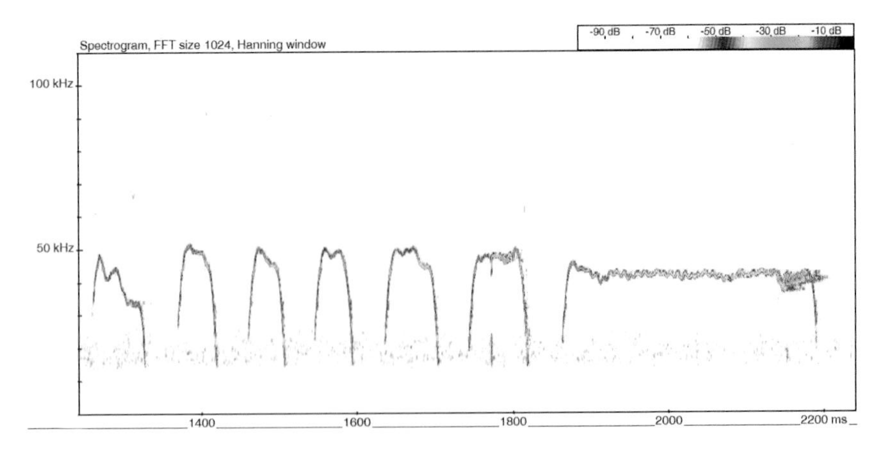

Figure 2.58 🔊 Hazel dormouse – variation (L Griffiths, 2018).

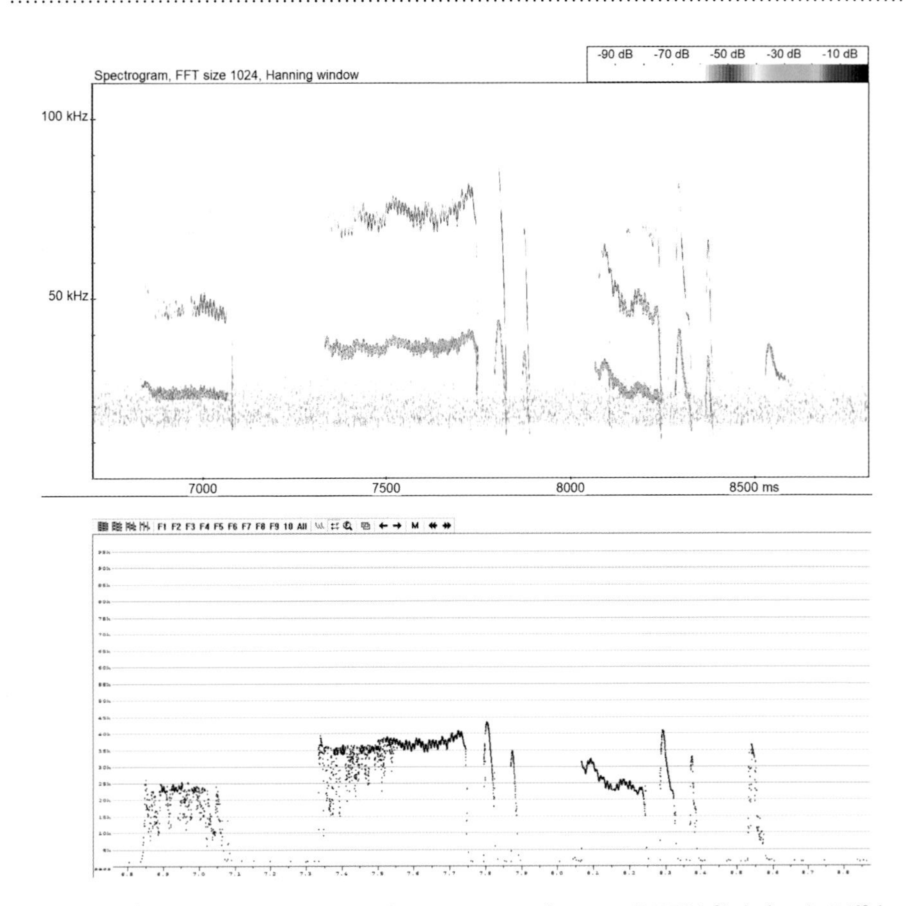

Figure 2.59 🔊 Hazel dormouse – complex sequence, with AnaLookW ZCA file below (L Griffiths, 2018).

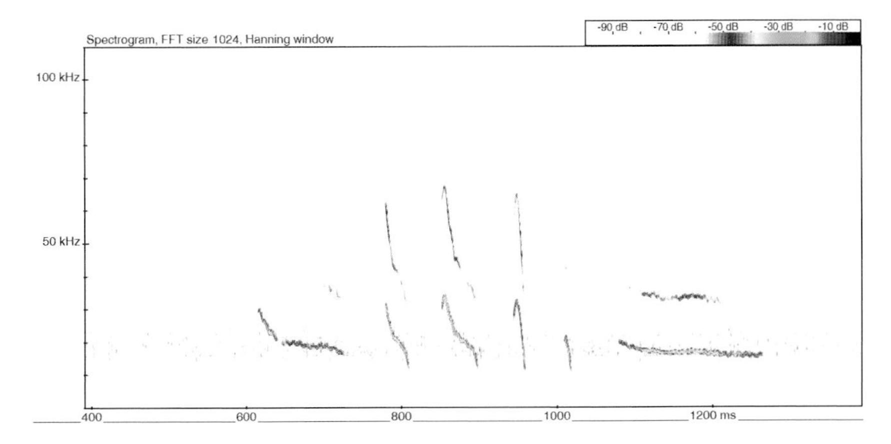

Figure 2.60 🔊 Hazel dormouse – complex sequence (L Griffiths, 2018).

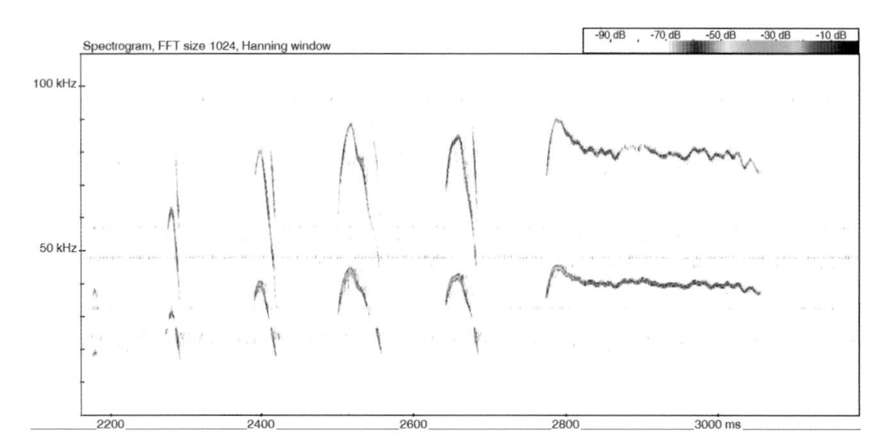

Figure 2.61 🔊 Hazel dormouse – baby during feeding time (K Bettoney, 2018).

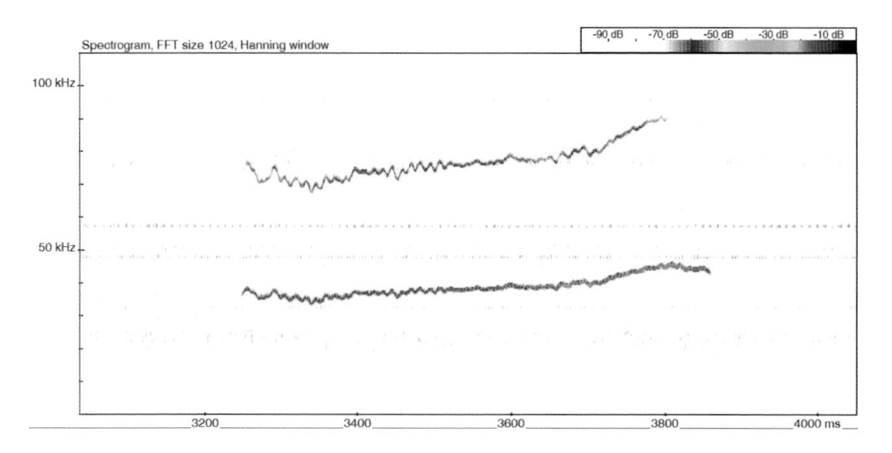

Figure 2.62 🔊 Hazel dormouse – baby during feeding time (K Bettoney, 2018).

Edible dormouse

Research into the acoustic vocabulary of edible dormouse (*Glis glis*) appears to be fairly limited. Goedele Verbeylen (Natuurpunt) very kindly assisted with obtaining some recordings for me to refer to, although due to copyright considerations I have been unable to provide these as sound files within this study.

However, one study that I was able to access involved captive animals in a stable colony (Iesari *et al.* 2017) where seven types of audible call were distinguished (ultrasonic vocalisations were not studied), these being referred to as: chatter, chirp, hissing, squeak, twitter, volley and whistle. The 'chirp' call (Figures 2.63 and 2.64) was described as being 'high-pitched, acute, short and recurring sounds with constant frequency modulation' delivered in a 'long series'. The context of these 'chirp' calls is thought to be related to agonistic territorial behaviour.

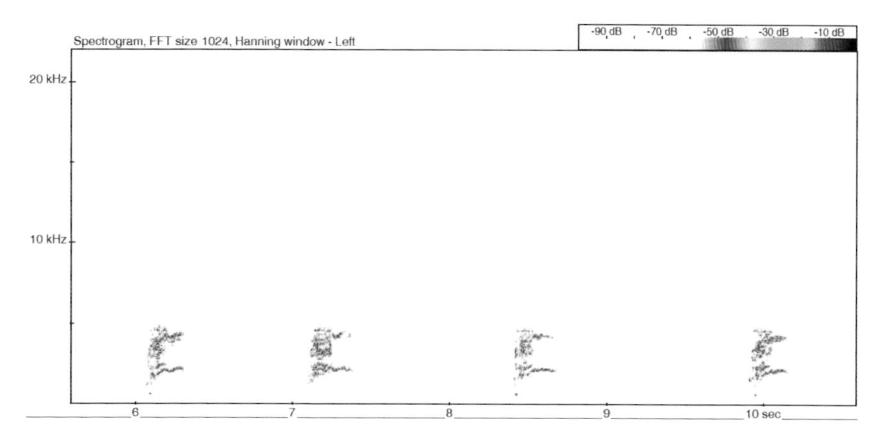

Figure 2.63 🔊 Edible dormouse – chirp call (Iesari *et al.* 2017; frame width 5000 ms, frequency scale 0–22 kHz).

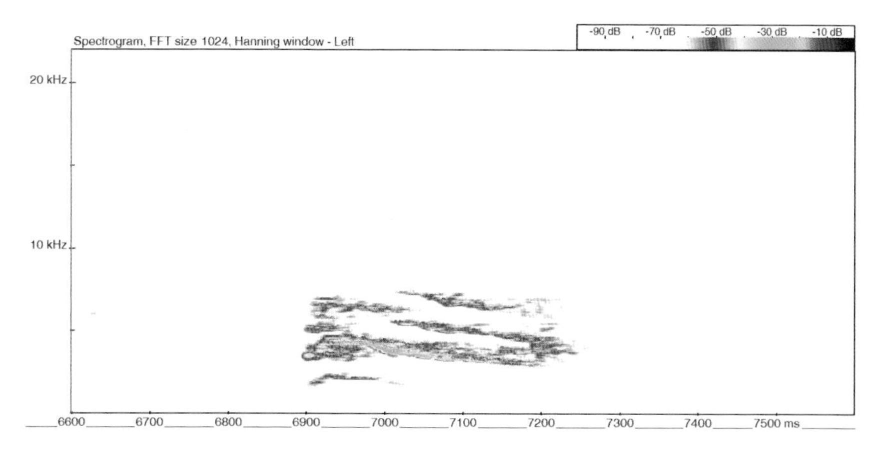

Figure 2.64 Edible dormouse – chirp call (frequency scale 0–22 kHz).

Hints and tips

Risks to bat survey methods and analysis

While recording many of the small terrestrial mammals for this publication, it was noted that they often showed inquisitive behaviour relating to new items placed within their environment, which of course included recording devices and microphones. It would therefore not be unreasonable to assume that remotely placed bat detectors, used for long-term monitoring purposes, could easily attract curious small mammals. In fact, within the bat worker world, we have regularly heard of microphones and/or their wind mufflers being nibbled by small mammals (and much larger animals as well, such as sheep, cows, deer etc.). With this in mind, I would suggest we have two fairly important things to consider. First, there is the occurrence of individual sounds being emitted by these small terrestrial mammals that could potentially be mistaken for bat activity. Secondly, there is the known risk that equipment could be damaged, potentially not only creating additional cost but also impacting upon survey deliverables. Tackling these issues, amongst others, Table 2.9 offers some potential solutions worthy of consideration.

Table 2.9 Consideration of small terrestrial mammals relative to bat surveys.

Subject	Problem	Potential solutions
Equipment damage	Small terrestrial mammals causing damage to equipment (e.g. wind muffler or microphone)	Create a small metal gauze or plastic cage for the microphone. The gauze should not touch the microphone or the wind muffler. Adopting approaches described below, under 'Inadvertently recording', may also help.
Inadvertently recording	Small terrestrial mammals being recorded inadvertently and potentially impacting upon sound analysis process	Other small mammals are more likely to be recorded when in close proximity to the microphone. The more distance between the microphone and any potential encounter (e.g. as high as possible from ground level) the better. Place the entire bat detector in such a way as to prevent a small mammal making progress towards the microphone area. Place the microphone/bat detector on a smooth-surfaced pole, reducing the chances of something being able to climb up. Alternatively, position a large smooth cone, open side pointing downwards, beneath where the microphone/detector is placed.
Noise filtering process	Applying a noise filter doesn't remove small mammal noise	A noise filter removing noise at levels below a frequency of 10 kHz should remove most audible sound, without having an impact on bat calls. Carry out a manual audit of filtered noise to ensure nothing of interest has been misallocated. The main issue potentially arises with ultrasonic noise, because much of the sound created falls within a similar range of frequencies to bat sound.
Bat classifier error	Small terrestrial mammals being mistaken for bats	Carry out a manual audit of calls that have been subjected to an automated classification process.
Human error	Person doing bat call analysis mistaking other mammals for bats	With experience many of the calls are less likely to be misidentified as bats. Some, however, are very similar to sounds created by bats. Appendix 2 offers some thoughts to bear in mind when faced with something unusual.
Anything occurring immediately around or beside the microphone could impact upon how it performs or how recorded sound behaves (e.g. sound echoing off a smooth flat surface close to the microphone). Therefore, any alterations made to a microphone system, or its immediate surroundings, that could impact upon recorded sound should be tested ahead of being used on case work.		

Following on from Table 2.9, I cannot stress enough the potential for the application of filters and the use of automated bat classifiers to cause misidentification of a small terrestrial mammal as a bat. This is not the fault of any one manufacturer as far as I can see, and it is a problem that potentially is inherent in the process because of similar frequencies etc. occurring in this group and in bats. In Table 2.10 examples are provided of small terrestrial mammal calls that have been put through two well-known and commonly used bat classifiers. All too often, non-bat noises, such as the examples shown, do not end up in noise folders, but are allocated to bat species, albeit at a wide range of confidence levels. To be fair, a trained user, upon seeing low confidence levels or multiple species being suggested within a single sequence, should know to be cautious with such calls during an audit. The calls referenced in Table 2.10 relate to figures used earlier in this chapter, so that you can refer to these and more easily appreciate why a classifier has reacted the way it has.

As can be seen from Table 2.10, the issue is quite wide-ranging. The potential for inappropriate or misguided mitigation/compensation is definitely there. Imagine not checking all those noctule calls for that wind farm application. At the time you thought it was a bit odd to have so many. Rats! I am going to spare you lots of other potentially disturbing scenarios. Within the table are two bat species (common pipistrelle and greater horseshoe). They have been included first of all to act as a control for the process. However, you will see from the table that in some instances the classifiers are demonstrating a higher degree of confidence that something that isn't a bat, is a bat, compared to an actual bat of a species that would normally be deemed to be a fairly safe bet. Yes, read that again.

In the bat world we regularly mention that, in most respects, classifiers don't cater well for bat social calls. For example, sometimes a social call is identified as echolocation from a totally different species (see Chapter 1). Now we have something else to consider, that it is possible (hold on a minute – not just 'possible', it is 'definite', as has been demonstrated here) for other animals that aren't bats to be identified as bats, and at times to a high confidence level. It doesn't help that when we look at a spectrogram it looks like a bat, we play it and it sounds like a bat, and then the automated classifier verifies the misguided initial instincts, in telling you that it is a bat. However, the issue doesn't just lie with the classifier, as it is always a human who ultimately decides, even if the decision is not to check. The classifier is there to give an opinion, but the technician ultimately is in control. So we need to be totally on our toes, thorough in our assessments, and not afraid to just step back, use our common sense, and say 'No, experience is telling me that it just isn't quite right', either at an individual call level or when looking at large dataset outputs.

The point that we really need to remember, when using a classifier, is that it can only generate outputs for the things that it has been asked to consider in the first place. So if the developer of a classifier has only used bat calls within the call reference library, then anything that looks like one of these calls can only be classified as a bat. However, if as time goes by we are able to add more than bats to a classifier, the classifier will have more options from which to hopefully make a stronger assessment.

Table 2.10 Examples of calls processed through automated bat classifier software programs.

Confirmed species	Figure	Automated bat classifier suggestions	
		Software package 1	Software package 2
Common pipistrelle	1.4	72% – Common pipistrelle	98% – Common pipistrelle
Greater horseshoe	1.7	87% – Greater horseshoe	97% – Greater horseshoe
Brown rat	2.1	94% – Noctule/Leisler's/serotine	96% – Noctule/Leisler's/serotine 85% – Common pipistrelle
Harvest mouse	2.11	80% – Soprano pipistrelle 72% – Common pipistrelle	97% – Common pipistrelle 95% – Soprano pipistrelle
Harvest mouse	2.12	72% – Nathusius' pipistrelle 67% – Common pipistrelle	97% – Common pipistrelle
Wood mouse	2.22	70% – Nathusius' pipistrelle 51% – Noctule/Leisler's/serotine	100% – No suggestion
Yellow-necked mouse	2.29	65% – Nathusius' pipistrelle 60% – Noctule/Leisler's/serotine	95% – Common pipistrelle
Shrew sp.	2.34	41% – Noctule/Leisler's/serotine	97% – Noctule/Leisler's/serotine
Bank vole	2.38	34% – Nathusius' pipistrelle	86% – Common pipistrelle
Water vole	2.46	84% – Noctule/Leisler's/serotine	98% – Noctule/Leisler's/serotine
Hazel dormouse	2.52	83% – Noctule/Leisler's/serotine	99% – Noctule/Leisler's/serotine
Hazel dormouse	2.60	65% – Noctule/Leisler's/serotine	98% – Noctule/Leisler's/serotine
Hazel dormouse	2.61	63% – Common pipistrelle 54% – Nathusius' pipistrelle	93% – Greater horseshoe 73% – Noctule/Leisler's/serotine 70% – Common pipistrelle

Only species occurring in the British Isles were considered in classifier output results.

The examples used are not considered to be the only examples of small terrestrial mammal calls that would produce similar results.

All scores with a degree of confidence greater than 60% are shown. In any example where a degree of confidence has not exceeded 60%, then the next highest degree of confidence is shown.

Where confidence is high (> 70%), giving a correct result.

Where confidence is high (> 70%), giving an incorrect result.

Where confidence level is lower (< 70%), in anticipation that the file would be dismissed when undergoing an audit.

Setting out to record small mammals

Initially when I set out to record small terrestrial mammals myself I thought that using a bat detector, with recording parameters appropriately set, seemed like a good idea. I was pleased to discover later, while carrying out the literature review, that I had not been the first person to do this (deliberately), and in fact many research studies have used either time expansion or full spectrum bat detectors to record small mammals.

The bulk of my recordings were made using a Batlogger M. While at the British Wildlife Centre (East Grinstead), I used five of these machines simultaneously to accumulate a large reference library very quickly. In effect, for each overnight period the machines were linked up to a separate glass enclosure in a way that would ensure, as far as possible, that no noise leakage would occur from neighbouring species, and also that the occupants wouldn't destroy the microphones (especially the ones I had borrowed). I have included some pictures to give you a better idea.

(*Top left*) Good to go for water vole. (*Top right*) Analysis station. (*Bottom left*) Typical glass enclosure set-up. (*Bottom right*) Don't nibble the microphone, with bald bloke in reflective mode.

Author's diary note

So how difficult can it actually be to pick up small mammals on a bat detector? Well I initially reckoned that it wouldn't be that difficult, as a good number of bat workers had said to me that they have definitely recorded rats and shrews etc. while carrying out surveys. But going out specifically to record 'other stuff' over a short period, as opposed to incidentally recording the other stuff over a lifetime of bat surveys, turns out to be far more challenging.

First of all, once you have a recording, how do you know which species it actually came from? I went on the basis that just saying *'it's a mouse'* wouldn't cut it with many of my batty peers. I can hear it now, and quite right too. *'Is it really a mouse, Neil?' 'How do you know it's not a vole, or a shrew?'* So getting it to species level, or at the very least genus, had to be the goal.

To catch a rat (or anything else for that matter), you need a trap. So I used Longworth traps. I sought mice, voles and shrews. *'Hey matey, hold the batty bus a minute. Aren't shrews protected?'* Yep, I had to get a shrew licence (thanks SNH Licensing Team). All seemed reasonably simple, and very quickly I started to catch stuff. First up was a wood mouse. I was well chuffed. Well, for at least three and half minutes. The little beggar escaped! OK, not to worry, there is something else in one of the other traps. A bank vole. This one was put into its temporary enclosure (at location of capture) and the recorder switched on. I watched as it scurried about. I thought to myself, there is no point in you looking for an escape route little fella (said in the style of Keith French), the enclosure is tight to the ground. Some 30 minutes later, still no noise recordings. Then I glanced at the ground next to the trap, only to see another bank vole staring up at me, which it did for a good 15 seconds – by which time I realised it was my no longer, ceased to now be, captive. As Lindsay 'Jaws' MacKinlay once announced during the Threave Bat Project, 'We are going to need a bigger boat' (thank goodness sharks aren't on the list!), or at least one with fewer holes. A more robust and longer-term solution would need to be created. Why create something when it already exists? The 'Pets At Home' hamster cage (with a nice grill on top for the microphone) will do the job, and be less stressful all round. At least, that is, until a mouse chews its way through the plastic base (Stuart Newson, personal correspondence).

Whatever I did, I would be restricted by geography and time as to how much could be caught and the diversity of species. I had a wish list that included everything small enough to fit in the cage, and had some success in my local area, catching local stuff. I was, however, beginning to feel that having a single animal of one species wouldn't really provide much vocalisation, in that it would be more likely to be communicating more often and naturally if it was in familiar surroundings, within its normal colony or family group. I regularly caught small mammals, but more often than not, one wood mouse, one field vole etc. I quickly considered the value of the limited data being gathered against the chaos imposed on these poor creatures, and thought there must be another way. Then in stepped Matt Binstead (Head Keeper, British Wildlife Centre). To be clear, he is too big to fit inside my cage, but where he did fit in very nicely was to allow me access to a wide range of captive small mammals at their excellent site in Surrey. Established populations, within enclosures, and all already accurately identified and behaving as close as you could get to natural conditions (made my hamster cage feel grossly inadequate). Anyway, it all worked out considerably better than I could ever have anticipated.

CHAPTER 3

Birds

Who gives a hoot?

By far the species group that you are most likely to encounter during bat surveys, apart from the bats themselves, would be birds. Within the British Isles we have over 200 breeding species (Svensson *et al.* 2010), which includes our summer visitors. This number does not, however, allow for the many other species that appear only during the winter or while migrating through.

Because there are quite a few species that could be encountered audibly while carrying out work at dusk or dawn (as well as overnight), the main focus of this chapter is going to be on what a surveyor is likely to hear, unaided by a bat detector. Every bird within the British Isles produces audible sound, and most of this, especially territorial singing, can be safely assigned to species – with training and experience. Birds are probably the most studied species group on the planet, with an impressive amount of published academic research (> 23,500 papers) having been carried out over a considerable length of time (Gasc *et al.* 2017). Added to this is the vast amount of resources available to those interested in the subject, be that books, websites, CDs, DVDs, software and apps. And the number of people interested in birds is also vast, with the Royal Society for the Protection of Birds (RSPB) alone having in excess of one million members. And, of course, the RSPB isn't the only organisation that people interested in birds can join. Some prefer to choose other organisations that reflect their specific interests, while many more, including 'garden' birdwatchers, are not members of anything.

People interested in birds describe themselves, or are described by others, in different ways, including 'birdwatcher', 'twitcher', 'birder' or 'ornithologist' (Oddie 1995). This may surprise a non-birding person (to you we are probably all 'twitchers'), as each of those descriptions, along with a few others, carries a very different meaning to those passionate about the subject. It may also surprise you to know that many people who are extremely good at bird identification rely almost entirely on what they can see (i.e. visual ID features), as opposed to identifying birds by sound. So, as with bats, when sound can be overlooked by some (e.g. during the analysis process), it is considerably more so the case, for many, when it comes to identifying birds. This is a shame, as there are many species of bird that are much easier to identify by sound than by sight. This approach (identification by sound) not only adds to confidence in the accuracy of the encounter, but also improves the speed with which the bird is identified.

Good knowledge about bird sound is especially beneficial when in a woodland setting, for example while carrying out a breeding bird survey. It would be impossible, and certainly extremely time-consuming, to do this job based on visual identification alone. In fact, this is a scenario best avoided by an 'ornithologist' without any knowledge of bird songs and calls. In practice, based on my own experience in this area, I would say that it would not be unusual for *c.* 75% of all data gathered during such a survey to be based on hearing alone (i.e. without a visual of the bird in question). To spend five minutes with your binoculars trying to get a sight of every singing blue tit, robin, wren and chaffinch would mean that you probably wouldn't be able to complete the survey on time. In any case, there will be many birds you just won't be able to get a visual on at all, due to vegetation etc. Also, if you were doing such a survey properly, you would normally find yourself within the woodland before sunrise. This would render any optical equipment useless at that level of light, especially so within a habitat that can be fairly dark even in the middle of the day. More often than not, an ornithologist with good knowledge of bird songs and calls, encountering bird noises within such settings, would immediately be able to identify them to species level and move quickly on.

How do you get to this level of knowledge if you are starting from scratch? Well, I would suggest that you start off slowly and that you base your learning around habitats, as opposed to trying to learn everything at once. It's a classic problem-solving technique. When faced with a massive problem or a complex task, how do you deal with it? You break it down into smaller components, and progress one step at a time. To help you with this, there are lots of resources available for learning or checking bird sound, as well as giving guidance as to recording them, should this be something you wish to do. The resources that I, personally, have found most useful during my own bird-related activities are listed in Appendix 3. However, there are substantially more out there should you wish to investigate for yourself, or take advice from others. These additional options would include numerous apps that can be downloaded onto your mobile phone or tablet and referred to in the field if needs be. Be careful, however, when you do this, as on occasions playing certain bird calls and songs in the field could cause unintended distress or artificially induced behaviour by the real birds within earshot. Apart from that, it can also be confusing to your co-workers!

I recall one occasion when I was co-delivering a bird sound course with my good friend David Darrell-Lambert (Bird Brain UK Ltd). For one of the sessions we split the course delegates into two groups, and we each went our separate way through a woodland. I was explaining to my group the difference between garden warbler (*Sylvia borin*) and blackcap (*S. atricapilla*) songs, this being a classic area of confusion for many. We had just heard a garden warbler. So, for comparison, I switched on my iPad with an external speaker and played a pre-recorded blackcap to my group. Quite a few times actually, as I attempted to allow them to hear the point I was making. Sometime later the two groups came together and exchanged notes, with David announcing that they had heard some brilliant classic blackcap singing. We very quickly established that it had in fact been my iPad, and at that point we had been close to David's group, albeit we could not see each other through the woodland cover.

To be a little more focused regarding advice to those wishing to start learning bird calls and songs, I would suggest as follows. It is best, for most people, to first of all learn the regular birds that they will most often hear, but which are not necessarily always easily seen while they are calling or singing. These will primarily be birds that occur in gardens, parkland or woodland. The added advantage with this group is when you are lying in bed, early in the morning during spring and summer, you can hear how much you are remembering. This helps to reinforce your knowledge, and you may also be able to impress your partner with your ornithological expertise. It is not unusual for me to enthusiastically ask my wife Aileen, first thing in the morning, 'Did you hear the dunnock?' And after years of this, she quite often is able to tell me what the birds are herself when I quiz her. If she gets three right in a row, I might even reward her with breakfast in bed. Seriously, though, many of these garden birds will also be the ones encountered during urban, parkland and woodland surveys. So, wherever you go you are able to reinforce what you have been learning, as well as being well on the path to becoming a breeding bird surveyor for those habitats.

Once you have a small number of these calls and songs sorted in your head and you are gaining confidence, you can then begin to slowly add others to your bank of knowledge. So, which ones should you learn first? I would say that anything big that is often out in the open is more likely to be identified visually, and therefore sound is often not your first point of reference. On the other hand, anything smaller that likes to be in amongst vegetation or tree foliage is going to be far harder to see, but easier to hear. In fact, often the only reason you know it's there in the first place is because you heard it. If you are reading this as a more experienced birder please bear in mind that I am currently thinking of someone who is setting out to learn from scratch, and is daunted by the vastness of what appears to be before them. Table 3.1 gives my opinion on the bird songs and calls to learn first ('The Do-able Dozen'). These are species regularly encountered during woodland/parkland/urban breeding bird surveys, and they can often be identified far more quickly by sound than by sight. You will find that a large percentage of the birds you hear calling or singing in your garden, or during a woodland visit, will belong to one of these species. Some of those listed in Table 3.1 can also be heard during darkness, at the very least at dusk or dawn.

Table 3.1 The 'Do-able Dozen' – bird songs and calls that can be most useful to learn first.

Robin (*Erithacus rubecula*)	Song thrush (*Turdus philomelos*)	Blue tit (*Cyanistes caeruleus*)
Wren (*Troglodytes troglodytes*)	Chaffinch (*Fringilla coelebs*)	Coal tit (*Periparus ater*)
Dunnock (*Prunella modularis*)	Greenfinch (*Chloris chloris*)	Chiffchaff (*Phylloscopus collybita*)
Blackbird (*Turdus merula*)	Great tit (*Parus major*)	Willow warbler (*Phylloscopus trochilus*)

Calls and songs

At this stage there is one thing that may, for some at least, be worthy of clarification. What is the difference between calls and songs? Birds produce sounds based on the scenario in which they find themselves, for example flight calls, alarm calls, contact calls, juvenile begging calls. All of these are correctly described as 'calls', but in addition we also have, for some species, more complex sounds relating to breeding activity or territorial defence, which are termed 'songs'. Taking account of all of the sounds made by birds, it should be borne in mind that individual calls and songs may have multiple purposes depending upon the context in which they are produced, and the message received by a listener (Constantine & The Sound Approach 2006). If we take the example of 'song' alone, this would be a description of sound made by a bird advertising for a mate during the breeding season. But it may also, simultaneously, be received by competing birds as a warning to them to stay away. So the same noise in effect has potentially the opposite effect depending on the recipient. Also, at other (i.e. non-breeding) times of the year some species (e.g. robin) can produce song, irrespective of any realistic expectation of mating.

In almost all species of songbird it is the male that sings, as he advertises for a mate and defines his territory. And then we have a range of song-like behaviours, attributable to mating-related activity, whereby the bird makes sounds that are not vocal. Examples would be the drumming made by many species of woodpecker, as well as another totally different drumming sound (this time made by feathers) which is produced by snipe advertising as they fly over their territory. Calls, on the other hand, are all the other noises that could be produced, including alarm calls and contact calls. In these respects, generally speaking, calls may be emitted by either a male or a female. Added to this would be calls that are made specifically by females, and those produced by fledglings and juveniles, which as they develop into adulthood will cease to be made.

Audible noise made by birds at night

Given the right circumstances pretty much any bird, including many of those listed in Table 3.1, has the potential to make noise during the dark hours. You are carrying out a transect, and you startle a bird and it kicks off into an unexpected racket or explodes into flight (e.g. woodpigeon, *Columba palumbus* 🔊). Simultaneously you are also launched into mid-air, leaving your walking boots fixed to the muddy ground, in the panic of the moment. You then quickly realise that it's only a bird. Here we go, put your boots back on and lace up tightly. '*It's only a bird*' – REALLY! The woodpigeon example is a classic, which I am sure you will have experienced on numerous occasions, but perhaps without realising what is usually responsible.

In this section I am going to give some guidance to some of those '*it's only a bird*' moments, as to what species it might actually be. At the very least this may satisfy curiosity. For some, hopefully, it may also help to create more of an interest in this area, ultimately improving upon the interpretation of what may be occurring at a site, other than bats. This may also have an impact upon assessments and the like. All of this, of course, with the corresponding benefits as outlined already in Chapter 1.

To pick up, again, on one of the points made in Chapter 1, I feel it is worth remembering that as a group of people studying the natural history of an area, bat workers are on site at times when most others are at home watching *Better Call Saul* with a nice cup of Tetley tea and a KitKat. At least this is what I think of, when I am out on cold, dark, quiet nights. So while we are out there, toughing it out, in an area where humans wouldn't normally go at that time of the night, we have the potential to encounter stuff that might have been missed, or that was just not possible to see or hear during the day. Some of these encounters will be of little or no relevance, granted, but others may be of high importance. The point is, that when you don't know what other things sound like (you are unlikely to get a good look – it's dark!) you are resigned to *only* being able to say 'It's only a bird'. What you have just heard, however, could in fact be considerably more important than all of the bats you are recording that night.

It is, however, unrealistic to know all of these sounds, especially if you are coming from a place where you currently don't know that many, if any at all. Bear in mind that many good ornithologists (and in some habitats you are only a good ornithologist if you know your sounds!) have spent a lifetime acquiring their knowledge about many hundreds of species. Bird songs and calls are not something you can learn on a one-week course, no matter how good the tutors. It takes time and experience, as well as the commitment and desire to want to learn, as opposed to it being something you have been told to do.

You do, however, almost certainly, have in your hands or in your pocket the ability to record the sound with a view to playing it back to a more experienced person later. Your bat detector may have a voice commentary function; this can be activated to record any other audible sound also. Failing that, your mobile phone probably has a recording function that can be used for the same purpose. And if you don't know where that is, most of us know how to record video on our phones. OK you are not going to see much on the video (it's still dark!) but the function also records the sound you are hearing. The point that I am emphasising (I am aware that we discussed this previously in Chapter 1) is that sometimes you will have the means to capture the sound that you are hearing so it can be referred back to. And so what if it turns out to be an alarm-calling blackbird, or an unseasonal robin? If it is something common and uninteresting to others, at the very least you have added to your experience and learned something new, which puts you in a far better place next time it occurs.

We are now going to consider different groups of birds that can be expected to be heard during darkness while carrying out bat surveys. Do bear in mind, however, that these birds make numerous other sounds over and above those provided here. What I am seeking to do, as far as possible, is to provide you with what you are most likely to hear most often, and what is most distinguishable audibly.

Spectrograms in Chapter 3

Spectrograms in this chapter use the following scales, unless otherwise indicated in the figure legend:

Time (x-axis): 5 seconds (5000 ms)
Frequency (y-axis): 0–22 kHz

When a figure legend includes the 🔊 symbol, this means that the figure has been created from a file in the Sound Library. The figure number matches the file number there. For more information about how to access and download files from the Sound Library please see pages xv–xvi.

Diurnal birds of prey

These are raptors that are mainly active during daylight hours. As such, due to their behaviour they are less likely to be encountered by bat workers during the hours of darkness. A number of species may occasionally be heard at night, however, including common buzzard (*Buteo buteo*) (Figure 3.1), red kite (*Milvus milvus*) (Figure 3.2) and hobby (*Falco subbuteo*) 🔊.

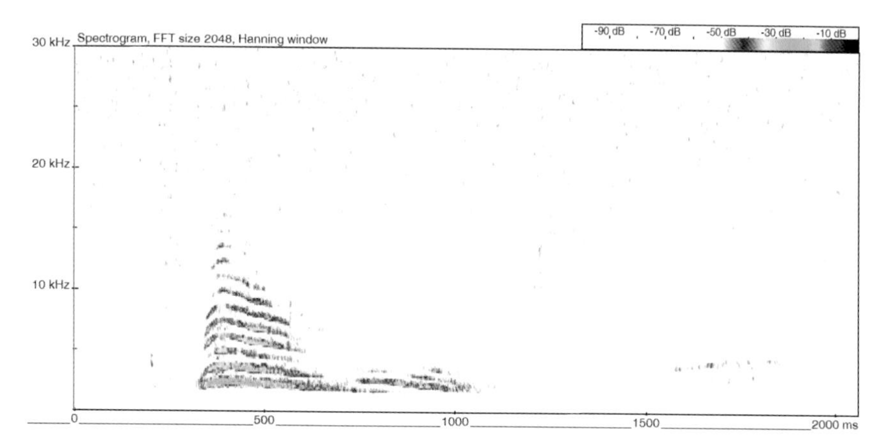

Figure 3.1 🔊 Common buzzard – call (frame width 2000 ms, frequency scale 0–30 kHz).

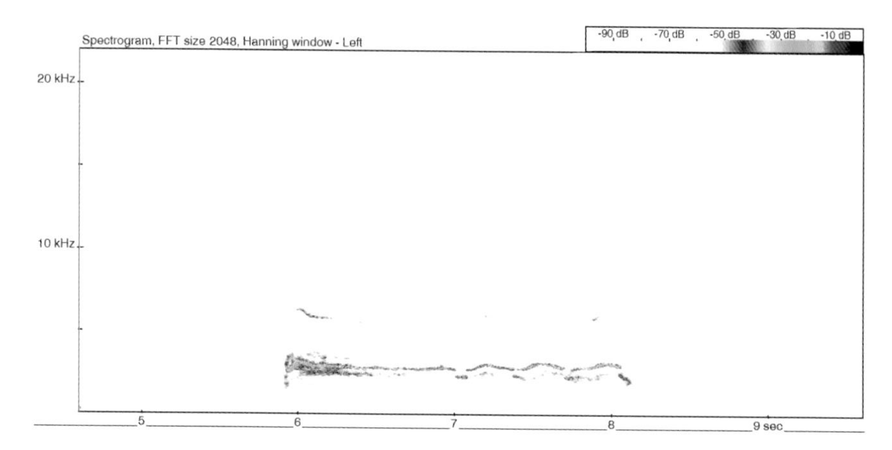

Figure 3.2 🔊 Red kite – call (D Darrell-Lambert, 2017).

Nocturnal birds of prey (owls)

To those who are not experienced with bird sound it may be of interest to learn that not all owls hoot. In fact, in the British Isles we have at least one species of owl (barn owl) whose screeching call is as far away from a hoot as you are ever likely to get. On the other hand, the classic owl 't'wit, t'woo' sound that we all grow up hearing during supposedly tense moments in thrillers belongs to the tawny owl, or to be more precise, two tawny owls. This classic call is in fact a combination of two separate calls, and it is quite common not to hear them together. The 't'wit' call 🔊 is a contact call that can be made by either a male or a female owl, while the drawn-out, longer 't'woo' 🔊 is usually attributable to the male's territorial hooting, although females make a similar sort of sound, albeit inconsistently. Both males and females may also produce a quiet bubbling call.🔊

When it comes to our other three species, first of all we have little owl, which produces a much higher-pitched territorial call, which I am sure would not be considered to be hoot-like by most of us. Then, finally, we have long-eared owl and short-eared owl. Learning these is quite easy, in that the long-eared owl produces long, slow-paced (i.e. well separated in time) hoots, whereas short-eared owls produce shorter and quicker-paced hoots. Their names help you to remember this.

Table 3.2 focuses mainly on the male calls made during the breeding season, as well as some of the noises made by young birds. A range of other calls also exist, including contact calls and alarm calls. Figures 3.3 to 3.11 show spectrograms of various calls from a range of owl species, with additional examples also being available in the Sound Library.🔊

Table 3.2 An overview of owl calls and behaviour in the British Isles (Hardey *et al.* 2009).

Species	Territorial and courtship behaviour	Nest location/fledgling calls
Tawny owl *Strix aluco*	Absent from Ireland. Vocal on their breeding territories throughout the year, but mainly from October through to April. The classic 't'wit, t'woo' call is actually two separate calls, these being a contact 'kwik' call 🔊 (the 't'wit') and the male's 't'woo' territorial call.🔊	🔊 Usually nests in holes and cavities within trees. Juveniles will produce single-note calls repetitively.🔊
Barn owl *Tyto alba*	Has been recorded breeding in the British Isles in every month of the year. Typically, however, breeding occurs during the period February to April. This species does not give a classic owl hoot. Males produce a screeching call,🔊 which is emitted usually in the vicinity of a nesting or roosting location.	🔊 Usually nests within structures, in hay bale stacks, or in large cavities in trees. Young in nest will make screeching begging calls.🔊

(cont).

Table 3.2 (cont).

Species	Territorial and courtship behaviour	Nest location/fledgling calls
Long-eared owl *Asio otus*	Usually breeds during the period March to April. Relatively speaking, owing to the short breeding season, territorial calls are not encountered often. Long, low-frequency hoots are made, these being repeated at a slow repetition rate.🔊 Wing clapping can also be heard during courtship behaviour.🔊	🔊 Usually nests in conifer trees. Once the young leave the nest they will locate themselves nearby, producing distinctive 'squeaky gate' begging calls.🔊
Short-eared owl *Asio flammeus*	More active during daylight hours than our other owl species. Usually breeds during the period March to April. Male calls are not often heard. In comparison with long-eared owl calls, short-eared owls produce short hoots, repeated at a faster rate. Wing clapping will also occur during courtship behaviour.	Typically nests at ground level, on moorland and heathland for example.
Little owl *Athene noctua*	Absent from Ireland. Usually breeds during March to April. Little owls are quite vocal during the breeding season, producing long high-pitched calls, which are repeated regularly. Also produces distinctive contact calls 🔊 and alarm calls.🔊	🔊 Often nests within trees or built-structure cavities.

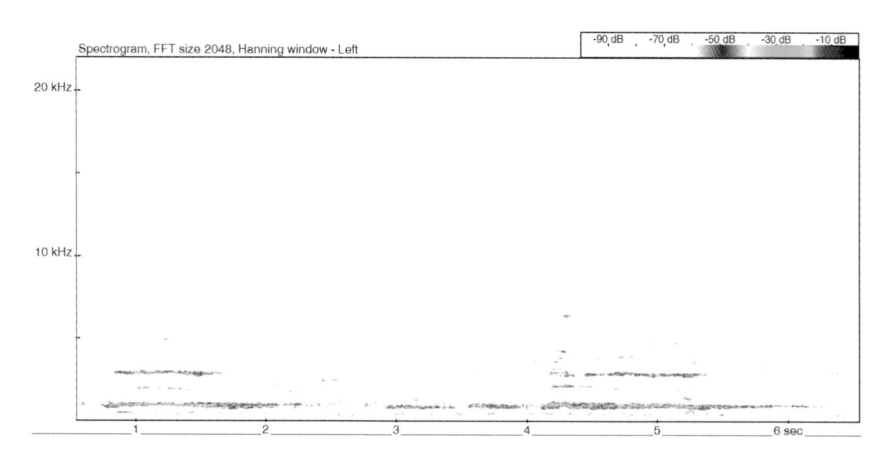

Figure 3.3 🔊 Tawny owl – male territorial hoot (D Darrell-Lambert, 2016).

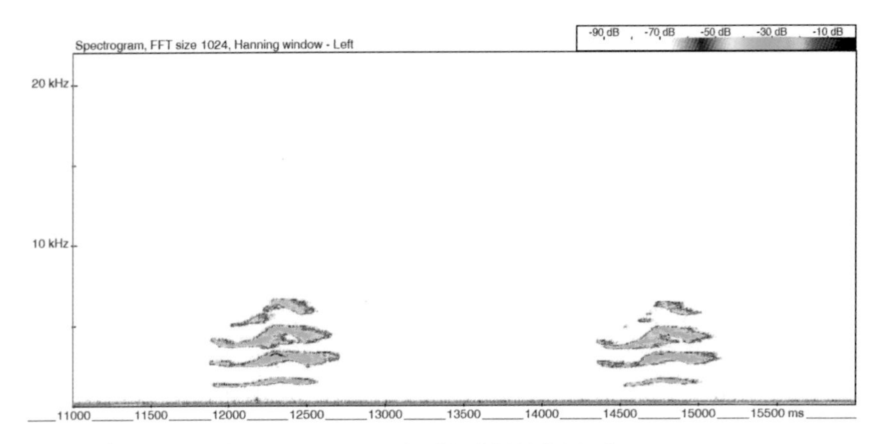

Figure 3.4 Tawny owl – contact 't'wit' or 'kwik' call (N Hull, 2019).

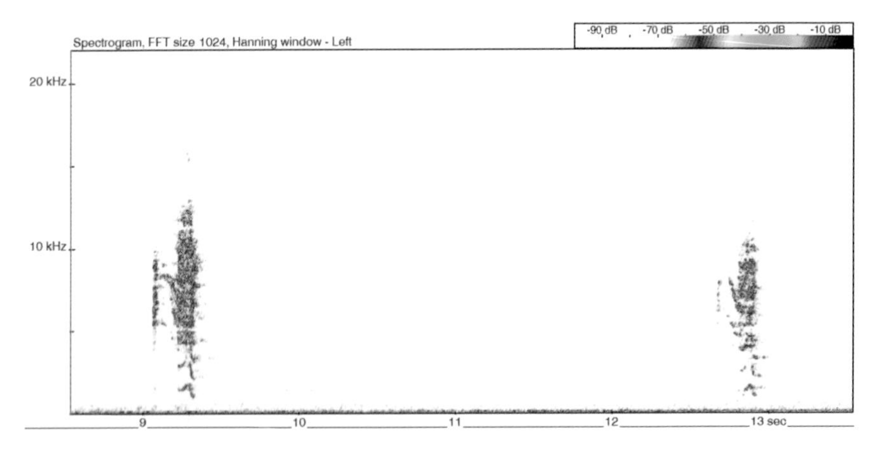

Figure 3.5 Tawny owl – juvenile single-note call (D Darrell-Lambert, 2008).

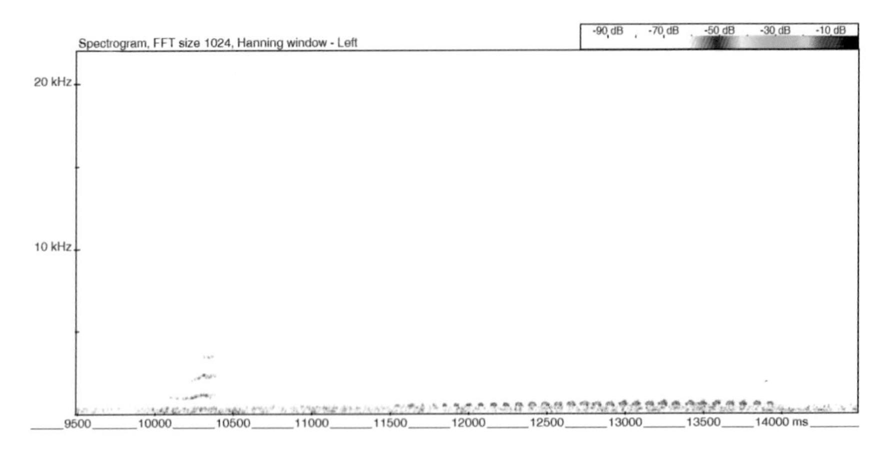

Figure 3.6 Tawny owl – bubbling call, preceded by a contact call (D Darrell-Lambert, 2016).

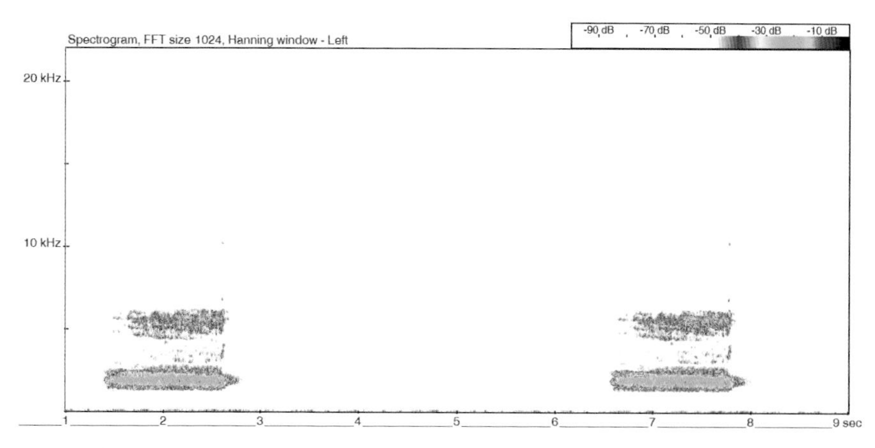

Figure 3.7 Barn owl – screeching call (N Hull, 2019; frame width 8000 ms).

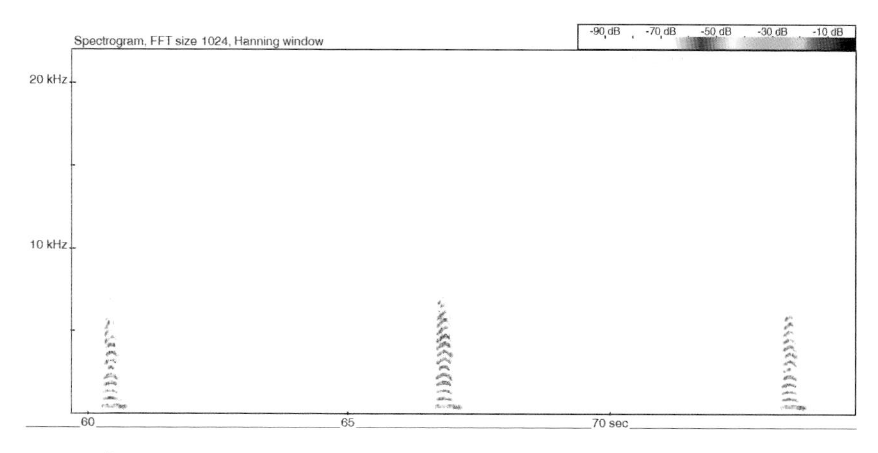

Figure 3.8 Long-eared owl – long hoot (H Lehto, 2015; frame width 15 seconds).

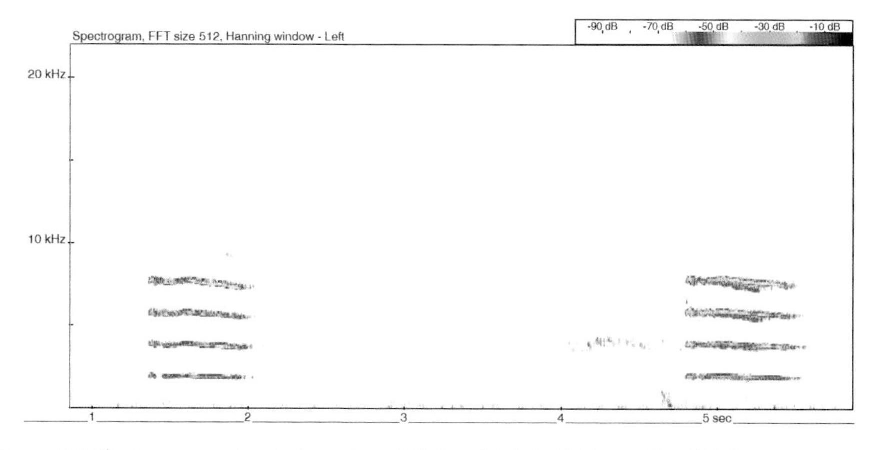

Figure 3.9 Long-eared owl – juvenile call (R Specht, Avisoft Bioacostics, 2019).

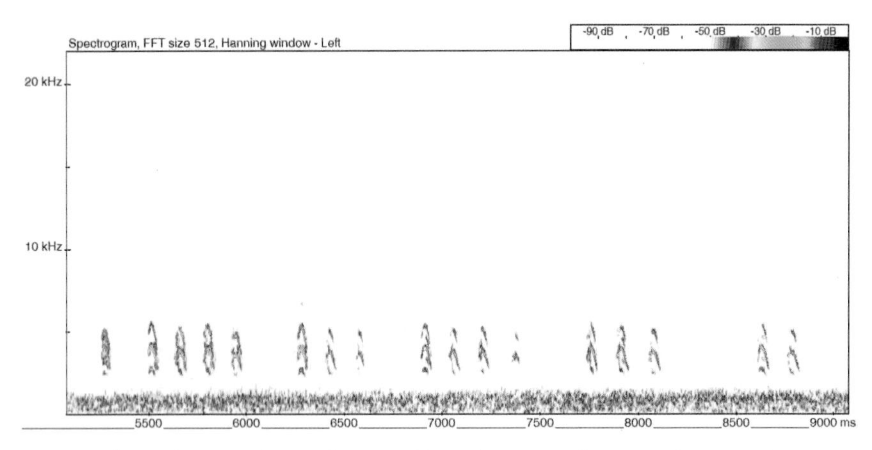

Figure 3.10 Little owl – alarm call (D Darrell-Lambert, 2013; frame width 4000 ms).

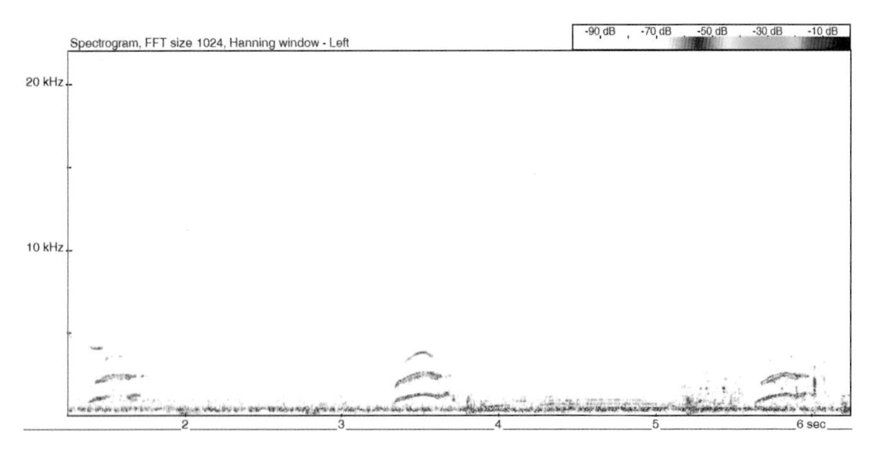

Figure 3.11 Little owl – contact call (D Darrell-Lambert, 2017).

Nightjar

Nightjar (*Caprimulgus europaeus*) has always had my attention. The first time I ever saw one was some 35 years ago when I was carrying out voluntary work for the RSPB at Minsmere, Suffolk. Ever since then, every time I have an opportunity to encounter these amazing birds I grab it with both hands. Such an opportunity arose in 2018 when David Darrell-Lambert suggested we go to Chobham Common (Surrey Wildlife Trust). 'What will we record there?' I asked. He read out a site list, and as soon as the word 'nightjar' was uttered – 'I'm in. Where will we meet?' Figures 3.12 and 3.13 are from recordings made that evening.

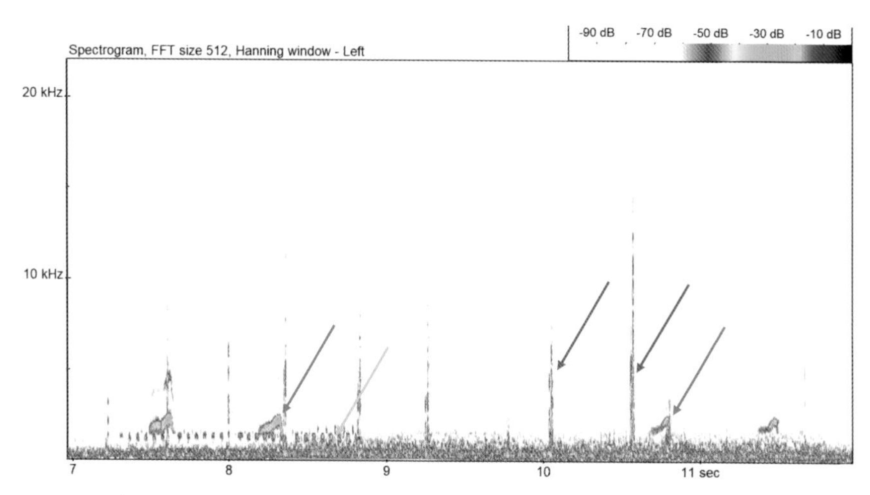

Figure 3.12 Nightjar – flight call (blue arrows) with wing clapping (red arrows), and another bird in the background churring (green arrow) (D Darrell-Lambert, 2018).

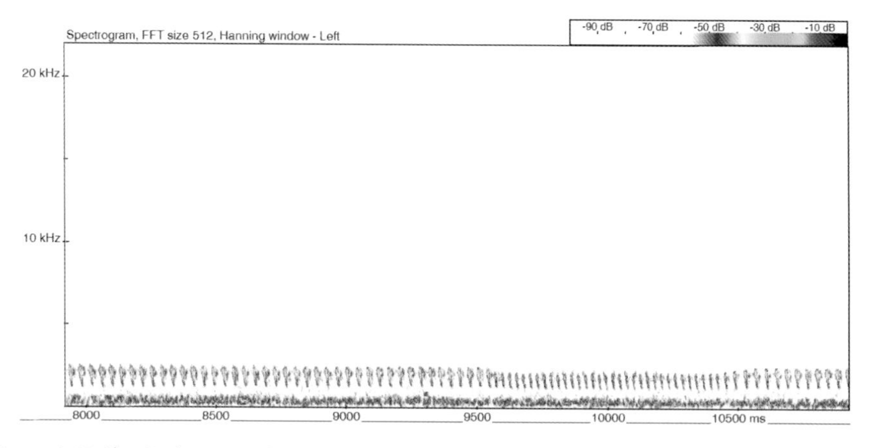

Figure 3.13 Nightjar – male churring (D Darrell-Lambert, 2018; frame width 3000 ms).

The churring call (Figure 3.13), produced as darkness falls, is not unlike the territorial calls of the natterjack toad (see Chapter 4, Figures 4.3 and 4.4). Both of these sound sources can be heard from some distance away, but the toad is far more irregular, and has particular habitat requirements which in many respects would not tie in with where you would typically expect to find a nightjar. It is also worth bearing in mind that nightjars are summer visitors to the British Isles, and don't usually arrive until well into May. Natterjack toads, on the other hand, will have been calling for many weeks beforehand.

Waders

Certain wader species can be quite vocal during darkness. As well as this, many often migrate at night, with flight calls being heard as they pass overhead. We will discuss migration separately, later on. Typically, waders are most often heard when they take off, or while in flight (i.e. flight calls). Your unexpected presence on a site may in fact have been the reason for the bird to take to the air in the first place.

Some of these vocal waders would be regarded as well sought-after species on many birder's lists, including stone-curlew (*Burhinus oedicnemus*), little ringed plover (*Charadrius dubius*) and golden plover (*Pluvialis apricaria*). Examples of typical calls produced by these species are included in the Sound Library.

The species that are more widespread, and therefore more likely to be heard across much of the British Isles, are described in Table 3.3. During summer months each of these can, on many occasions, be ruled in or ruled out based on the habitat in which a sound was encountered. Table 3.3 gives a broad description of typical breeding habitat and flight calls for these species, but please do bear in mind that the examples given are typical, as opposed to exhaustive.

Table 3.3 Widespread breeding waders within the British Isles – typical breeding habitats and call descriptions.

Species	Typical breeding habitat and flight call descriptions
Oystercatcher *Haematopus ostralegus*	Nests usually at ground level, often on farmland, on bare ground or in short vegetation. Also nests on flat roofs, including in more urban settings. Has various calls, the most distinctive being loud and high-pitched piping sounds, often heard in flight, and also at night during the breeding season.
Lapwing *Vanellus vanellus*	Often found nesting on farmland, fields and meadows, especially when habitat is damp and/or adjacent to wetland or moorland. Call can be described as a 'peewit' sound, which is also the name given to this bird in some parts of the British Isles.
Curlew *Numenius arquata*	Found nesting across a range of habitats, but in particular rough grassland, moorland and bog. The flight call is reminiscent of the bird's name, and during the breeding season you can hear its drawn-out bubbling song produced also while the bird is in flight.
Woodcock *Scolopax rusticola*	Usually nests on ground within woodland, especially associating with damper areas. When startled, will explode into the air, often also causing the surveyor to be startled. A much bigger bird than a snipe, and often seen flying at dusk, sometimes giving an inexperienced observer (from a distance) the impression of a very large bat. Its display flight, at dusk, is called 'roding'. It is accompanied by a higher-pitched call (likened to a blown kiss), often interspersed with a series of much lower-frequency grunts.

(cont).

Table 3.3 (cont).

Species	Typical breeding habitat and flight call descriptions
Snipe *Gallinago gallinago*	🔊 Found nesting in wetland, bogs and wet meadows, and can be quite active at dusk or dawn. When startled it will quickly take off and call harshly once or twice.🔊 Often heard at dusk and dawn, as well as overnight. It can produce a 'ticka' type advertisement song from a stationary position. Its aerial display noise (known as 'drumming') is created by the bird's outer tail feathers vibrating. This sound has been described by some as a high-pitched bleating.🔊
Redshank *Tringa totanus*	🔊 Nests on ground in wetland habitat, bog or moorland. Has a distinctive flight call, of two or three notes, described as 'tee-hoo' or 'tee-hoo-hoo'.🔊
Common sandpiper *Actitis hypoleucos*	🔊 Nests on the ground, in habitat associated with rivers and bodies of fresh water. Often active at night. High-pitched calls may be continuously uttered as the bird flies low over water or close to shore.🔊
Ringed plover *Charadrius hiaticula*	🔊 Usually nests on open bare ground and in the vicinity of water. Its call is a short two-syllable whistle, with more emphasis on the second syllable.🔊

Figures 3.14 to 3.24 show spectrograms relating to some of the more regularly heard calls for the species discussed in Table 3.3. In addition, examples of other calls are provided in the Sound Library.🔊

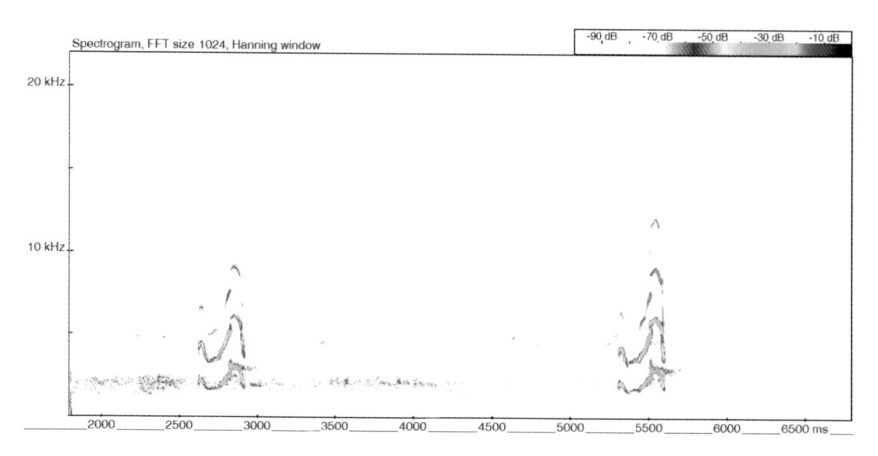

Figure 3.14 🔊 Oystercatcher – flight call.

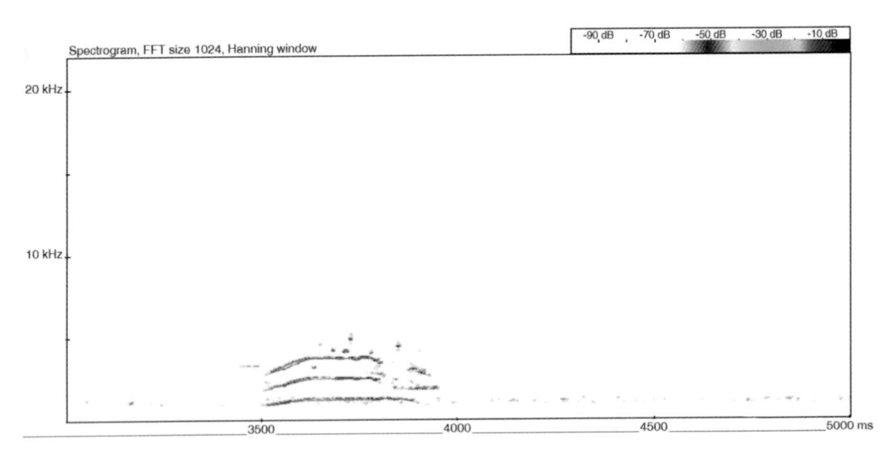

Figure 3.15 Lapwing – flight call (frame width 2000 ms).

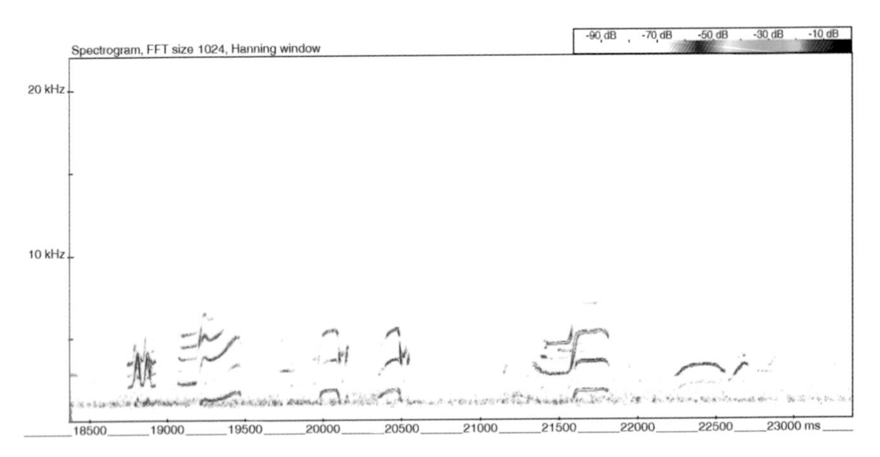

Figure 3.16 Lapwing – display call.

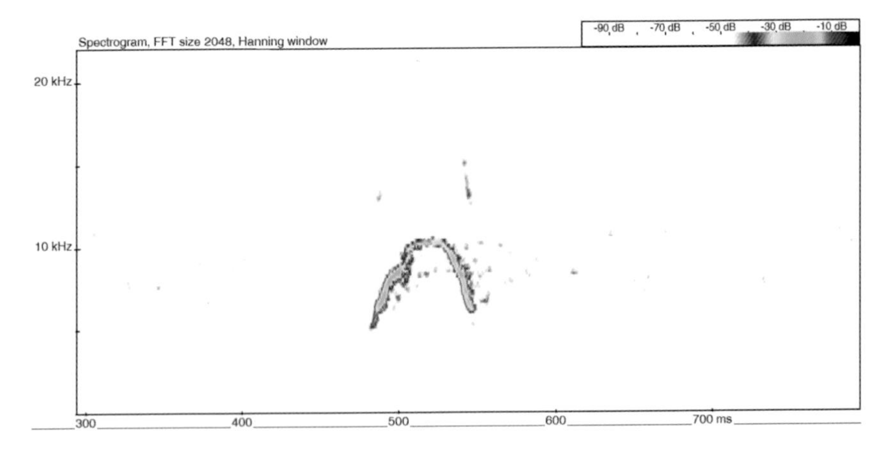

Figure 3.17 Woodcock – flight call while roding (frame width 500 ms).

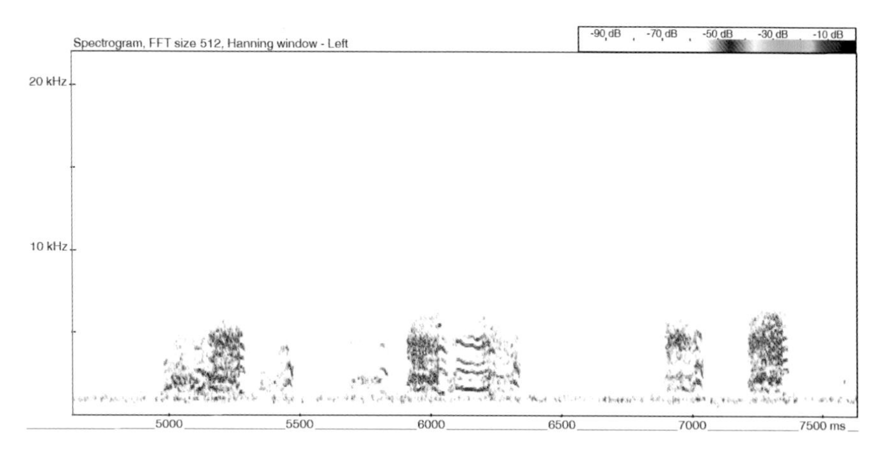

Figure 3.18 Snipe – flight call, with drumming (D Darrell-Lambert, 2018; frame width 3000 ms).

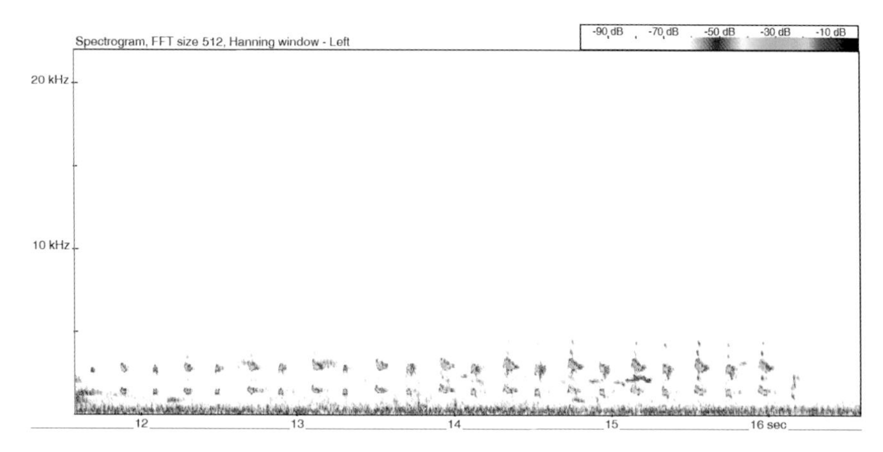

Figure 3.19 Snipe singing (D Darrell-Lambert, 2018).

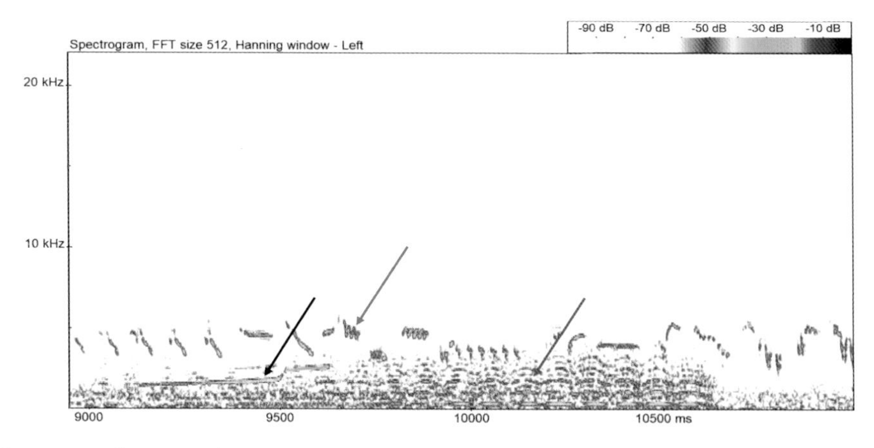

Figure 3.20 Snipe drumming (red arrow) with curlew (black arrow) and skylark (blue arrow) also present (D Darrell-Lambert, 2018; frame width 2000 ms).

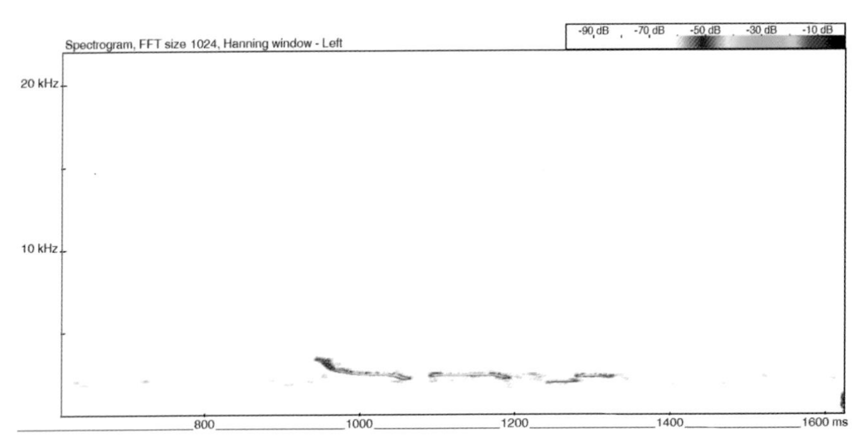

Figure 3.21 Redshank – flight call (D Darrell-Lambert, 2017; frame width 1000 ms).

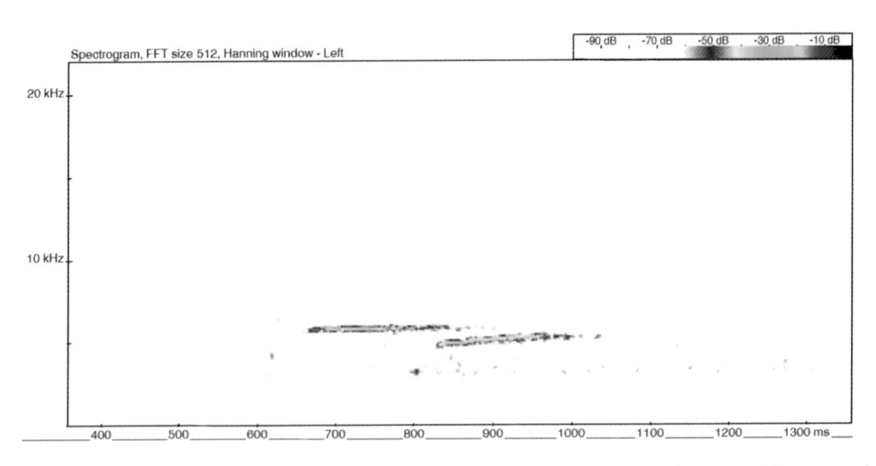

Figure 3.22 Common sandpiper – flight call (D Darrell-Lambert, 2018; frame width 1000 ms).

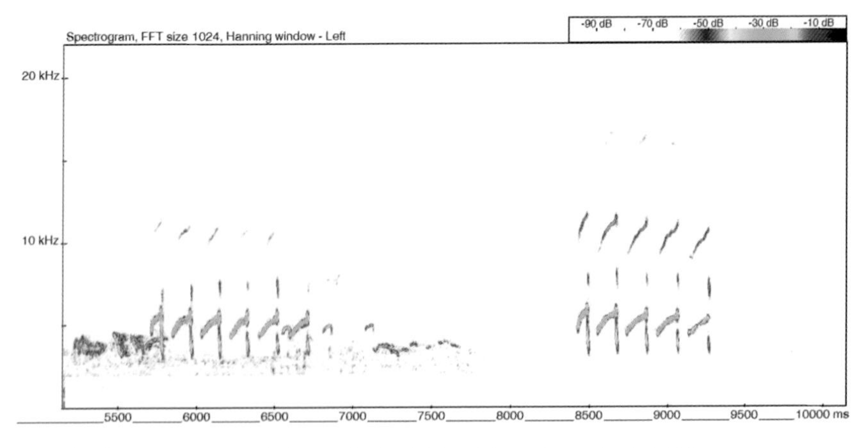

Figure 3.23 Common sandpiper – display flight (D Darrell-Lambert, 2017).

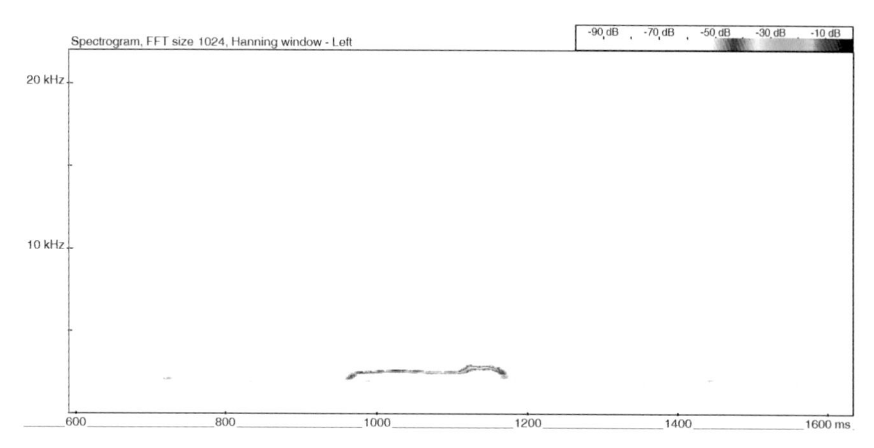

Figure 3.24 Ringed plover – flight call (D Darrell-Lambert, 2018; frame width 1000 ms).

Crakes, rails and coots

Birds in this group are usually found either within or in association with water bodies, or wetland areas. One exception to this is the corncrake (*Crex crex*), which is usually heard (not very easy to see!) on drier land, for example in fields and meadows hidden within ground vegetation. Corncrake is a rare species, as is spotted crake (*Porzana porzana*). Both of these are more often known to be present from their distinctive calls, and encountering either would make an experienced birder very happy on any day of the week. Examples of their calls are contained in the Sound Library.

The species that we are most likely to encounter, wherever we are in the British Isles, are coot (*Fulica atra*) and moorhen (*Gallinula chloropus*) (Figures 3.25 to 3.28). These species are not only widely distributed within our area, but also common. In addition to these, water rails (*Rallus aquaticus*) are widespread, albeit far less common and considerably harder to see, owing to their typical skulking nature within reedbeds. Water rail calls have been described as a '*variety of squeals, grunts and whistles*' (BWP*i* 2009), some of which, once known, allow an experienced listener to confidently identify its presence. Figures 3.29 and 3.30 provide some spectrogram examples, with additional recordings also being available in the Sound Library.

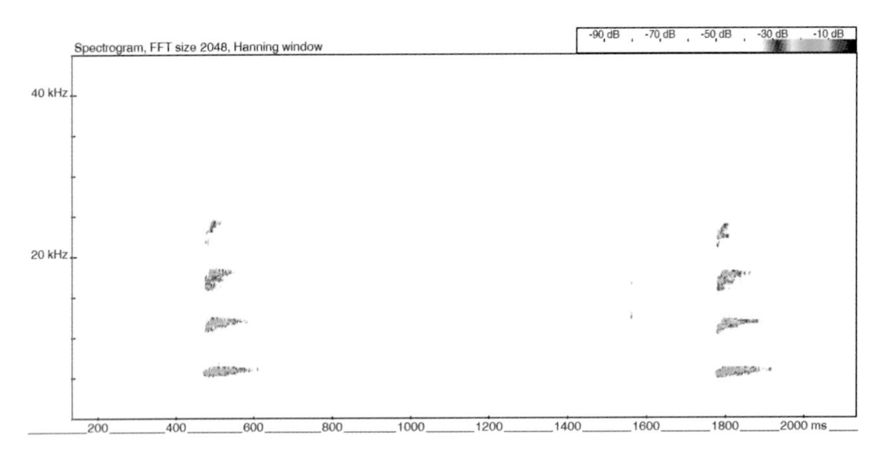

Figure 3.25 Coot – high-pitched call (frame width 2000 ms, frequency scale 0–45 kHz).

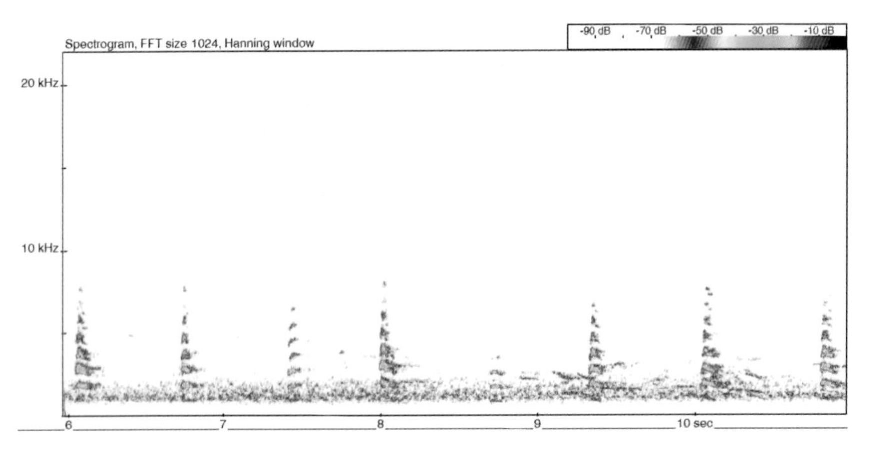

Figure 3.26 Coot – call.

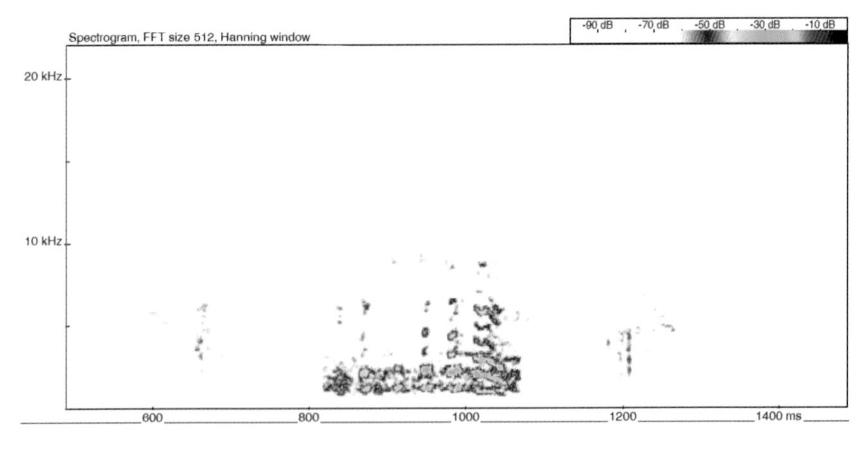

Figure 3.27 Moorhen – alarm call (frame width 1000 ms).

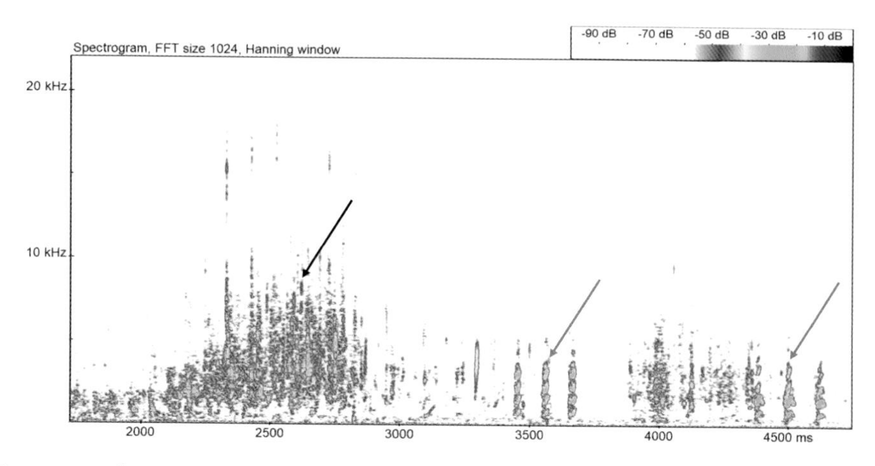

Figure 3.28 Moorhen – call (blue arrows), preceded by splashing (black arrow) after the bird had been disturbed (frame width 3000 ms).

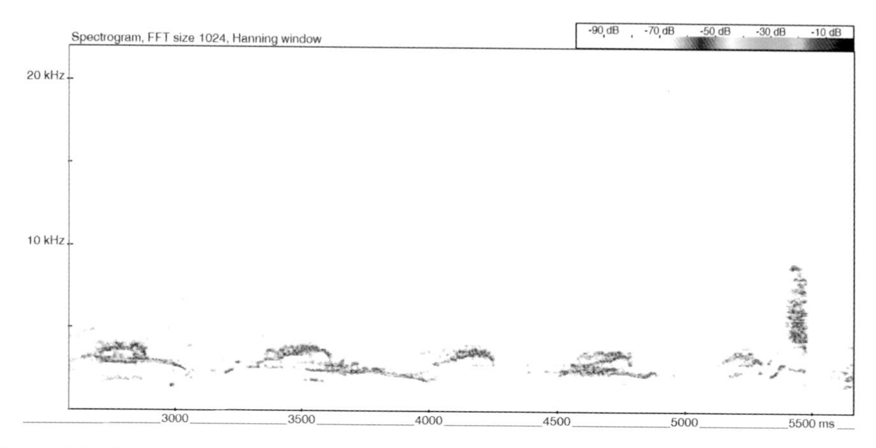

Figure 3.29 Water rail – squealing call (frame width 3000 ms).

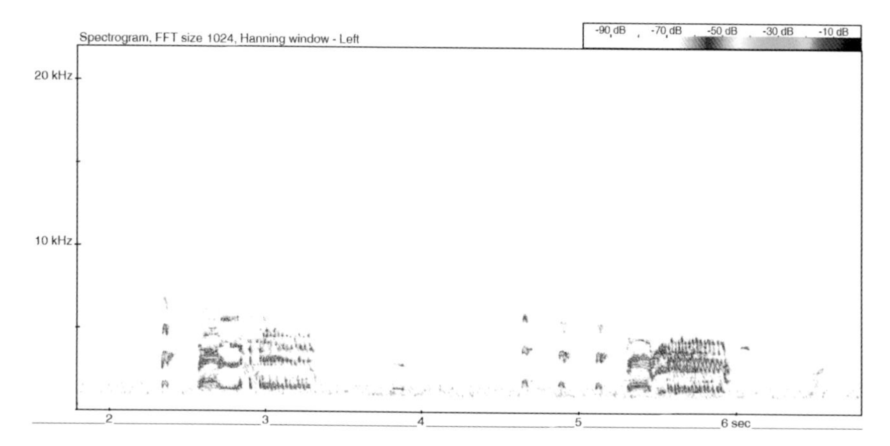

Figure 3.30 Water rail – displaying (D Darrell-Lambert, 2017).

Grey heron

Grey heron (*Ardea cinerea*) is widespread across the British Isles, showing a strong association with water bodies, rivers and wetlands. Typically, it can be found nesting in trees (heronries), and in large groups. In the Sound Library there is a recording of birds present in a heronry containing approximately fifteen nests, recorded during the day and before the young had hatched.🔊

Herons can move around in flight during darkness, and on a number of occasions I have jumped out of my skin when one has flown overhead unnoticed until it loudly announced its presence. Figure 3.31 relates to such an occasion.

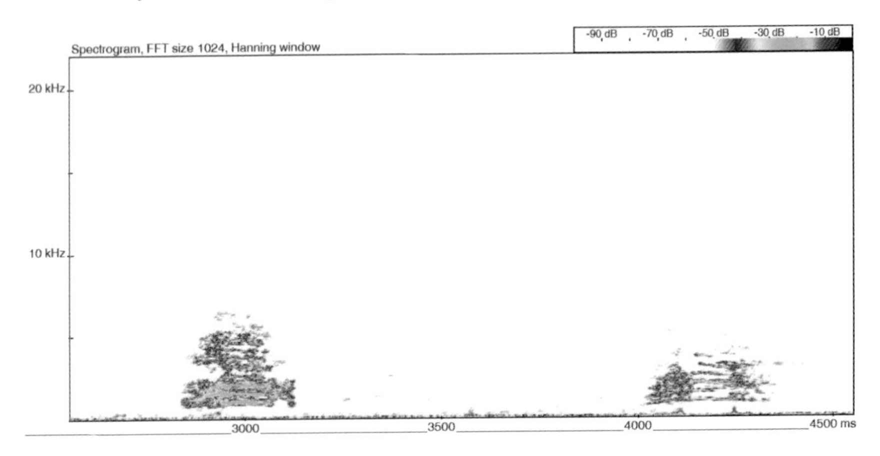

Figure 3.31 🔊 Grey heron – flight calls (frame width 2000 ms).

Kingfisher

Kingfishers (*Alcedo atthis*) call regularly in flight, and one is often heard to be coming before it is seen. They can also produce territorial or mating-related calls during darkness. Figures 3.32 and 3.33 provide examples.

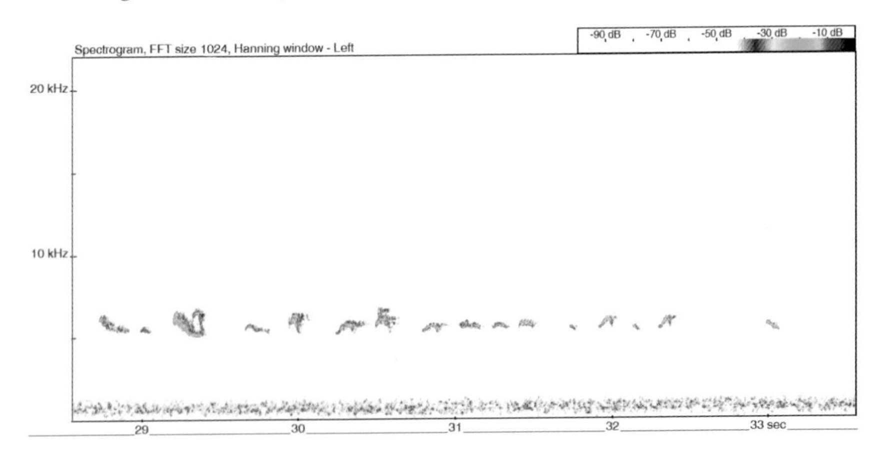

Figure 3.32 🔊 Kingfisher – call (D Darrell-Lambert, 2015).

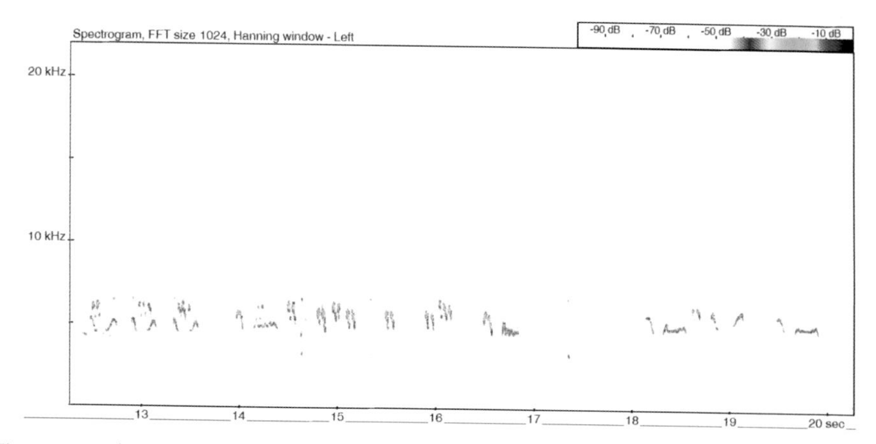

Figure 3.33 Kingfisher – call (D Darrell-Lambert, 2016; frame width 8000 ms).

Swift

Swifts (*Apus apus*) are summer visitors to the British Isles, being one of the last migrants to arrive and one of the earliest to depart. Their screaming is distinctive, and they will often be active right up until darkness, being one of the last daytime 'flight foraging' birds to go to bed. I have, on many occasions, seen swifts out foraging during the period when our early-emerging bats (noctules) may also be active. Figure 3.34 shows a typical example of swift 'screaming' calls.

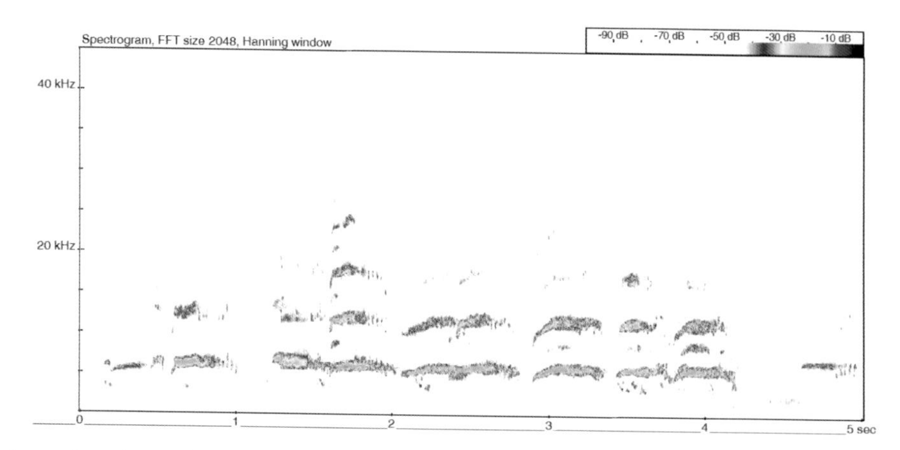

Figure 3.34 Swift – flight screaming calls (frequency scale 0–45 kHz).

Daytime passerines – night-time shenanigans

The term 'passerine' is a collective name given to perching birds, most of which are relatively small and sing. I am now going to discuss a small selection of these, choosing species that you are more likely to encounter during bat surveys in the British Isles.

Warblers

Many of our warblers are summer visitors to the British Isles. They can be a very difficult group to identify visually, partly because of similarities between certain species, and partly because many are particularly difficult to see as they skulk within vegetation. Either that or they are obscured from view and silhouetted high up in a tree canopy. Thankfully, identifying them by sound is often far easier. If there is one group of birds for which knowing their songs is going to be hugely beneficial, then it is this one. Once learned, many of what previously would have been documented as LBJs (little brown jobs) suddenly become easy to give names to. The first summer I had after learning my warbler songs was a fantastic experience, and I remember at the time feeling that I had immediately gone up to a whole new level of competence and confidence. If you could only just learn two, then chiffchaff 🔊 and willow warbler 🔊 (see Table 3.1) are easy to remember, and although visually they are very similar, in full song they couldn't be further apart. They are also the two most widespread and regularly encountered warbler species across the British Isles.

One other warbler species worth knowing, which is also a summer visitor, and fairly widespread, is grasshopper warbler (*Locustella naevia*). It can be particularly vocal at dusk and dawn. As the name suggests, its song is quite similar to what you would expect a grasshopper to sound like. Figure 3.35 shows a typical example, and it would be quite at home perched in amongst many of the calls within Chapter 5 (*Insects*).

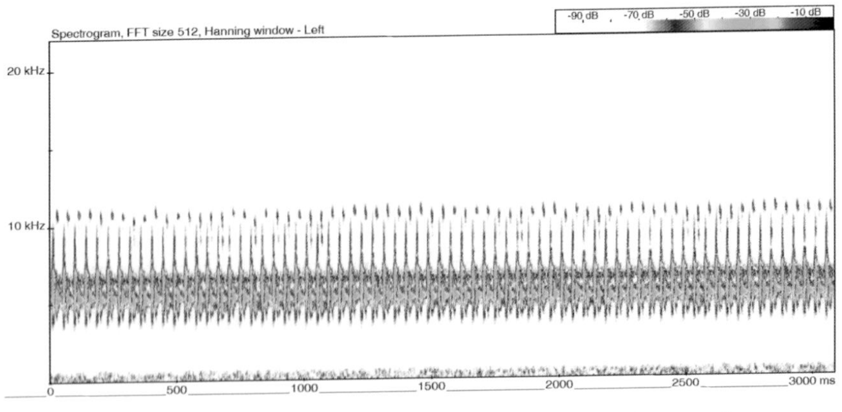

Figure 3.35 🔊 Grasshopper warbler – song (H Lehto, 2014; frame width 3000 ms).

As well as grasshopper warbler, other warbler species may also be heard singing during darkness, especially early in the breeding season (David Darrell-Lambert, personal correspondence). The more widespread of these 'offenders' include sedge warbler (*Acrocephalus schoenobaenus*) 🔊 and reed warbler (*A. scirpaceus*).

Wren

Wren (*Troglodytes troglodytes*) is one of the commonest breeding birds in the British Isles. For one of our smallest birds, it is surprising how loud a singing wren can be. '*What's making that noise?*' '*It's a wren.*' '*No way! They can't be that loud, can they?*' This species will often be heard at dusk and dawn, and can also be picked up on bat detectors. In fact it is the bird that most regularly gets sent to me as a 'query' bat social call.

The song usually has a distinctive trill element to it (Figure 3.36). Figure 3.37 shows an alarm call, which audibly can be described a series of loud 'teks'. People can sometimes get this confused with the 'tick' calls of a robin (Figure 3.39). An easy way to remember which is which is that wrens 'tek' and robins 'tick'. OK, so you are now maybe asking, '*Why is that easy?*' It's simple: wren has the letter 'e' in its name, and robin has the letter 'i', each of which corresponds to the vowel in the call description.

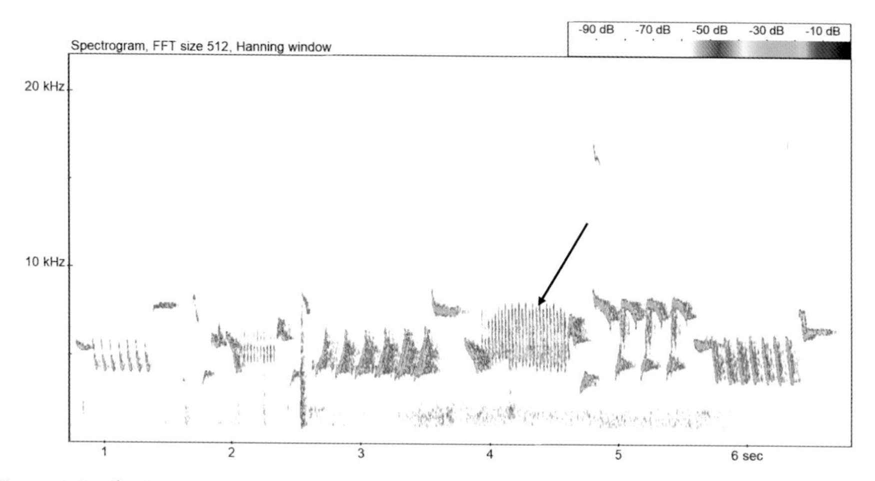

Figure 3.36 🔊 Wren – song, with a trill element (black arrow) (frame width 6000 ms).

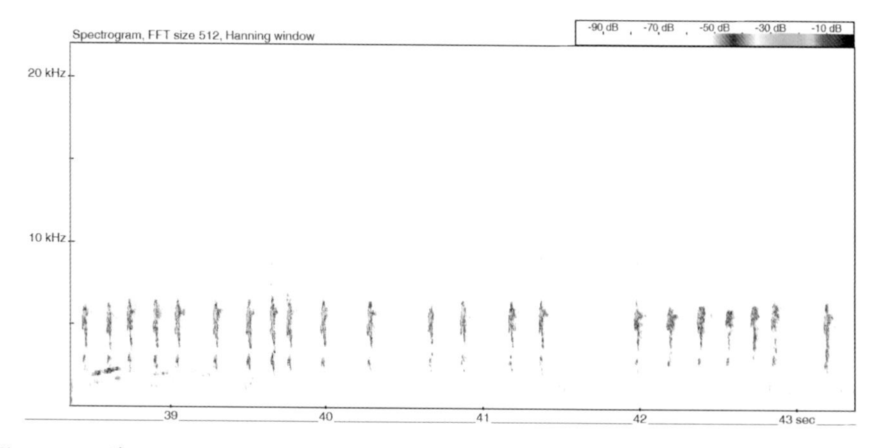

Figure 3.37 🔊 Wren – tekking alarm call.

Robin

Similar to wren, a robin (*Erithacus rubecula*) will often make noise at dusk and dawn. They will also sing during complete darkness. The robin's song (Figure 3.38) is quite thin (i.e. high-pitched) with a slow pace and deliberate pauses between each phrase. In some respects it has the style and pace of a blackbird, but at a higher pitch. The alarm call is a repeated 'tick' (Figure 3.39), as opposed to the wren, which produces a rougher 'tek' noise (Figure 3.37). Advice on differentiating and remembering these calls has been given above.

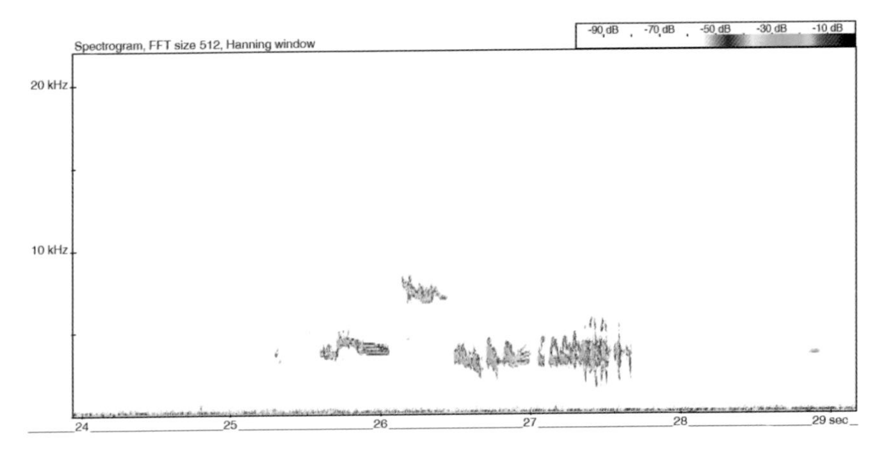

Figure 3.38 Robin – song.

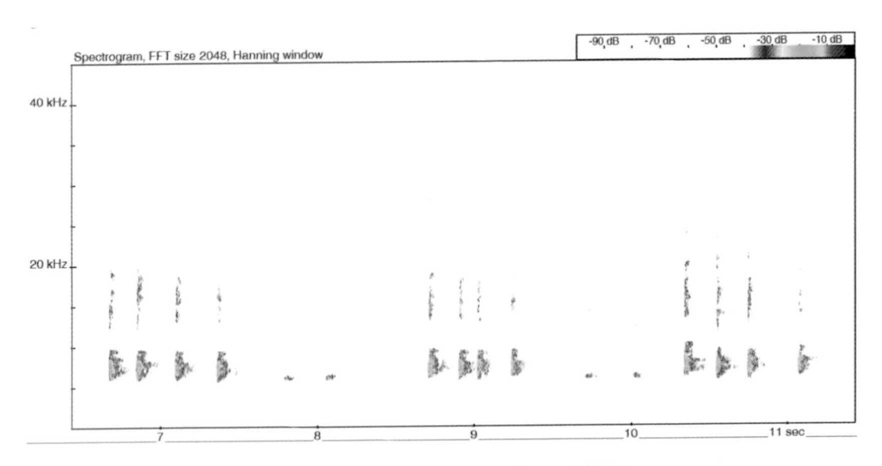

Figure 3.39 Robin – ticking alarm call (frequency scale 0–45 kHz).

Blackbird

The blackbird (*Turdus merula*) is a species of thrush. It is well known for making a racket at dusk, and there are three particular sounds to listen out for. The first is the song (Figure 3.40), which is made up of a longish phrase, followed by a pause of similar length, then another phrase, another pause, and so on. In terms of pace it is not dissimilar to a robin, but the pitch is far lower and more full-bodied, as opposed to the weaker, thinner-sounding robin. Blackbird song can also be confused with that of mistle thrush (*T. viscivorus*).🔊 With mistle thrush the song phrases are much shorter, with equally shorter pauses in between.

The other two sounds to listen out for each relate to distress behaviour. First of all you have the repeated 'alarm chinking' sounds that can often be heard (Figure 3.41), and then you have the clattering (Figure 3.42) created by a flushed bird, as it quickly heads off in the opposite direction (it was probably you!). When this flushed call happens, it wouldn't be uncommon for the bird to then start 'alarm chinking' once it lands. Both of these calls are often heard at dusk, either separately or in conjunction with each other.

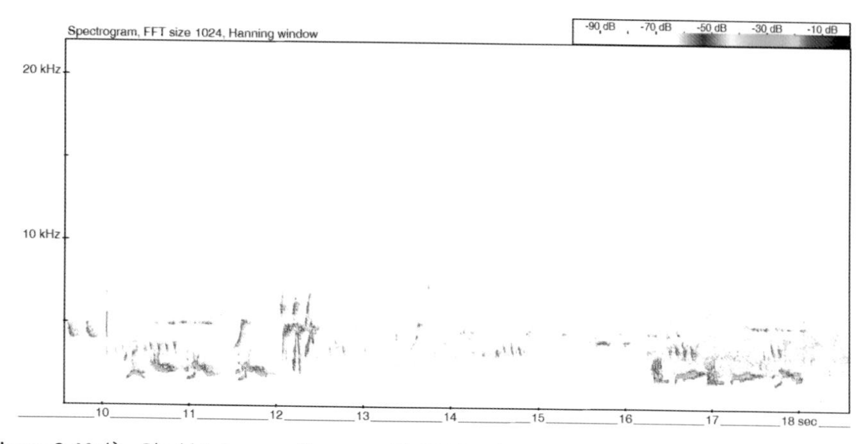

Figure 3.40 🔊 Blackbird – song (frame width 9000 ms).

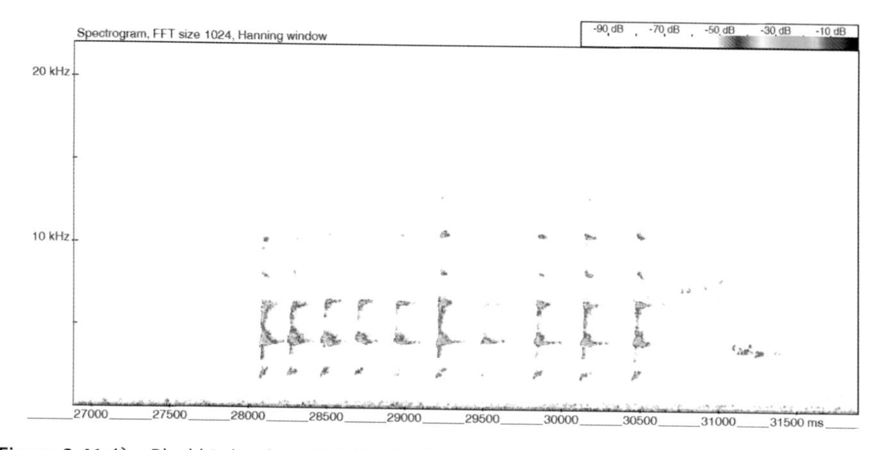

Figure 3.41 🔊 Blackbird – alarm 'chinking' call.

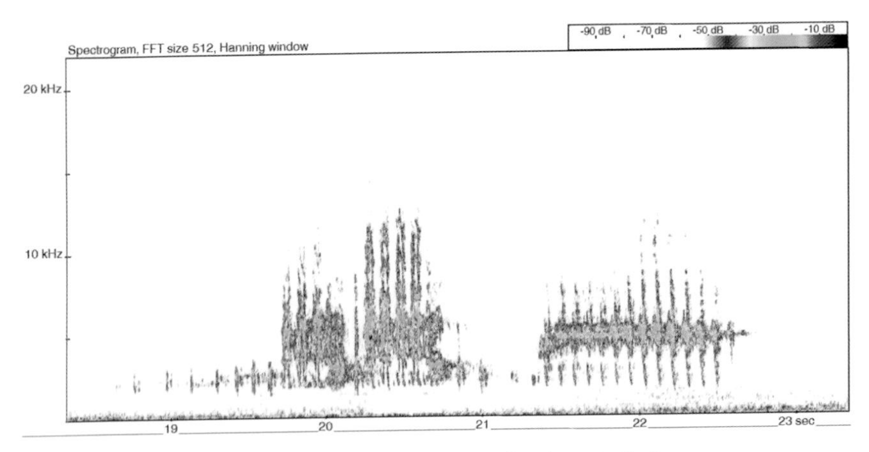

Figure 3.42 🔊 Blackbird – flushed clattering noise, made when startled.

Song thrush

Song thrush (*Turdus philomelos*) may also be heard at dusk, and occasionally well into darkness. The pitch of the song is quite similar to a blackbird, but the phraseology is very different. The songs tend to have longer segments and relatively shorter pauses in between. They will have elements that are quite random sounding, but the important thing to listen for is the repetition of phrases, which will always occur. After a short pause the bird will continue, but next time the repeated notes will usually be different. Figure 3.43 shows a good example of a song, with Figure 3.44 showing a ZCA example of a bird recorded by Sandie Sowler using an Anabat detector. Figure 3.45 is an alarm call, and 3.46 a flight call.

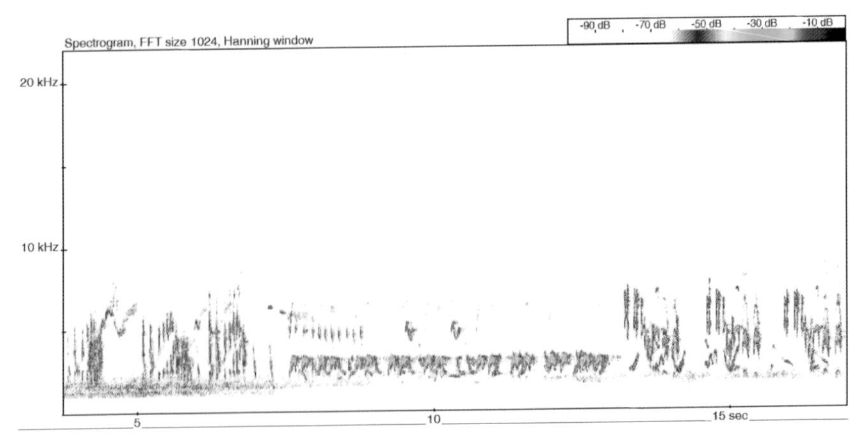

Figure 3.43 🔊 Song thrush – song (frame width 13 seconds).

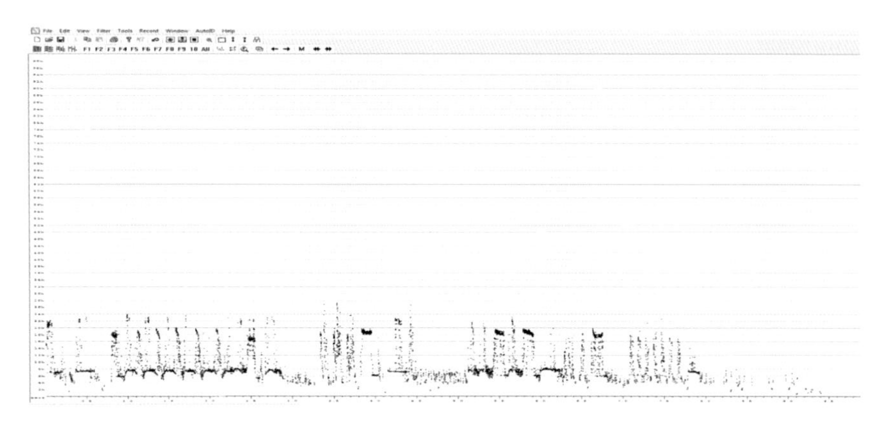

Figure 3.44 Song thrush – song, ZCA in AnaLookW (S Sowler, 2018; F3, uncompressed, *c.* 10 sec window).

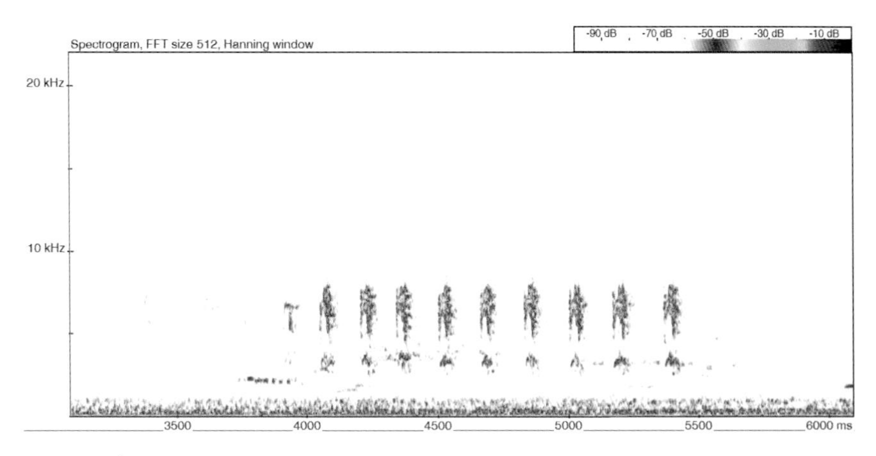

Figure 3.45 Song thrush – alarm call (frame width 3000 ms).

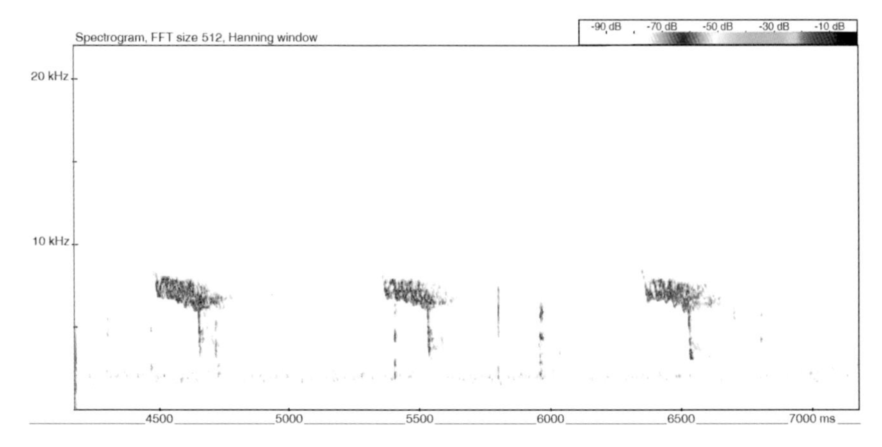

Figure 3.46 Song thrush – flight call (frame width 3000 ms).

Nightingale

I don't suppose I would be forgiven if I failed to include nightingale (*Luscinia megarhynchos*), bearing in mind that this summer visitor is well known as a night-time songster (it also sings during the day). Its distribution is restricted to the southern half of Britain. The song has lots of repeated notes, but different in delivery to that of a song thrush. Many people say that the nightingale produces the most beautiful song of all of our species. Personally, I disagree, feeling that the blackbird edges it. Blackbirds are common and taken for granted somewhat, I would suggest. Like all matters of taste when it comes to music, each to their own. Let's just agree they are both wonderful in their own way. Figure 3.47 shows a classic nightingale sequence.

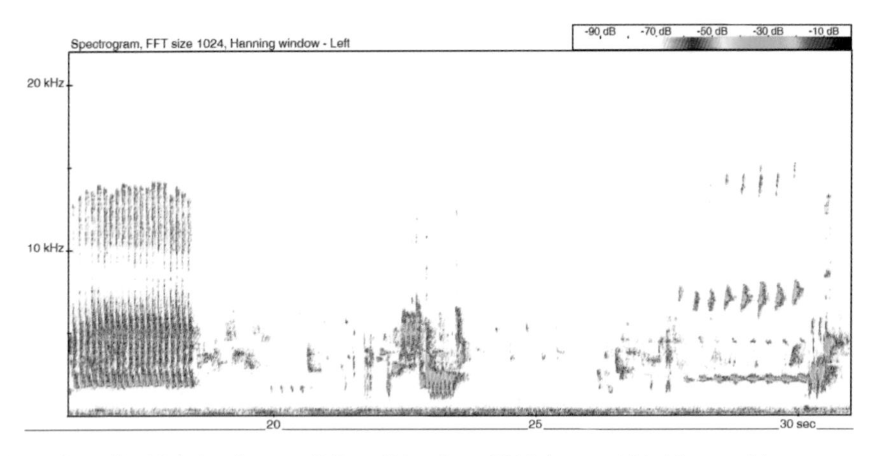

Figure 3.47 🔊 Nightingale song (D Darrell-Lambert, 2018; frame width 15 seconds).

Migrating birds

In the spring and the autumn, many bird species migrate during darkness, and while doing so can also be fairly vocal, producing flight calls in order to aid group cohesion when in a flock (Franke 2018). Birds that migrate singularly tend to be silent, or at the very least far less vocal, as there is no benefit to using up energy reserves calling out, when there are no other birds there to share information with.

Within the world of ornithology, with equipment and associated software becoming cheaper and more accessible, there has, in recent years, been a rapid growth in what is often described as 'nocmig' (from the words 'nocturnal migration'). Nocmig, or NFC (night flight call) recording is the nocturnal, audible equivalent of visible migration watching. To carry out this type of work the surveyor uses sound recording equipment to record the flight calls of birds migrating, unseen, overhead. Procedures for carrying out such surveys are described within a 'simple protocol to enable individuals or groups to conduct Standardised Nocturnal Flight Call Monitoring (SNFCM)' (Gillings *et al.* 2018).

Having recorded the sound, the analyst then identifies the birds to species level. This can, for many species in many scenarios, be regarded as diagnostic, as different bird species have their own specific flight calls. However, as in bats, there can in some cases be confusion calls which are more difficult to separate with high degrees of confidence. Occasionally, therefore, it is better to narrow calls down to groups, as opposed to individual species. Also, as in bats, automated classifiers are beginning to make an appearance. My suspicion is that surveyors wishing to identify bird calls will face challenges similar to those encountered when classifying bats, with at least some of the points raised in Figure 1.15 needing to be borne in mind. One other thing to consider, however, is that there are many more species of bird out there making noise. Added to this, each species has a wide repertoire, as well as differences that depend on age or sex, as well as geographical variations (e.g. local dialects). It is pleasing to see that in the protocol described earlier (Gillings *et al.* 2018) the authors say that

> identifying NFCs is often extremely difficult, and this is very much a collective work in progress. Recognising this, it is always better to err on the side of caution. Demand high standards of yourself. Identify to species where possible, but if in any doubt, use one of the species aggregates (e.g. 'flycatcher or chat spp.') rather than guessing.

Thankfully, from a bat worker's perspective, birds migrating at night do not impact as much as they could, as some migration is completed, or at least well under way, by the time that the April batting season commences. As we get closer to the autumn, migration for many species on the move doesn't commence until August, continuing through to October. Therefore, bird migration is clipping either side of our activity survey season for bats. During these periods all kinds of birds that you would not necessarily expect could be heard or recorded inadvertently, including thrushes, swallows, martins, other small passerines, raptors and geese. In addition, waders will often migrate at night and can be quite vocal when doing so. David Darrell-Lambert explains that

> migration isn't limited to when you would typically expect (e.g. spring and autumn). Some failed breeders may migrate south earlier (e.g. June onwards), and I have had whimbrel and greenshank as early as July in the Midlands. Spotted redshank, little ringed plover and common sandpiper can also be heard during the night in July and August.

Overall, where does that leave us? My suggestion would be that if you become aware of noise above your head, coming from birds flying over a site, it would do no harm whatsoever to record the sound and ask someone to attempt to give you an identification of the source. Likewise, a longer spell of regular bird sound picked up on a bat detecting system as an audible noise may very well be worthwhile investigating, especially if it is occurring during the migration period.

Recording birds

As discussed, and demonstrated, in Chapter 1 and earlier in this chapter, quite often it is possible to record birds, as well as other audible sound, using your bat detector (e.g. voice function) or your mobile phone. These techniques may produce good enough results if all that is being sought is a second opinion from someone else later on.

Birds are so accessible to us all, however, and if you wish to obtain better-quality results it is worth considering other recording options. These are quite different to what we are accustomed to within the bat world and the approaches I have described already.

The choice of microphone is a big consideration when recording any target species, and within this we have considerations such as omnidirectional, directional, sensitivity and frequency responsiveness, as well as overall performance across the frequency range involved. These features, and others, are taken on board by manufacturers when choosing the most appropriate microphones for their respective bat detectors. In the same way, those studying other taxa have to make similar judgements on the equipment they use, and different equipment may be needed. For a start, microphones that are suitable for recording bats will have a different specification to those used for recording lower-frequency, audible bird sounds, for example.

Web resources such as those listed in Appendix 3 give lots of advice as to what equipment is suitable for getting the best results. Included within these equipment options is an excellent piece of additional kit, which for sound falling within certain frequencies can be used to improve the quality of any recordings being sought. This item is a parabolic reflector, or dish, and it is used by many naturalists across a range of taxa, as well as for recording other sounds in completely different sectors.

Parabolic reflectors (binoculars for your ears)

As I have just mentioned, as well as microphones suitable for the job, and suitable digital recording devices, ornithologists can also use a parabolic reflector in order to acquire good-quality results. This piece of equipment needs to be of a specific diameter and depth in order to maximise its effectiveness within, broadly speaking, what we would describe as audible sound.

Using a dish has the effect of magnifying the sound, meaning that it is louder and clearer when being recorded. Most of the recordings used within this chapter, whether produced by myself or by David Darrell-Lambert, have been obtained using one of these in conjunction with a good-quality microphone. The equipment I use (see picture), is a Telinga (www.telinga.com) PRO-X system, which includes their stereo MK2 microphone and a foldable dish. I use this in conjunction with a Roland R-05 WAVE/MP3 digital recorder, with its settings adapted in order to power the microphone. David also uses a Telinga system, with the specifications differing slightly from my own.

When you start using a parabolic dish it can take a little bit of getting used to. You are accustomed to using your ears, whereby you hear something and you can tell the direction the sound is coming from. With a dish your whole sense of directionality diminishes. This is because instead of listening with two dishes, one on either side of your head (ears), you are now listening with one massive ear. As well as this, everything sounds so much closer, hence the expression 'binoculars for your ears'. So you are in the field, recording a sound. It's really

All ready to record birds – just add your own human! The Telinga parabolic dish, with MK2 microphone (far right) hooked up to a Roland R-05 digital recorder.

loud in your ears, and you think, 'Yes, I'm on it' – only to turn the dish in a different direction to find that you were nowhere close to 'on it'. Those using bat detectors will be accustomed to this up to a point, but when you hear the sound magnified it really does become considerably more evident.

Now one question, as a bat worker, you may now want to ask is, '*Why the funky pip don't we use parabolic reflectors for recording bat calls?*' Well some manufactures have done this to some extent, but you wouldn't notice the dish as easily, as they are considerably smaller, and referred to as horns or parabolas. These very directional microphone systems can have advantages for recording bats in certain scenarios, as there can be benefits from increasing directionality (as opposed to omnidirectional microphones) depending upon the specific data you are after. However, for the reasons we are about to discuss, parabolic dish specifications for recording bats are on a quite different, and in some respects unworkable, scale compared to what is beneficial for recording birds.

The challenge is all linked to the typical frequencies where bats make noise, the sound waves relative thereto and the corresponding dimensions of a parabolic reflector for it to be of any use to do the job. In short, the reflector would need to be very small, by comparison, to work, making it not very practical (David King, Batbox Ltd, personal correspondence).

Chris Corben (www.hoarybat.com) very kindly gave me some further explanations, as follows:

The physics are working against you when it comes to using parabolic dishes for bat work. You lose sound quality with distance (i.e. due to attenuation), and relatively speaking, compared to birds, bat sound doesn't travel that far. For the directionality to work at the frequencies emitted by most bats, the dish would need to be very small. For example, to get the same directional pattern for a bat at 40 kHz you would need one-20th of the dish aperture for a bird at 2 kHz. Both birds and bats vary greatly in frequency, so there isn't just one aperture which is good for bats and another which is good for birds. Rather, there is a lot of overlap depending on the species involved and what you are trying to achieve, but still, an aperture which is a good compromise for a wide range of birds just won't work for a wide range of bats.

Another complication is that birds are generally recorded while stationary or at least being seen by the surveyor, while bats are generally recorded unseen and in flight. A very narrow beam pattern would therefore be much harder to keep aligned to a bat than to a bird. When taking account of all of this, you are trying to find a bat in the dark, with a small dish. The bat will need to be very close and you would need to be pointing the dish exactly to where the bat is. Not easy, bearing in mind that the bat is moving at speed within a dark environment.

So, in conclusion, don't go buying a 'normal' sized parabolic dish thinking that it's going to work well, if at all, for recording bat echolocation.

Hints and tips

Field and analysis techniques

While in the field, carrying out a bat survey, if you are conscious that you are picking up bird-emitted sound on your detector, it would be helpful to make a note of this, so that it can be allowed for during the analysis phase. This is especially the case if the person doing the analysis is not the same person who was in the field recording the data in the first place.

During analysis, how do you know that it is definitely a bird you have recorded? Well, sometimes that can be relatively easy. First of all, looking at the call sequence, if it is a bird in song, as opposed to a flight call, contact call or alarm call for example, it will usually look complex and be occurring over a long period of time. The sound will normally be spread over many seconds, whereas with bats you tend to get lots happening within a couple of seconds, or even less.

Quite often when bat workers see bird-related noise on a spectrogram, their first thought is that it is a bat social call, as opposed to bat echolocation, or indeed a bird. I have on a number of occasions been sent a bat social call, only to discover that it was a bird. Wrens and robins are the most frequent offenders, this probably being due to how widespread and common each of these species is, meaning that they would be the more likely ones to cause issues in this respect. Also, they are both relatively vocal at dusk or dawn, and robins even overnight.

However, there are also, undoubtedly, occasions when it just isn't that simple (e.g. bird calls as opposed to more complex song). Taking this into consideration, I will now move on to a discuss sound analysis relative to bird-related noise in more detail. Some of what follows may also be worthy of consideration when looking at noise emitted from other sources, as discussed elsewhere in this book.

Are you seeing the whole picture?

All bird calls and songs within the British Isles are audible, albeit certain species may be more problematic for people with difficulty in hearing higher frequencies. But, despite them being readily audible to most of us, there are other things to consider relating to bat detector work and sound analysis.

A bat detector microphone's frequency responsiveness would normally be expected to avoid recording bird species at the lower frequencies (i.e. < 10 kHz) you would normally expect these to occur. However, you may pick up some birds calling at a higher frequency, for example goldcrest (*Regulus regulus*), which can produce frequencies above 8 kHz, and lesser whitethroat (*Sylvia curruca*), which is capable of achieving frequencies beyond 12 kHz (Constantine & The Sound Approach 2006). These species are not typically night-time birds, but activity close to dawn or dusk would certainly be feasible.

This becomes more confusing, as it means that your detector's microphone may not be picking up the complete story, in that it may only be recording the higher-frequency elements of a largely lower-frequency series of sounds. Therefore, what you are seeing is only the upper part of a picture. Secondly, harmonics can be quite common, and if present they will be closer to, or within the frequency range capable of being picked up by a detector (Harry Lehto, personal correspondence). In effect, you could have an audible sound that isn't registering strongly, if at all, and you are seeing just the harmonics that occur within the microphone's frequency responsiveness range. In theory, therefore, you could have something loud at a fundamental frequency (the first harmonic) of *c.* 8 kHz, with a second harmonic within the microphone's range at *c.* 16 kHz. All you see is the second harmonic – and you assume it can't be a bird, or anything else easily audible for that matter. This point was also highlighted by another contributor (Graham Sennhauser, Tetrix Ecology Ltd) when he had recorded oystercatcher on a site (using a static bat detector), with only the higher-frequency elements being present during analysis.

We therefore have a bit of a dilemma. On the one hand it is really useful when bat detector microphones don't pick up lower-frequency sound (note that some detectors pick up lower frequencies better than others). But on the other hand, if all of the sound was accurately recorded it would make some problem-solving scenarios far easier, in that we could more confidently rule in or rule out other sources of noise.

I am going to demonstrate this effect by using two random examples of bird calls picked up where harmonics were very evident. First of all oystercatcher (Figure 3.48), and then dunnock (Figure 3.49). In each case I would ask you to imagine what these spectrograms would look like if lower frequencies had not been recorded, or if so, only weakly. In fact, that's maybe too much to ask you to do, when I can easily demonstrate this to you. Look at the corresponding figures, 3.50 and 3.51. They look different to the originals, don't they? It's easy when you know what the originals look like, but now imagine that all you had was the second set of examples.

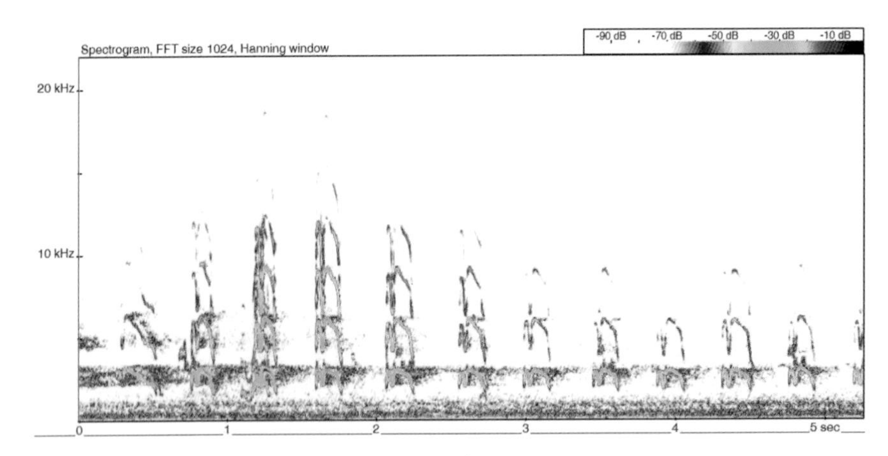

Figure 3.48 🔊 Oystercatcher – flight call (original).

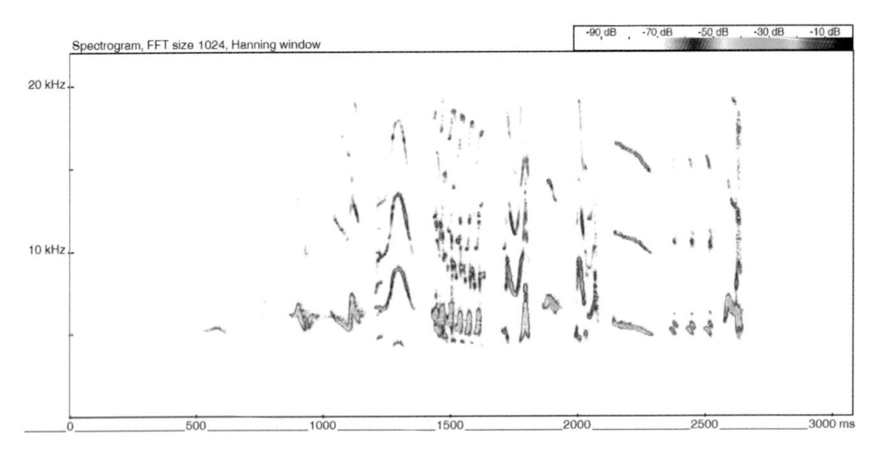

Figure 3.49 🔊 Dunnock – song (original; frame width 3000 ms).

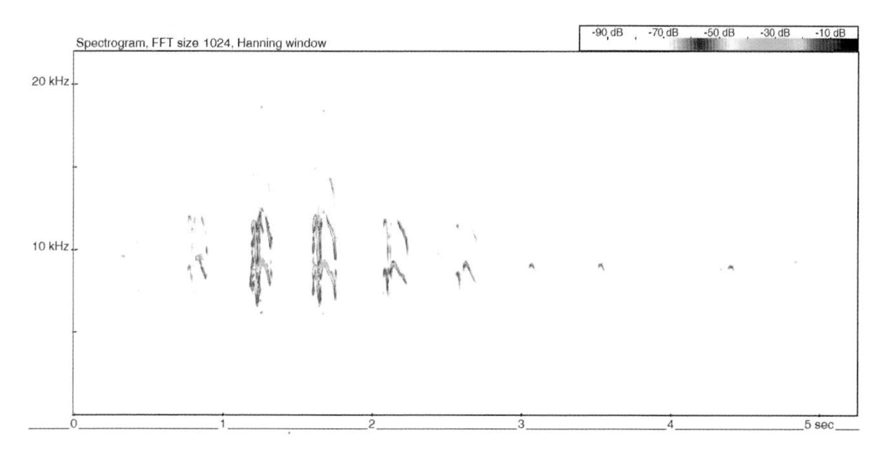

Figure 3.50 Oystercatcher – flight call (lower frequencies absent).

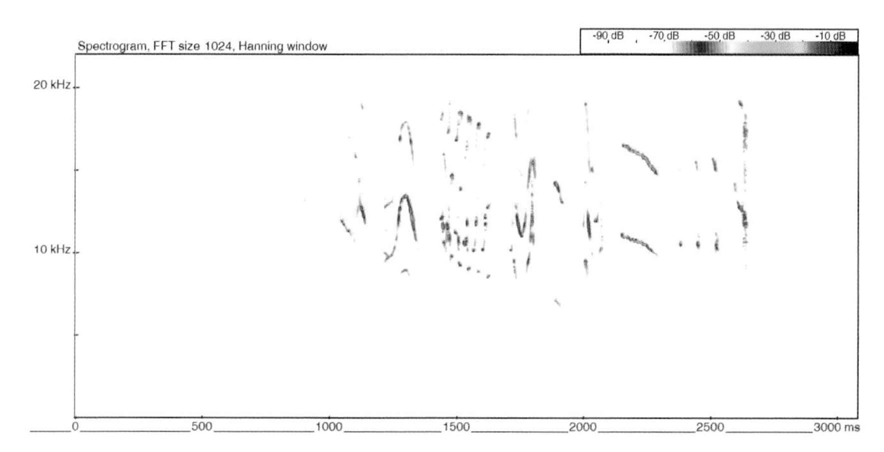

Figure 3.51 Dunnock – song (lower frequencies absent; frame width 3000 ms).

When faced with something like this, there may in fact be clues. First of all, look at the position of the harmonics in the second set of figures (3.50 and 3.51; also refer to Appendix 1, Figure A1.5). If we were seeing a true recording of the whole sound we would expect the second harmonic to be double the frequency of the first (or fundamental), as seen on the spectrogram. For example, in the oystercatcher example (Figure 3.50) the fundamental appears to be at a maximum frequency of *c.* 8 kHz, but the second harmonic, if accurate, is nowhere close to the expected *c.* 16 kHz. Something just isn't measuring up, and this should make an analyst suspicious, to say the least. Secondly, listening to the sound may help you decide that it is a bird (or something else) after all, albeit you may struggle as to its identity.

Use your ears

It would be unwise to base your judgement purely on what you are seeing. Remember to use everything at your disposal, including your ears!

If you are able to listen to the track in its original, real-time, format you should be able to hear if it's a bird. Often this will mean that either you or a more experienced person will be able to identify it, if not to species level, then at least to potential sources of the sound. If you haven't already listened to the examples shown in Figures 3.48 to 3.51, then now might be a good time to do so. With the lower frequencies removed, the oystercatcher no longer sounds like the same bird species. In fact it now sounds a bit like a dodgy goldcrest.🔊 Despite this, however, it does still sound like a bird.

There are a couple of options available to you in order to listen to the sound you are seeing on your software. The first obvious option is that you use the software to play the call in real time. Many of the bat analysis software packages allow you to do this. If you are unsure how to do it, then within the Sound Library I have created a pdf screen grab tutorial, *Listening to Your Software*, that walks you through the process for the most popular packages used by bat workers in the British Isles. Now, sometimes when you try to do it through the analysis software you either cannot hear anything, or it doesn't allow you to play the call. In either of those situations definitely don't give up, as there is a second option, as alluded to earlier. If you come out of your bat analysis software and browse for the sound file within its folder, and then click on it, it should open and play the sound using your PC's nominated media player (e.g. Windows Media Player). You may now very well find that you are able to hear the sound. Quite often this second approach has been the one that has established precisely what is happening. Whichever option you use, I would always recommend using headphones, as this enables you to be much more aware of any sound present, especially quieter and higher-frequency noise. As well as this, if you don't know what it is that you are about to be listening to, it possibly isn't wise to have it blaring across the office for all to hear!

Having listened to the sound in real time, if you are still not sure, or you don't hear anything, the next approach is to slow the sound down by a factor of 10 (i.e. time expansion × 10). What do you hear now? Quite often bats slowed down 10 times can sound very bird-like. And birds slowed down often sound quite whale-like, or at least what many of us think whales sound like based on what we hear during natural history programmes on the television (e.g. humpback whale, *Megaptera novaeangliae*, song).

Adopting an approach as described above, you have now listened to the call in real time to establish if it was audible, and if so what it sounded like. Then, if it wasn't audible, you have listened to it in time expansion to see if that has helped you to hear anything ultrasonic. There will be times, however, when it isn't going to be as easy as this. Appendix 2 provides the outline of a process that I use when trying to solve more complex sound mysteries. I would not suggest that my approach is perfect or the only way to do this, but it is a logical process (as opposed to just randomly doing stuff and pressing buttons) that more often than not works for me, and I am happy to share it.

Pulling it all together

Allowing for everything we have discussed in this section, Table 3.4 summarises many of the issues to be taken account of when potentially encountering birds during bat surveys or the sound analysis process.

Table 3.4 Consideration of birds relative to bat surveys.

Subject	Problem	Potential solutions
Equipment damage	Birds causing damage to equipment (e.g. wind muffler or microphone)	Create a small metal gauze or plastic cage for the microphone. The gauze should not touch the microphone or the wind muffler.
Inadvertently recording	Birds being recorded inadvertently and potentially impacting upon sound analysis process	All bird sound in the British Isles is audible and below the threshold for recording bat echolocation. The specifications of individual bat detector models and microphones will impact upon their susceptibility to pick up audible non-bat-related noise. A potential issue arises with higher-frequency sounds and harmonics, in that these may fall within the microphone's range of frequency responsiveness. In this respect caution is advised, including a proper manual audit of any calls recorded that have been subjected to a filtering or automated classification process.
Noise filtering process	Applying a noise filter doesn't remove bird noise	A noise filter categorising noise at levels below a frequency of 10 kHz should remove almost all bird-created sound, without having an impact on bat echolocation recordings. Carry out a manual audit of filtered noise, to ensure nothing has been misallocated. Also bear in mind that harmonics may not have been removed during filtering process.
Bat classifier error	Birds being mistaken for bats	Unlikely to occur with this group, unless higher-frequency harmonics have been recorded.
Human error	Person doing bat call analysis mistaking birds for bats	Listening in both real time and time expansion (× 10) would normally mean that it is unlikely that birds would be misidentified by an experienced technician as bat echolocation. It is more likely that bird sound could be mistaken for bat social-call sequences. Appendix 2 offers some thoughts to bear in mind when faced with something unusual.
Anything occurring immediately around or beside the microphone could impact upon how it performs or how recorded sound behaves (e.g. sound echoing off a smooth flat surface close to the microphone). Therefore, any alterations made to a microphone system, or its immediate surrounds, that could impact upon recorded sound should be tested ahead of being used on case work.		

Author's diary note

My earliest recollection of being interested in natural history is when, while at infant school, I returned home one afternoon and presented my mum with an earthworm which I had earlier carefully picked up from a puddle, and nurtured all the way back to the skyscraper where we lived. I was very excited as I got into the lift and progressed up to the fifth floor. My mum answered the door, and I announced, 'I am going to build a wormery.' Her look was nothing short of horrified, as she screamed and ordered me to go back down in the lift and dispose of the poor creature.

A couple of years later we moved house, and now, having a garden, I started to get into birdwatching. I guess my parents were quite pleased that at least there wouldn't be any creepy crawlies in the home. Although not interested themselves, they would take me somewhere most weekends to see different species, over and above my rapidly growing garden list, which was increasing in stature with every new feeder that was being added to the bird table.

Birdwatching had me hooked right from the 'get go', and I quickly moved on to being a birder (not that I knew that's what it was called at the time). My knowledge and ID skills by the age of 12/13 were impressive (less so now, I'm afraid to say). I owe a lot of this to having an excellent Young Ornithologists' Club (YOC) leader, Logan Steele, who is still very active in the Scottish bird world today. For those 'youngsters' out there, the YOC used to be the junior branch of the RSPB, and we all had our YOC embroidered kestrel emblems proudly stitched onto our rain jackets by our mums. I still don't know why I didn't get beaten up at school.

Almost 35 years later, Logan and I coincidentally ended up working for the same insurance company (Norwich Union, now Aviva), and, in particular, I recall a conversation we had one afternoon in Edinburgh. He didn't work from the same office as myself, so we rarely met. That day, I saw him in the distance, and I quickened my step in order to catch up, so that we could share the lift. I said good afternoon, and asked him how he was. I then proceeded to tell him about this excellent spot I had found for watching raptors (he was a member of his local raptor group, after all). As he listened attentively, I told him all about the sightings I had witnessed there. Anyway, it turned out that he was in the building to meet with our regional manager (Tracey), who, as it happens, was also my boss's boss. So, being helpful, I took Logan to Tracey's room, continuing to tell him more about my latest birding exploits.

Tracey welcomed him in by saying, 'Brilliant to meet you again John.' The guy I had been talking to wasn't Logan after all! Yes, I had just met some random stranger (considerably more senior than me in the business) in the lift and had swamped him with ornithological anecdotes. He didn't know me at all, but had remained polite throughout, and thought he had just met some nutter in the lift. I doubt that he would have ever been able to tell a kestrel from a griffon vulture, let alone a sparrowhawk from a goshawk.

And after Tracey had finished her meeting? Yes, I was called in to explain myself, as John had obviously told her all about it. Before long everyone on the floor (*c.* 40 people) as well as our entire Glasgow office (where Logan was based) knew what had happened, and quite rightly I was at the buzzard's butt of many jokes for some time.

CHAPTER 4

Amphibians

Listen before you leap

Within Britain and Ireland we have a selection of amphibian species (newts, frogs and toads), some of which are native, while others are non-native (i.e. alien). Newts don't make a noise, and therefore are not covered in this chapter. We are only ever likely to encounter sounds emitted by frogs and toads, which belong to the order Anura, with the individual species therein being described as anurans.

As bat workers we are often out at night during spring and summer, and if we are in the right habitat we have the opportunity to be pleasantly distracted by the diverse noises made by frogs and toads. During the mating season some of our species can be heard when it's light, but it is as dusk falls that all of

Table 4.1 Overview of native species of frog and toad occurring in the British Isles.

Species	Distribution	Notes on typical habitat and behaviour
Common frog *Rana temporaria*	Widely distributed throughout Britain and Ireland[1][2]	🔊 Usually nocturnal, but mating can also occur during daylight.[3] They are regarded as a generalist,[2] breeding in small to medium, shallow, vegetated ponds, with few or no fish. For most of the year, outside of the breeding season (normally concluded by late March) they are found in damp habitats, including lightly grazed pasture, scrubland, open woodland, parkland, gardens and moorland.[4] Ponds used by common frogs for breeding may also be used by common toads. However, there is usually a clear difference in timing, with frogs usually spawning before the arrival of toads.
Northern Pool frog *Pelophylax lessonae*	Reintroduced, with localised population in Norfolk[1] Sites are highly sensitive, with restricted access	A member of a group called 'green frogs' or 'water frogs'.[5] Visually, hard to distinguish from the non-native southern clade of pool frog, but genetic research and analysis of male advertisement calls helped to confirm that an isolated population in England, which became extinct in the 1990s, had in fact been a native species and related to those occurring in Scandinavia.[6] The northern pool frog is now the subject of a reintroduction programme in Norfolk.[7] They live close to water throughout the year, usually associating with smaller water bodies, ponds and ditches.
Common toad *Bufo bufo*	Widely distributed throughout Britain, but absent from Ireland[1][2]	🔊 Found in a range of habitat types, with breeding ponds typically medium to large, not necessarily well vegetated, and may contain fish. Breeding period typically up to 14 days.[8] Outside of breeding period they occur in lightly grazed pasture, scrubland, open woodland, parkland, gardens and moorland.[4] Ponds used by common toad for breeding may also be used by common frog. However, common toads tend to commence breeding about one month later than common frogs. Once in their ponds, mating activity is intense. Adults will only be present at their spawning pond sites for a short period of time (a matter of days).[3]
Natterjack toad *Epidalea calamita*	Rare and restricted to certain parts of Britain and Ireland[1][2]	🔊 Usually associated with coastal habitat, where during the breeding season they are found within shallow, small to medium-sized ponds (often of a temporary nature). Outside of their breeding season, typically found in sand dunes, saltmarshes and sandy, lowland heath.[4][9] The natterjack toad breeding season starts later than Common frog and Common toad, and continues for a longer period, through to July or early August.

Additional native species

Agile frog (*Rana dalmatina*) – only found on Jersey, with range restricted to the southwest of the island. Population in decline and heavily protected.[10][11]

References

[1] NARRS 2018; [2] McInerny & Minting 2016; [3] Gent & Gibson 2012; [4] Beebee 2013; [5] Inns 2009; [6] Froglife 2018; [7] Buckley & Foster 2005; [8] Wells 1977; [9] Speybroeck *et al.* 2016; [10] Buley *et al.* 2001; [11] J Horn, Nurture Ecology Ltd, personal correspondence, 2018.

Table 4.2 Overview of non-native species of frog and toad occurring in the British Isles.

Species	Distribution	Notes on typical habitat and behaviour
Marsh frog *Pelophylax ridibundus*	Southeast England, particularly in Kent and Sussex.[1][2]	🔊 One of the species belonging to a group of frogs referred to as 'green frogs' or 'water frogs'.[3] They live close to water, year round, in areas of low-lying pasture, open fields, parkland and gardens. In particular they tend to be found in the vicinity of, or within, medium to large unshaded ponds, well vegetated and often containing fish.[4] Will jump dramatically into water if disturbed.
Edible frog *Pelophylax* kl. *esculentus*	Localised in areas in southern and eastern England.[3]	A fertile hybrid resulting from breeding between marsh frog and pool frog.[5] As such, it belongs to the group of frogs referred to as 'green frogs' or 'water frogs'.[3] They live close to water, year round, in areas of low-lying pasture, open fields, parkland and gardens. In particular they tend to be found in the vicinity of, or within, medium to large unshaded, vegetated ponds, often containing fish.[4]
Midwife toad *Alytes obstetricans*	Relatively well established in Bedfordshire. Other localised populations elsewhere in England and Wales.[1][2][5]	🔊 Typically terrestrial, often associating with gardens, but in the vicinity of their breeding ponds – which do not usually contain fish.[4] Although a non-native, is not considered to be a threat to our native ecology. After mating, the males carry the females' egg strands with them, accumulated around their back legs. They do this until favourable conditions are found and the eggs are ready to hatch, at which point the male deposits them into water.[6]

Additional non-native species of note

Pool frog (southern race) (*Pelophylax lessonae*) 🔊– distinct clade from our native northern pool frog, with localised populations in southern and eastern England.[1][6]

Iberian water frog (*Pelophylax perezi*) 🔊– another species of 'green frog' or 'water frog' that may be encountered in southern England.[1][6]

North American bullfrog (*Lithobates catesbeianus*) – a large non-native species that has a negative impact on our native ecology. Isolated populations have been found in southern England,[2] usually preferring large and well-vegetated water bodies.[4] Very rarely encountered, but has bred in the past and has subsequently undergone control measures.

African clawed frog (*Xenopus laevis*) – a completely aquatic frog that poses a threat to our native biodiversity. Looks quite different to other species you are likely to encounter, with a flattened build and circular eye pupils on the top of the head.[5][6]

Fire-bellied toad (*Bombina* spp.) – a number of species found in localised sites, released as a by-product of the pet trade. If threatened they will play dead and expose the warning colours on their bellies.[5][6]

References

[1] Alien Encounters 2018; [2] NARRS 2018; [3] Inns 2009; [4] Beebee 2013; [5] Froglife 2018; [6] Speybroeck *et al.* 2016.

our Anura become vocal. Rather conveniently for us, and essentially for them, each species has its own distinctive advertisement call, which helps us greatly in establishing the presence of our native species, as well as providing more information about non-native populations which may need to be monitored and managed in order to help protect and conserve our own biodiversity. All of this, of course, can also inform developmental impacts relating to the project that we may be out doing the bat surveys for in the first place.

When it comes to monitoring amphibians in the British Isles there are various sources of reference, and a number of monitoring programmes that you can contribute to (see Appendix 3).

We have four species of Anura regarded as native (Table 4.1), two of which (common frog and common toad) are common and widely distributed, while the other two (natterjack toad and northern pool frog) are more localised, northern pool frog extremely so, with its current distribution being restricted to a couple of areas in Norfolk.

As well as our native species there are numerous non-native species occurring in the British Isles (Table 4.2). All of these are regarded as being either established as breeding in the wild, or at the very least known to be present with their self-sustainable status being uncertain. In some cases these species are carefully monitored, and when necessary controlled (Alien Encounters 2018).

Audible advertisement calls made by Anura

One of the easier ways to establish the presence of any of our frog or toad species is through the species-specific advertisement calls that they make during the breeding season (Wells 1977). The reader should, however, also be aware that the anurans are known to make a variety of other noises. Toledo *et al.* (2014) described the repertoire of calls covering all behaviours as falling into three main categories: reproductive, aggressive and defensive. These categories were then further divided into thirteen sub-categories. Within the reproductive category, for example, there are six sub-categories, namely advertisement, courtship, amplectant (i.e. during the mating embrace), release, post-oviposition male release and rain. Thus, it can be confusing for the non-specialist reviewing the literature on this subject, and a further source of confusion is that the term 'territorial call', which might be assumed to be the same as an 'advertisement' call, usually refers specifically to a different call altogether that can be made aggressively between males, as one encroaches into another's territory (Walkowiak & Brzoska 1982).

The bulk of studies relating to the acoustic behaviour of Anura relates to what are often described as the previously mentioned 'advertisement' calls (i.e. falling within the reproductive category), and not those specifically labelled as 'territorial'. These advertisement calls – which may also be referred to as mating calls (Bogert 1960) – are regarded as being species-specific, although the call parameters can exhibit changes influenced by ambient temperature and the size of the calling male (Schneider & Sinsch 2007). It is these calls that we are going to focus on within this section, as they are the calls we are most likely to encounter. In most species worldwide these calls are produced by males, during the breeding season, to establish territories and attract females towards a chorus of males, and then to a specific calling male. They are also thought to enable males to space themselves out from each other (Toledo & Haddad 2005). The more recognisable and regularly heard calls are produced by the males (though in some species females may also produce responsive mating calls) when they are advertising territory in this way during the main mating/spawning period (Erik Paterson, personal correspondence, 2018). Some species are what would be regarded as 'explosive breeders', meaning that the breeding period for a discrete population of a particular species may last only a couple of weeks, or even just a few days. Other species are regarded as prolonged breeders, indicating a breeding period in excess of a month (Wells 1977).

Our two common and widespread native species typically breed earlier in the season in the south and west of Britain, and later in the north and east. The breeding period may also be later at higher altitude (Gent & Gibson 2012). Focusing on advertisement (mating) calls, Table 4.3 shows the months when these are most likely to be encountered for each of the species described. Bear in mind, however, that seasonal variations in weather, as well as geographical location and altitude, can all have an impact on the typical breeding season.

Table 4.3 Mating/spawning period (shaded green) for frogs and toads (adapted from Inns 2009).

	Jan	Feb	Mar	Apr	May	Jun	Jul	Aug	Sep
Common frog									
Common toad									
Natterjack toad									
Northern pool frog									
Marsh frog									
Edible frog									
Midwife toad									

Species-specific examples

We are now going to consider the advertisement calls made by Anura on a species-by-species basis, incorporating notes, spectrograms etc. as appropriate. Please bear in mind that what follows are examples, and not considered to be the entire repertoire for any one particular species. We are focusing purely on advertisement (mating) calls, but these animals are known to also emit other sounds, as discussed earlier in this chapter.

Spectrograms in Chapter 4

Spectrograms in this chapter use the following scales, unless otherwise indicated in the figure legend:

Time (x-axis): 5 seconds (5000 ms)
Frequency (y-axis): 0–11 kHz

When a figure legend includes the 🔊 symbol, this means that the figure has been created from a file in the Sound Library. The figure number matches the file number there. For more information about how to access and download files from the Sound Library please see pages xv–xvi.

Toad species

Common toad

Common toad (*Bufo bufo*) males can be heard advertising on their breeding ponds, mostly at night, but they may also call during the day (Gent & Gibson 2012). The call is quiet, relatively high-pitched and rough, described as a repeated 'qwark' (Froglife 2018). The call has also been likened to calls made by coot 🔊 (Speybroeck *et al.* 2016). The sound frequencies of mating call sequences have been measured as occurring typically at 4 kHz, with the duration for individual calls being 400–1200 ms, produced at a repetition rate of 12–32 pulses per second (Walkowiak & Brzoska 1982). 'Short territorial call' sequences and 'long territorial call' sequences have also been described. When handled, common toads can emit a noise that sounds similar to their mating calls (Erik Paterson, personal correspondence, 2019). This call might be noticeable, for example, as a result of humans on Froglife's 'Toad Patrol' assisting the toads to make it safely across busy roads, as they head towards their breeding areas each spring.

Figure 4.1 shows an oscillogram and a spectrogram of typical advertisement calling, and Figure 4.2 shows a call emitted by a male while paired with a female during spawning. In the latter figure, common frog is also calling at a lower frequency, as marked by the black arrows.

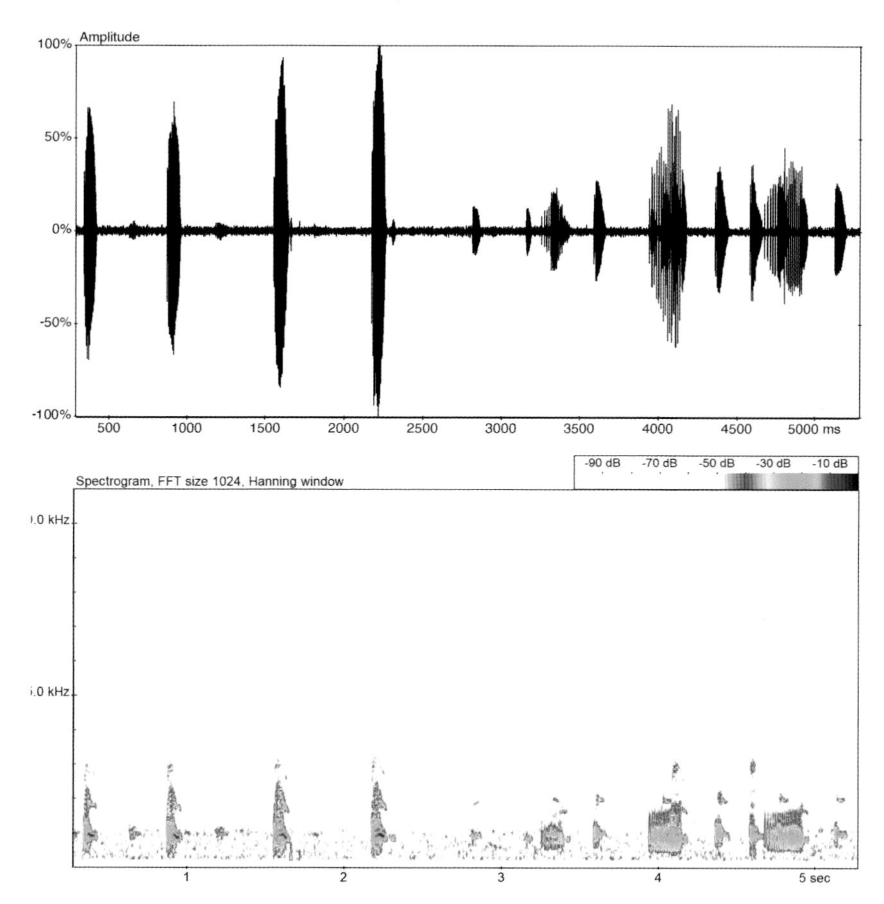

Figure 4.1 🔊 Common toad – advertisement call. Oscillogram (above) and spectrogram (below) (R Specht, Avisoft Bioacoustics, www.avisoft.com).

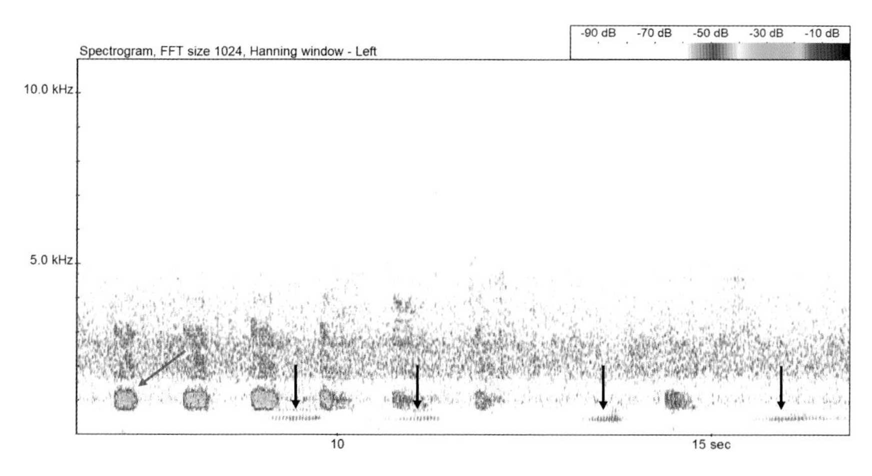

Figure 4.2 🔊 Common toad – spawning calls (red arrow) with common frog (black arrows) (R Jones, 2019; frame width 10 seconds).

Natterjack toad

Natterjack toads (*Epidalea calamita*) produce a very distinctive advertisement call, which is often the easiest way to detect their presence. The call is relatively loud, and far-carrying, compared to common frog and common toad. Males produce a long series of audible loud, 'errr, errr, errr' calls to attract females during the breeding season (Froglife 2018). The peak frequency often appears to be in the region of 2 kHz or lower. On still nights they can be heard from over a kilometre away. The call reminds me, as it does others, of nightjar,🔊 which produces its long drawn-out churring call from dusk onwards (see Chapter 3, Figure 3.13). This toad and this bird are each important species to identify accurately, so be careful not to get them confused, audibly, from a distance. One differentiator is that you would not expect to hear nightjar, a summer visitor to the British Isles, until well into May.

Figure 4.3 shows an oscillogram and a spectrogram of the same individual toad calling. Figure 4.4 shows a spectrogram, within which a power spectrum is displayed showing a peak frequency of 1.7 kHz. Note that in this second example there are in fact two toads calling simultaneously, with one much louder than the other, and therefore strong harmonics are being picked up for this louder individual.

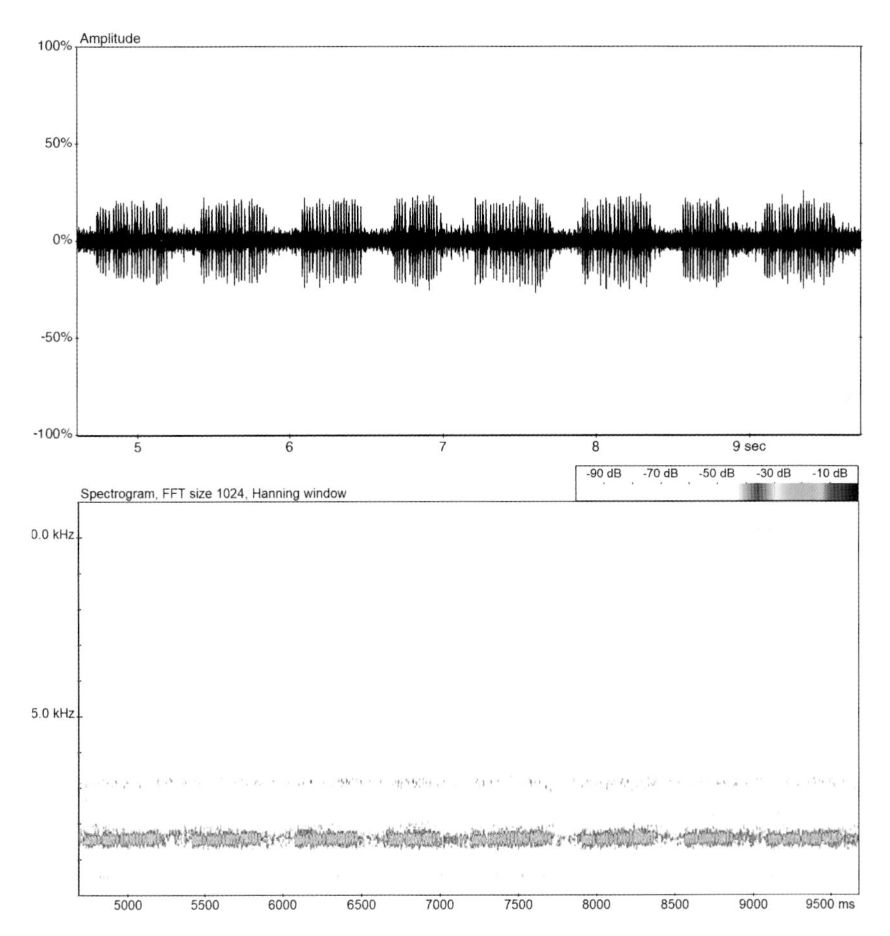

Figure 4.3 🔊 Natterjack toad – advertisement call. Oscillogram (above) and spectrogram (below).

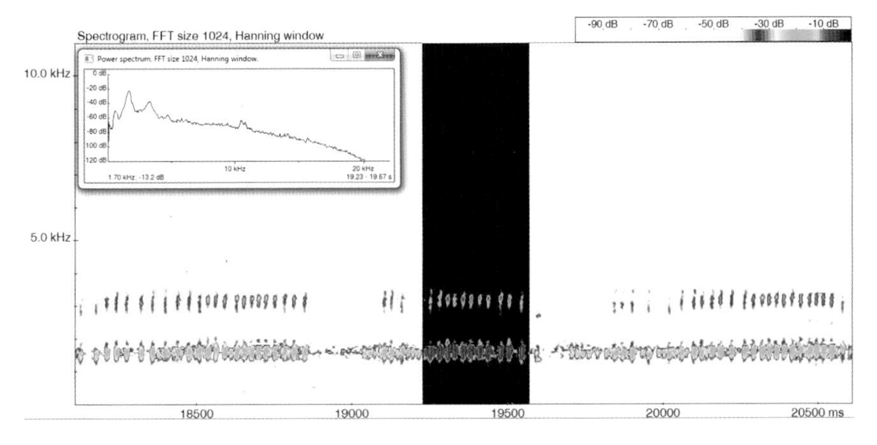

Figure 4.4 🔊 Natterjack toad – advertisement calls, two individuals, with power spectrum displaying a peak frequency at 1.7 kHz for the fundamental part of the calls.

Midwife toad

The electronic beeping sonar-like call of midwife toad (*Alytes obstetricans*) is very distinctive, and does not immediately sound like what an inexperienced person would expect from a toad or a frog. This species typically calls from dusk onwards, on warm summer nights (Helen Muir-Howie, personal correspondence). It can be confused, on the continent, with scops owl (*Otus scops*). This is unlikely to be the case in the British Isles, as scops owl would be regarded as a very unusual occurrence. For those accustomed to using frequency division bat detectors, the audible call of this toad is not hugely dissimilar to what an echolocating *Nyctalus* species of bat would sound like on a frequency division bat detector.🔊

Both male and female midwife toads produce mating-related calls (Figure 4.5), with the female responding with reciprocation calls to the advertisement calls of males (Heinzmann 1970). Often it calls from land-based (as opposed to aquatic) hiding places, which may not necessarily be located near to standing water (Beebee 2013).

The calls alter in frequency ever so slightly, for example as alternate notes, as this toad produces a long series of calls which are somewhat musical to the ear by comparison to other frogs and toads that you are likely to encounter within the British Isles. Typical peak frequency would be *c.* 1.5 kHz, with individual call durations being *c.* 120 ms.

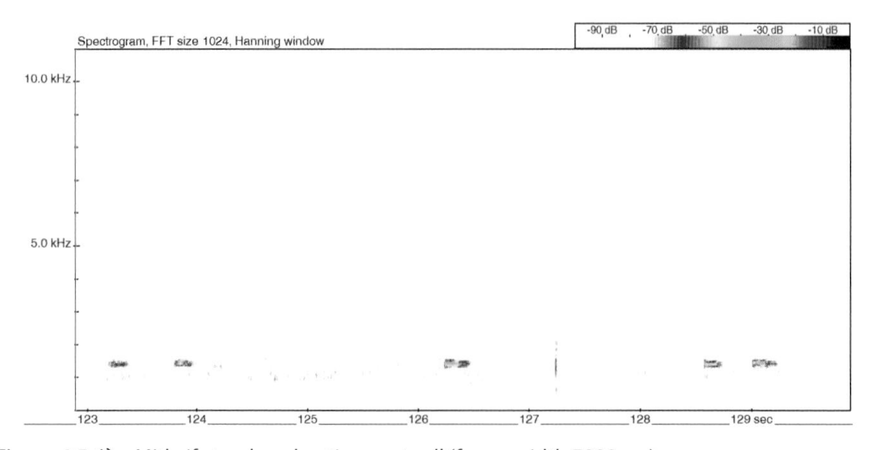

Figure 4.5 🔊 Midwife toad – advertisement call (frame width 7000 ms).

Frog species

Common frog

The common frog (*Rana temporaria*) is usually our earliest anuran to commence breeding activity, as early as February. Adults will only be present at their spawning pond sites for a short period of time, often only a matter of days (Wells 1977, Gent & Gibson 2012). Males call with a quiet repetitive croak (Figures 4.6 and 4.7), often produced as a chorus by a number of animals at their breeding ponds. These calls usually occur at dusk and through the night, in order to attract females for mating purposes (Beattie 1987). Occasionally, this species may also be heard calling during the day, and during rain, as per the call within the Sound Library kindly provided by Rogan Jones.🔊

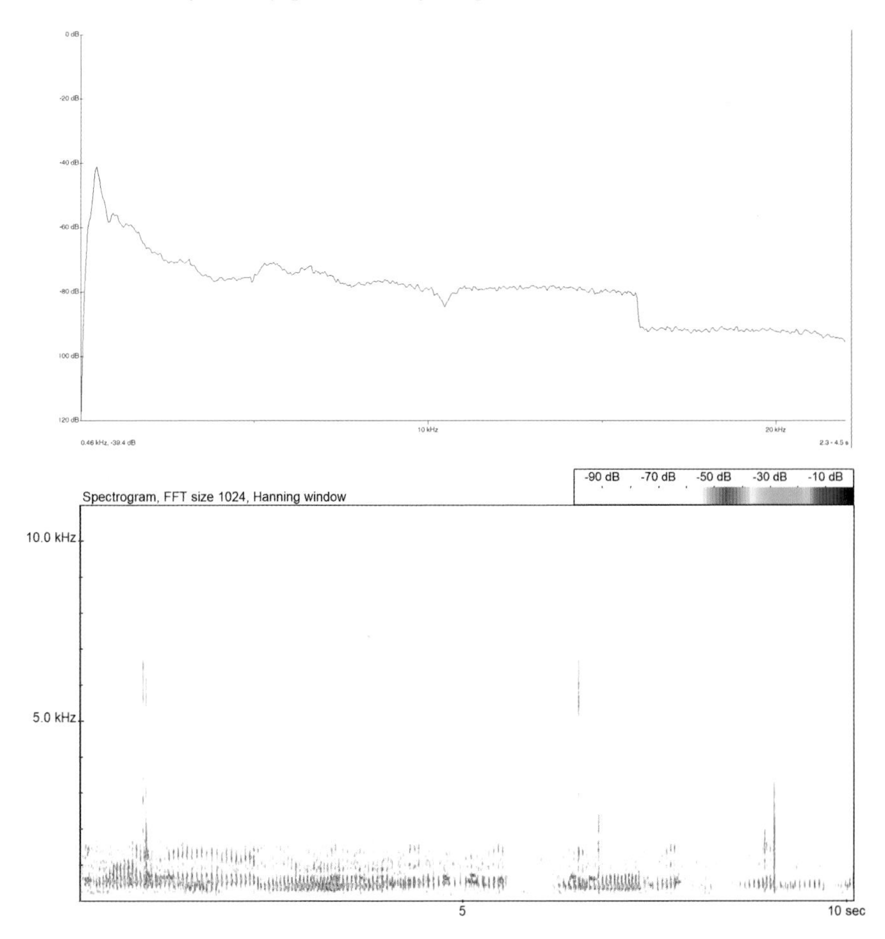

Figure 4.6 🔊 Common frog – advertisement call. Power spectrum (above) and spectrogram (below) (T Clarke, 2019; frame width 10 seconds).

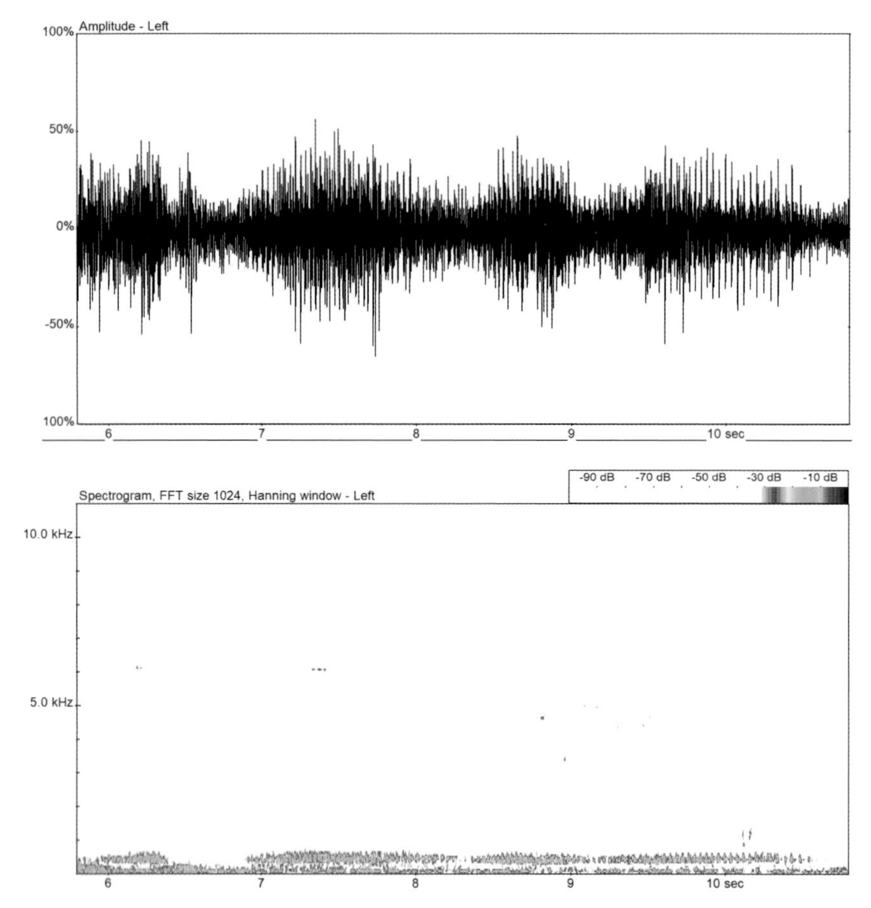

Figure 4.7 Common frog – advertisement call. Oscillogram (above) and spectrogram (below) (N Hull, twoowlsbirding.blogspot.co.uk, 2019; frame width 5000 ms).

Water (or green) frogs

The following five species belong to a group called 'water' or 'green' frogs. As a collective they are relatively difficult to tell apart visually, further complicated by the occurrence of hybrids and introduced aliens across their overall range. They are strongly aquatic in their preference of habitat and are 'sun worshippers', often observed basking in the vicinity of their vegetated and unshaded pools (Inns 2009, Speybroeck *et al.* 2016). Their calls are species-specific and therefore can be an aid to accurate identification, but establishing precise species from the calls does take relatively more experience.

Marsh frog

Marsh frogs (*Pelophylax ridibundus*) produce a loud and fast series of short metallic notes (Speybroeck *et al.* 2016). The calls (Figure 4.8) can also be mistaken for ducks quacking (been there, done that – Middleton, you plonker!) or the sound of rapid laughter. They are particularly vocal and will call during the day as well as at night.

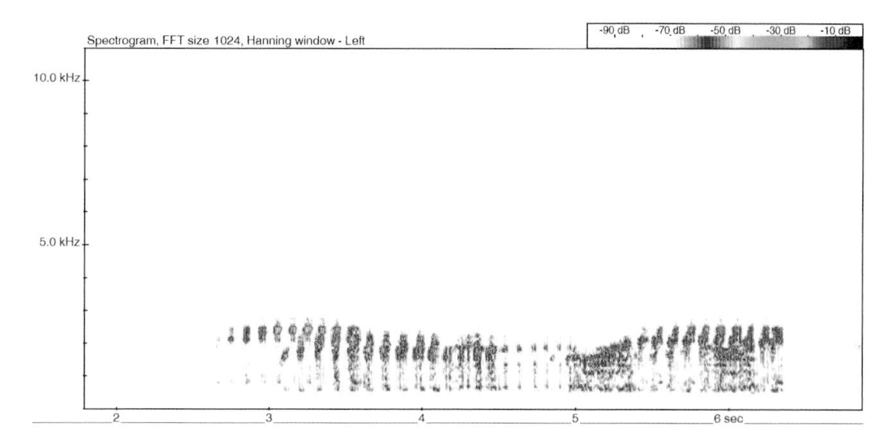

Figure 4.8 Marsh frog – advertisement call (R Specht, Avisoft Bioacoustics, www.avisoft.com).

Edible frog

Calls from edible frogs (*Pelophylax* kl. *esculentus*) may remind you of a duck quacking, with calling taking place during the night or daytime. The call is loud and rattling and can last several seconds (Speybroeck *et al.* 2016). It has been described as being not dissimilar to Baillon's crake (*Porzana pusilla*), which would be classed as a very rare bird within the British Isles (Martin Scott, HiDef Aerial Surveying Ltd, personal correspondence).

Iberian water frog

Figure 4.9 shows a typical call from Iberian water frog (*Pelophylax perezi*), which can occur within southern England.

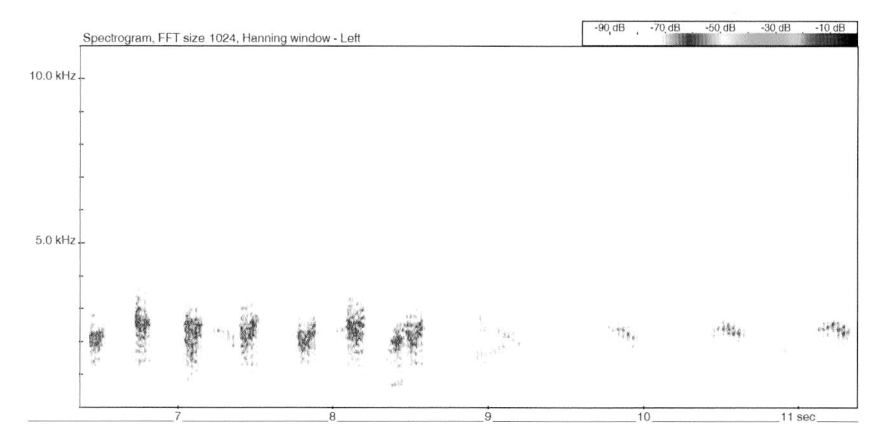

Figure 4.9 Iberian water frog – advertisement call (R Specht, Avisoft Bioacoustics, www.avisoft.com).

Northern pool frog

Regarded as native to Britain, the northern pool frog (*Pelophylax lessonae*) belongs to a distinct and rare northern clade of pool frog, separate from the southern pool frog. It is particularly vocal, calling at night and also during the day. It is often seen basking when it's sunny, and has a long rattling, duck-quacking type of call (also described as rapid laughter) that may last several seconds (Speybroeck *et al.* 2016).

Because the sites where this species occurs have restricted access, it has not been possible to obtain recordings of this native animal.

Southern pool frog

Figure 4.10 shows a typical call from the southern clade of the pool frog, which is closely related to, but distinct from, the rare and native northern pool frog. Southern pool frog has a wider range within Britain, occurring at localised sites across southern and eastern England.

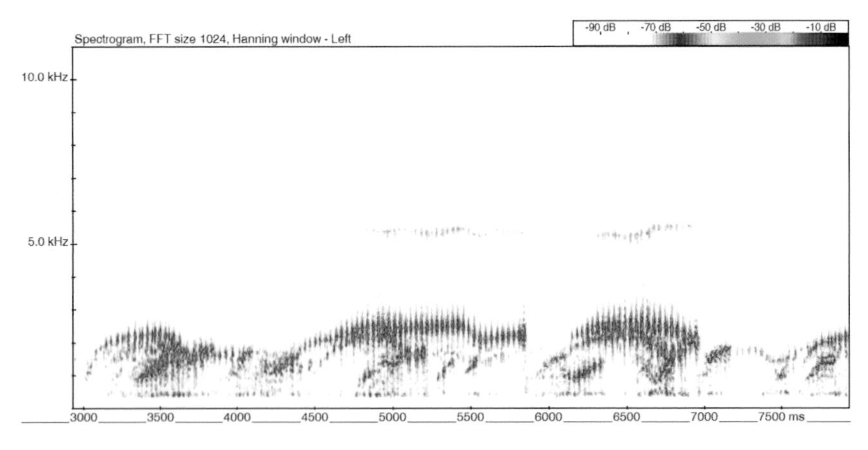

Figure 4.10 🔊 Southern pool frog – advertisement call (R Specht, Avisoft Bioacoustics, www.avisoft.com).

Hints and tips

Risks to bat survey methods and analysis

It is unlikely that calls made by frogs or toads would interfere with the analysis or results of a bat survey, and as such there is not much need to consider positioning of equipment, from this perspective, when it comes to such surveys. Bat detector microphones would normally have their frequency responsiveness set to a range that avoids recording the low frequencies at which you would expect frogs and toads to produce noise. Consequently it would also not be anticipated that any bat filters or classifiers would misidentify these calls as bats, and noise filters set appropriately should be able to easily remove such sounds during the filtering process.

Setting out to record amphibians

Acoustic monitoring and recording of frogs and toads is not usually carried out in the British Isles for consultancy purposes, as our more common species are relatively quiet, with short breeding seasons (Beebee 2013). In some parts of the world (e.g. North America), however, acoustic surveys for establishing the presence of frogs and toads can be relatively common, and for some species would be regarded as reliable (Pellet & Schmidt 2004), depending on the resources available and the data sought.

In the British Isles, one important consideration from a native species' perspective is the rare natterjack toad. Its calls, which are far-carrying and produced over a longer breeding season, can be very useful for identification purposes in order to establish overall presence/absence, as well as the numbers present in an area (Sewell *et al.* 2013). Also, bearing in mind their heavily protected status, acoustic surveys would normally be deemed preferable, as this approach causes less disturbance. Being aware of, and recording, audible sound can also be very useful in establishing the presence of our non-native and more localised species (see Table 4.2) including the 'water' frogs and midwife toad. Studying and recording acoustic behaviour within the Anura could therefore be of great interest from a developmental impact perspective, as well as for academic research and/or the monitoring and control of alien species.

Various bioacoustics methods have described suitable equipment for recording these species, as well as the call features deemed most useful to measure during analysis (Schneider & Sinsch 2007). In addition to using spectrograms, researchers studying the acoustic behaviour of frogs and toads have, in particular, paid more attention to the oscillograms produced by these species, as an oscillogram '*reveals the precise temporal structure*' of call sequences (Schneider & Sinsch 2007). Microphones performing effectively within the range 20 Hz to 15 kHz have been recommended as appropriate. Having said all this, the use of any normal microphone system would at least be an aid to verification of species in many instances. For example, as with other species groups discussed in this book, using the voice-note feature on your bat detector or the video tool on your mobile phone can often be good enough for *ad hoc* field recordings.

It should be borne in mind that within Britain and Ireland, natterjack toad is given strong protection against disturbance etc. (Gleed-Owen *et al.* 2013), and although northern pool frog is not afforded any special protection currently (Buckley & Foster 2005), the reintroduction sites for this species are highly sensitive, with restricted access.

Author's diary note

I didn't know that much about frogs and toads before I started this project, but I knew a man who did. Erik Paterson was the group secretary of Clyde Amphibian and Reptile Group (CARG), and very knowledgeable about these amazing animals. I really wanted to record natterjack toad (see picture), as it is one of our rarest native species. So off I went with my wife Aileen (if it's rare, she's there!) one Friday evening in May 2018 to meet up with Erik and some other members of CARG at RSPB Mersehead

(Dumfries & Galloway). I wasn't sure what to expect, as I had shamefully done no research into what they actually sounded like prior to being there. Well, I was in the hands of experts (including licence holders), and anyway I wanted Erik to tell me all about it first-hand. After a short walk along the beach, it wasn't too long before we were in natterjack territory and I was obtaining my first ever, deliberately intended, amphibian recordings. My word, they sound just like hesitant nightjars.

Natterjack toad, RSPB Mersehead, May 2018.

A few months passed, and then I reacquainted myself with the amazing sonar-like calls of midwife toad. I had fond memories of encountering these in the past while on holiday in France, and have now given myself a good kicking for not having recorded them at the time! I could listen to this sound forever and not get close to bored, especially with some nice French cheese and a glass of Syrah. I wonder if the rock band Genesis was inspired by these calls when they wrote the keyboards for their hit 'Tonight, tonight, tonight'. Or is that just me? Go on, have a listen and see what you think.

CHAPTER 5

Insects

Anyone for cricket?

By and large if you go out batting in Scotland (where I spend most of my time), it is a rare event to encounter any insect-generated sound. This certainly makes life much easier from an acoustic perspective. However, before you start thinking 'lucky beggar', remember that up in this part of the British Isles we have more than our fair share of midges. This means that the occasional screams of frustrated and agitated bat surveyors can add to the local soundscape, and make bat work at times very uncomfortable. I am now feeling quite itchy all over from the very thought of it, and it's only January.

When deciding what to include and what to leave out from this chapter, I had to draw a line somewhere. A line beyond which I decided that I could not step, mainly for fear that it would open the door to a never-ending labyrinth of corridors that I just didn't have time to travel, or, once travelled, space to fill. Early on someone asked me, 'Are you going to include wasps, bees, bluebottles and mosquitoes?' I politely said 'No, definitely not', while inside my head I thought, 'Oh Neil, what have you let yourself in for?' Even allowing just for the insects that are noisier at night (and therefore more likely to be picked up by a bat detector) was going to be a task that couldn't be properly achieved within a single chapter. So once again, please bear in mind the overarching spirit of this work, in that it aims to open your ears and introduce you to what's out there, as opposed to being a full-on encyclopaedic account of everything that ever buzzed in Basingstoke, or anywhere else for that matter. As you will see, this chapter is dominated by bush-crickets, although there will be short diversions into other species groups. As I am about to do now, in Figure 5.1 – which shows the buzzing noise made by a species of hoverfly. I am introducing this early on, to give you an example of what 'buzzy' things look like on a spectrogram. This means that you can compare this low-frequency, near-CF call, showing multiple harmonics, with the other groups we will discuss shortly. The hoverfly buzzes are indicated with blue arrows, and a great tit was also present, singing (red arrows).

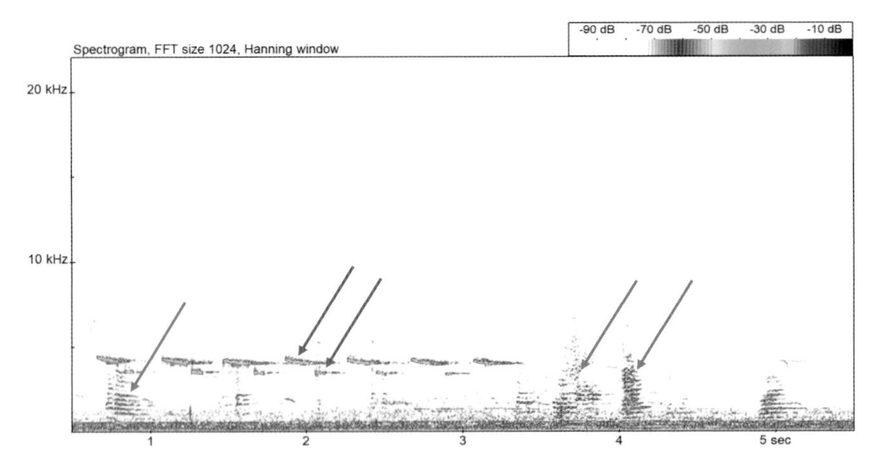

Figure 5.1 Hoverfly species (blue arrows), with singing great tit (red arrows) (frame width 5000 ms, frequency scale 0–22 kHz).

There are many reasons why it is useful to know about insect sound as recorded on bat detectors, and occasionally these sounds can be confused with bats (e.g. type B social calls, type C social calls), as well as other things (e.g. idling car engines). But it's not all about the bats. Knowing about bush-cricket sounds, and other insects for that matter, can actually help immensely with their conservation and our overall understanding of their distribution and abundance. In addition, they are proving to be excellent indicators of healthy environments, as well as being 'sensitive indicators of climatic change' (Natural England 2015).

For many years now, researchers have been studying insect sound. When we consider the work that has gone into grasshoppers, true crickets and bush-crickets as examples, many of those involved have been using our bat detectors to locate and identify these creatures in the British Isles (Lee 2004) as well as overseas, for example in France (Penone *et al.* 2013). In fact the relatively recent developments in bat detecting equipment have had huge knock-on benefits for the study of insect sound, as well as other taxa (e.g. small terrestrial mammals – see Chapter 2).

Now, I am not sure if any of these 'insect people' ever asked permission from anyone at BCT to use our bat detectors for this purpose. Indeed, they may even have been thinking that BCT was a TLA for the Bush-cricket Trust. I mean, it's not as if we would ever use one of their butterfly nets to catch bats! TLA, BTW, is an abbreviation for 'three-letter abbreviation', and BTW, 'by the way'. Get back to the point, FGS man!

As bat workers we can play a huge part in contributing towards the knowledge of insects in the locations where we are doing our bat work. An excellent example of this is already well under way. As a result of a citizen-science project undertaken by the British Trust for Ornithology (BTO) using static bat detectors (Norfolk Bat Survey), a substantial 'bycatch' of bush-cricket calls has been analysed, and this is contributing towards a greater understanding of the species involved (Newson *et al.* 2017, BTO 2019). This project should be seen as an inspiration to bat workers everywhere, as to what is possible when there is a willingness to diversify into other areas. I will touch upon this project again later in the chapter.

In order to make you aware of some of the many additional resources that are available on the subject matter of this chapter, Appendix 3 provides a good selection, which are all well worth a visit.

'But it's not our data – it belongs to the client'

In the world of ecological consultancy we hear this a lot, and of course it is absolutely correct to point this out, and to be cautious if you were using data gathered as a 'bycatch' from paid work for a client employing you to be on their site for another purpose altogether. So, with a strong emphasis on caution, I am going to give a perspective based on numerous years of working on ecology projects on behalf of clients.

There are some clients who quite rightly will not wish their data to be used for another purpose. There may be all kinds of valid reasons for this, and it is not for me to judge whether the client is right or wrong. In fact, in the world of business, 'the customer is always right.' Albeit, they may be open to being gently persuaded otherwise, especially if there is something in it for them. Over the years I have known quite a few developers and the like who would enthusiastically participate in anything that helped to show them in as good an environmental light as possible. It may help them, in that it could counteract the public perception of the negative impact their work is having on the environment. And the cheaper the better. A conversation along the lines of the following might just do the trick.

> Hi Client. You know how you are paying us to gather all of this bat data for your site. You maybe don't appreciate that simultaneously the equipment is recording bush-crickets. These are an important indicator species, and in the British Isles we are trying to understand more about their behaviour and distribution. On the basis that we won't charge you any more, how would you feel about your bush-cricket records being sent off to an organisation to contribute towards their wider conservation objectives? And it might also be a good PR for yourselves if you wanted to include an article in your customer newsletter, demonstrating a positive environmental output from the development.

I know many client types who would be receptive to this sort of approach, and allow the consultant to proceed. It's all about the timing of your conversation, the strength of the relationship you have with the client, and the benefits for them potentially outweighing any perceived negatives. Think about your client base. Clients come in all shapes and sizes. Land managers, historic building preservation, conservation bodies, renewable energy providers. At the very core of some of these clients' values, matters such as conservation and the environment may be key. So, for some of these clients, contributing towards conservation goals of species groups as a whole is something that they will at the very least consider. The widely reported recent publication about the worrying state of insect communities on our planet (Sánchez-Bayo & Wyckhuys 2019) is an ideal platform on which to build a potentially positive conversation about the whole matter. And obviously this whole approach doesn't relate just to insects – the discussion could be broadened out to other subjects.

Spectrograms in Chapter 5

Spectrograms in this chapter use the following scales, unless otherwise indicated in the figure legend:

Time (x-axis): 2 seconds (2000 ms)
Frequency (y-axis): 0–110 kHz

When a figure legend includes the 🔊 symbol, this means that the figure has been created from a file in the Sound Library. The figure number matches the file number there. For more information about how to access and download files from the Sound Library please see pages xv–xvi.

Orthoptera

Orthoptera is an order that includes a large group of insects. In particular, the ones that may be more of interest to us, because of their acoustic behaviour, are the Tettigoniidae (bush-crickets), Acrididae (grasshoppers), Gryllidae (true crickets) and Gryllotalpidae (mole-crickets). Between these four families there are almost 50 species known to occur within the British Isles, of which 30 are native (GRIRS 2019). In almost every instance each one of these insects produce species-specific sound. This means that the use of bioacoustic monitoring equipment to gather field data is an extremely useful, non-invasive approach from which to monitor population densities, as well as geographical distribution, environmental health and the climatic effects upon the natural world.

As mentioned earlier, an excellent example of this already occurring in the British Isles is the project being carried out by the BTO whereby huge amounts of bush-cricket data have been gathered, simultaneously with static bat detector surveys, in Norfolk and further afield (Newson *et al.* 2017, BTO 2019). The project has not only collected useful data, but has also demonstrated that the data analysis, which many would have considered an impossible task, can be carried out efficiently. The use of an automated classifier (developed by Yves

Bas) has achieved high accuracy rates (> 85%) in identification for at least some species of bush-cricket.

Before we consider the bush-crickets any further, it will be beneficial to discuss, albeit briefly, the other members of this order (crickets, mole-crickets and grasshoppers). Many of these 'other' species do make noise, and potentially could be encountered during bat surveys, especially bearing in mind that some survey methods require bat detectors to be activated before dusk and after dawn (i.e. during daylight conditions). In addition to this, a number of these species are rare, nationally scarce or localised. By highlighting this, it may bring to the forefront of some people's minds the importance of sharing unusual records with other interested parties.

First, however, I feel it would be useful to discuss how calls are described in this chapter. Within the academic study of Orthoptera, the descriptors used by entomologists to describe calls (Ragge & Reynolds 1998) are completely different from those used within the world of bats (see Appendix 1, Figure A1.6 and Table A1.1). But I have taken the decision, throughout this book, to adopt the same terms that I would use when describing bat social calls (Middleton *et al.* 2014), as to introduce you to a completely new range of terms in each chapter could confuse matters. So in this chapter I am not using the same terms as the entomologists, and Figure 5.2 provides guidance as to the terminology used in this chapter. Referring to that figure, we have one call comprising many sequences. In the case shown (great green bush-cricket), the sequences are doubled up (i.e. paired). Within each sequence there are multiple components which would be seen even more clearly if you zoomed in on the oscillogram. Think about the call as being a paragraph and each sequence as a sentence, within which you have numerous words (components). If we were able to study the components in detail, we might find that each 'word' could be single-syllabled or multi-syllabled.

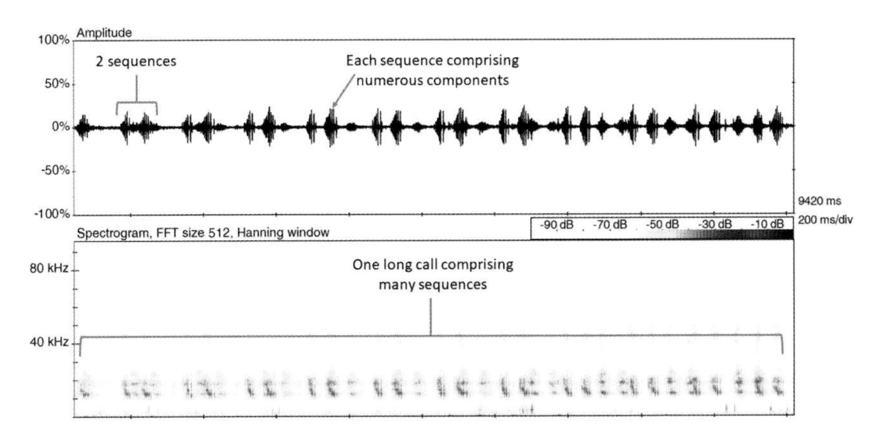

Figure 5.2 Descriptors used in Chapter 5. Great green bush-cricket (C Nason, 2018) (adapted from Middleton *et al.* 2014).

Gryllidae (true cricket) and Gryllotalpidae (mole-cricket) species

Not all crickets are bush-crickets. In the British Isles we also have true crickets (three native species), as well as our native and very rare single species of mole-cricket. Table 5.1 describes the ones that are more likely to come up in conversation. Note that true crickets and the mole-cricket usually sound more musical, albeit squeak-like, and their calls are well within our audible range. On the other hand, bush-crickets and grasshoppers tend to sound harsher and not musical to the ear. Among the true crickets that occur in the British Isles it is usually only the male that makes a sound, whereas in mole-cricket females can also stridulate (Ragge & Reynolds 1998). Figures 5.3 to 5.7 provide examples of calls emitted by the species listed in Table 5.1.

Table 5.1 Overview of true cricket and mole-cricket species occurring in the British Isles.

Species	Distribution	Notes on typical habitat and behaviour
Field cricket *Gryllus campestris*	Southern England, and southeast Wales[1]	🔊 A very rare species within the UK. Can be heard from May through to August,[2] usually calling during daytime when temperatures are > 13 °C. May also be heard in the early evening.[1] Calls are continuously repeated in grouped sequences, usually of four, with the initial call within each grouped sequence being quieter than the others.[2]
House cricket *Acheta domesticus*	Recorded occasionally throughout Britain and Ireland[1]	🔊 A non-native species. Nocturnal, and usually encountered within or near structures. Can be active and heard calling at any time of the year.[1] Calls are continuously emitted in grouped sequences of three or two, each call being of similar amplitude.
Wood cricket *Nemobius sylvestris*	Restricted in range to the far south and west of England[1]	Nationally scarce, occurring within woodland edge or scrubland areas. Adults are active from June through to the end of October.[1] This species is relatively quiet, in amplitude, compared to other species, so often goes unnoticed, unless in a chorus comprising a group of males.[2]
Scaly cricket *Pseudomogoplistes vicentae*	Localised populations in Dorset, Devon and Pembrokeshire[1]	Extremely rare and localised. Adults are active from August through to October. Does not stridulate.[1]
Mole-cricket *Gryllotalpidae gryllotalpidae*	Southern England[2]	🔊 Very rare. Utilises loose, damp soil, often at the edge of a wetland. Calls are produced from within a burrow, and are usually made during the evening and at night. The sound has been described as being similar to nightjar churring,🔊 at a frequency of < 2 kHz.[2]
References [1] GRIRS 2019; [2] Ragge and Reynolds 1998.		

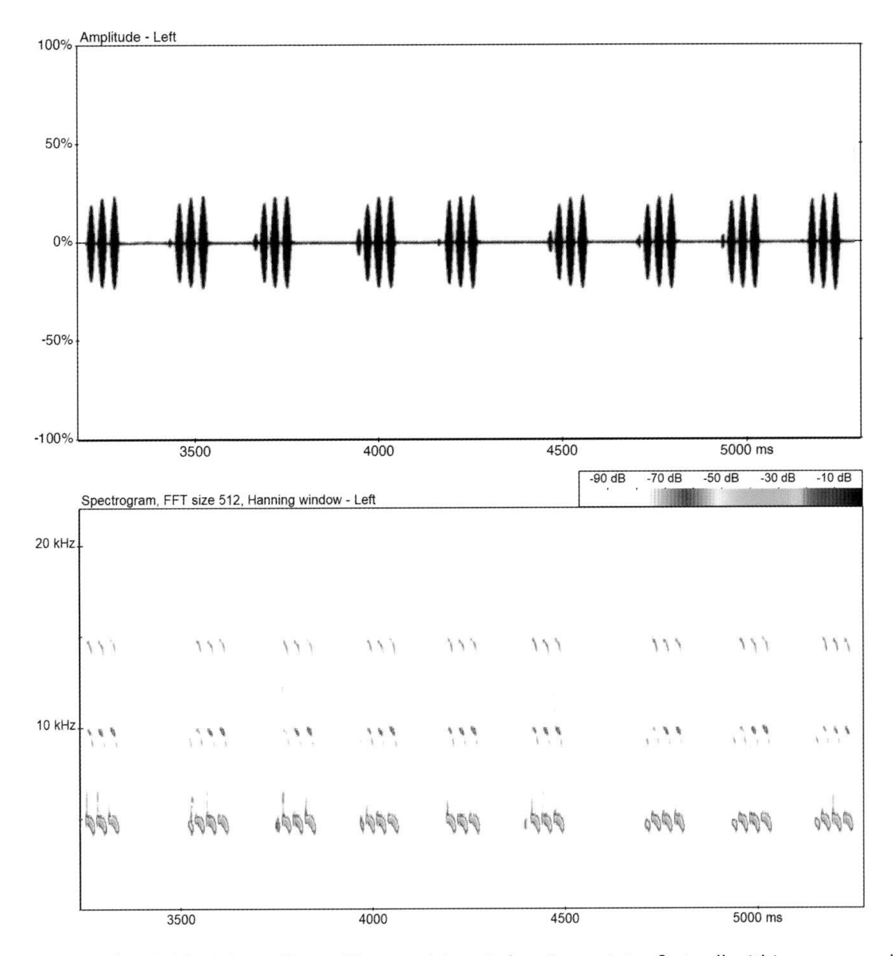

Figure 5.3 Field cricket call – oscillogram (above) showing quieter first call within sequenced groups of four, and spectrogram (below) (R Specht, Avisoft Bioacoustics, www.avisoft.com).

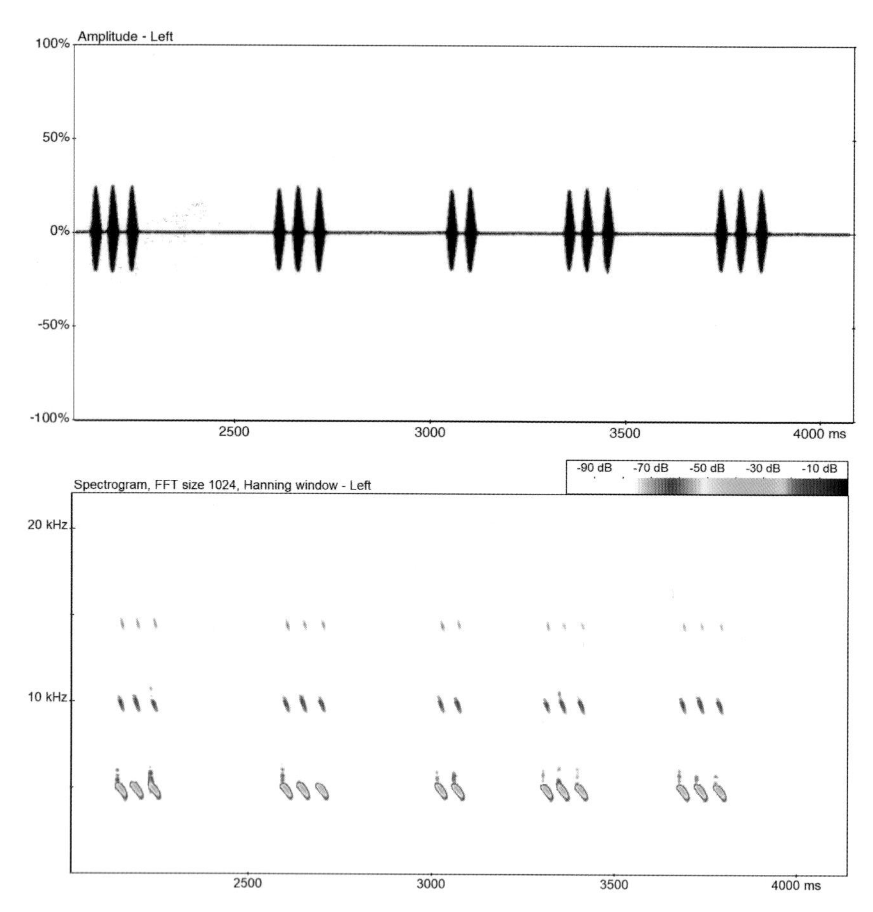

Figure 5.4 House cricket call – oscillogram (above), and spectrogram (below) (R Specht, Avisoft Bioacoustics, www.avisoft.com).

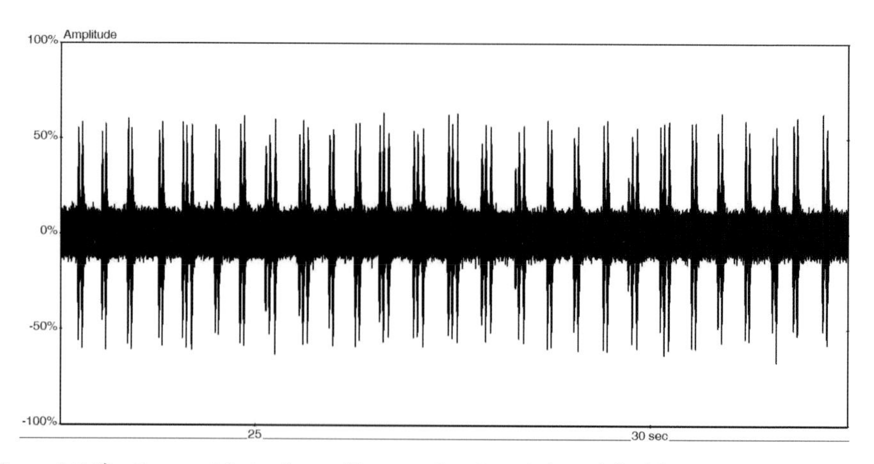

Figure 5.5 House cricket call – oscillogram showing triple and double grouping over longer time period (M Hughes, 2019; frame width 10 seconds).

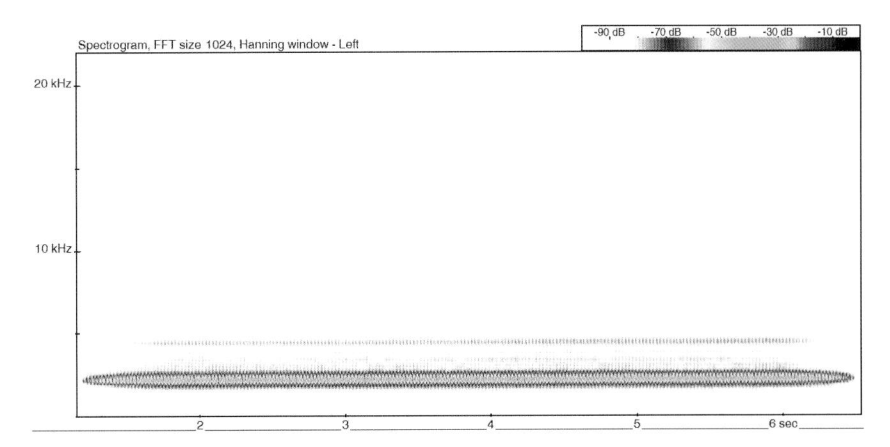

Figure 5.6 Mole-cricket, showing a full sequence of calls (R Specht, Avisoft Bioacoustics, www.avisoft.com; frame width 5000 ms).

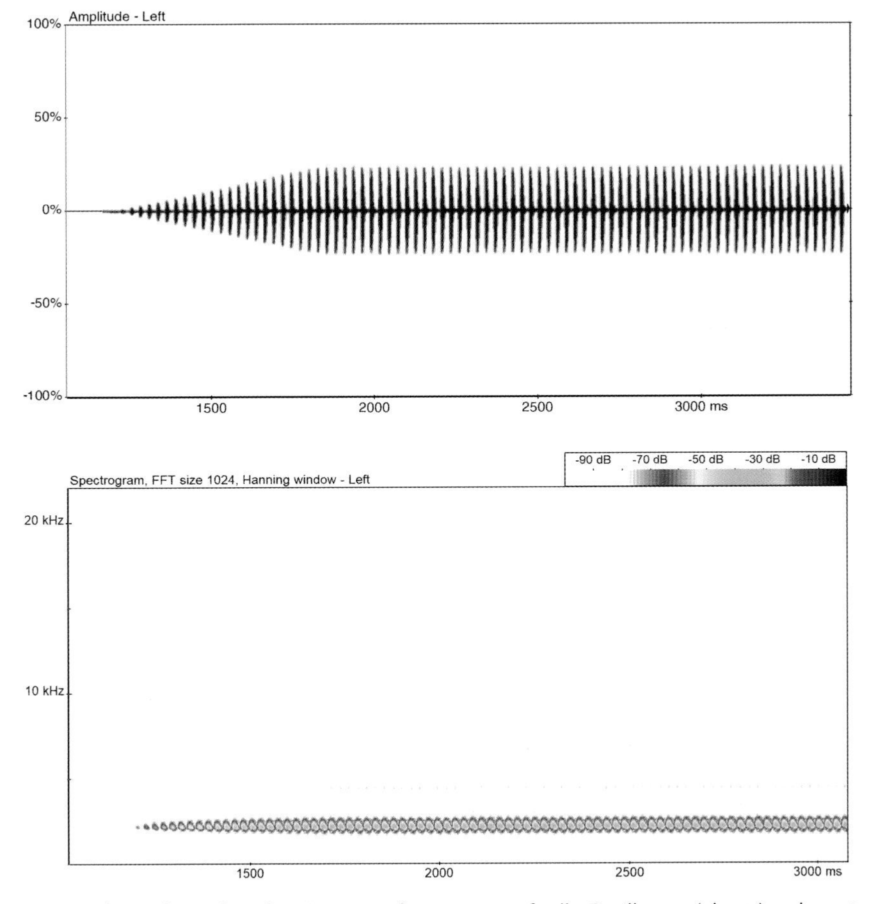

Figure 5.7 Mole-cricket, showing part of a sequence of calls. Oscillogam (above) and spectrogram (below) (R Specht, Avisoft Bioacoustics, www.avisoft.com; frame width 2000 ms).

Acrididae (grasshopper) species

Twenty species of the family Acrididae occur in the British Isles, thirteen of which are regarded as native. By and large grasshoppers call during the daytime, although for some this behaviour can go on into the early evening. Unlike true crickets they do not sound musical to our ears, but more buzzlike or hissing in nature (Ragge & Reynolds 1998). Generally their calls are well within our audible range, therefore making them more noticeable and easier to hear than many bush-crickets.

As well as what is described as 'normal' song, males can also become involved with other males in what is called a 'rivals' duet'. Separate to this scenario, when a receptive female is in his vicinity, a male will start producing a 'courtship' song which may be followed by an 'assault' song immediately before mating, and then, finally, a 'copulation' song during the mating process (GRIRS 2019). In some species of grasshopper receptive females are known to call in response to a singing male.

Owing to their mostly diurnal behaviour and lower calling frequency, and depending upon the microphone frequency responsiveness of a bat detector, they may not be picked up that loudly, or at all, during bat-related work. In Table 5.2 I have included our native species, with a particular emphasis on those which are more common and widely distributed. Figure 5.8 shows an example of meadow grasshopper calling. Due to constraints in time and space it isn't possible to include more examples of grasshopper sounds, but I would refer you to the referenced material, as well as the resources described in Appendix 3, should you wish to learn more about their calls and behaviour.

Table 5.2 Overview of native grasshopper species occurring within the British Isles.

Species	Distribution	Notes on typical habitat and behaviour
Common green grasshopper *Omocestus viridulus*	Occurs throughout Britain and Ireland[1]	🔊 Damp, unimproved grassland areas. Adults are active from July through to November. This is a relatively loud species, with a rapid series (e.g. 15–20 pulses per second) of clicks that slow down in repetition rate towards the end of a sequence. Sequences typically last 10–25 seconds. The call starts off quietly and then quickly increases in volume, before coming to a sudden end.[1][2]
Field grasshopper *Chorthippus brunneus*	Occurs throughout Britain and Ireland[1]	Short, dry vegetated areas, with good sunny conditions. Adults are active from June through to early December.[1] The calls are a series of short, abrupt chirps (each lasting between 120 and 250 ms), typically repeated at intervals of 1–5 seconds.[2]
Mottled grasshopper *Myrmeleotettix maculatus*	Occurs throughout Britain and Ireland[1]	Dry habitat, exposed to sunshine. Adults are active from June through to October.[1] Calls are a sequence of 10–25 short buzzes, usually lasting 8–15 seconds. Calling begins quietly, building up to a climax.[1][2]

(cont).

Table 5.2 (cont).

Species	Distribution	Notes on typical habitat and behaviour
Woodland grasshopper *Omocestus rufipes*	Southern England[1]	Woodland rides and clearings, as well as grassland areas or heath close to scrub or woodland. Adults are active from June through until October.[1] A similar call to common green grasshopper, but tends not to last as long (5–10 seconds).[2]
Stripe-winged grasshopper *Stenobothrus lineatus*	Southern England and East Anglia[1]	Grassland with chalk or limestone substrate, heath and dune systems. Adults are active from July through to October.[1] The call is relatively slow,[2] sounding like a short, pulsating, metallic rasp.[1]
Meadow grasshopper *Chorthippus parallelus*	Occurs throughout Scotland, England and Wales. Not present in Ireland[1]	🔊 Coarse grasses in a wide range of habitats such as sand dunes, saltmarshes, woodland rides, roadside verges, waste ground, valley wetlands and wet, grassy moorland. Adults are active from June through to November. Buzzlike call, usually repeated at intervals of 3–5 seconds.[1][2]
Lesser marsh grasshopper *Chorthippus albomarginatus*	Southern and eastern England, southern Wales and western Ireland[1]	Sand dunes, saltmarshes and shingle banks, low-lying pastures and grassy slopes. When present inland, tends to associate with rough grassy areas, roadside verges, waste ground, parkland, set-aside arable land and in damp woodland clearings. Adults are active from July through to October.[1] Calling is usually a series of 3–4 sequences, at intervals of 1.5–5 seconds.[2] Courtship calls sound like a clock being wound up.[1]

Additional native species of note

Large marsh grasshopper (*Stethophyma grossum*)
Heath grasshopper (*Chorthippus vagans*)
Rufous grasshopper (*Gomphocerippus rufus*)
Lesser mottled grasshopper (*Stenobothrus stigmaticus*) – Isle of Man only
Blue-winged grasshopper (*Oedipoda caerulescens*) – Channel Islands only
Jersey grasshopper (*Euchorthippus elegantulus*) – Jersey only

References

[1] GRIRS 2019; [2] Ragge & Reynolds 1998.

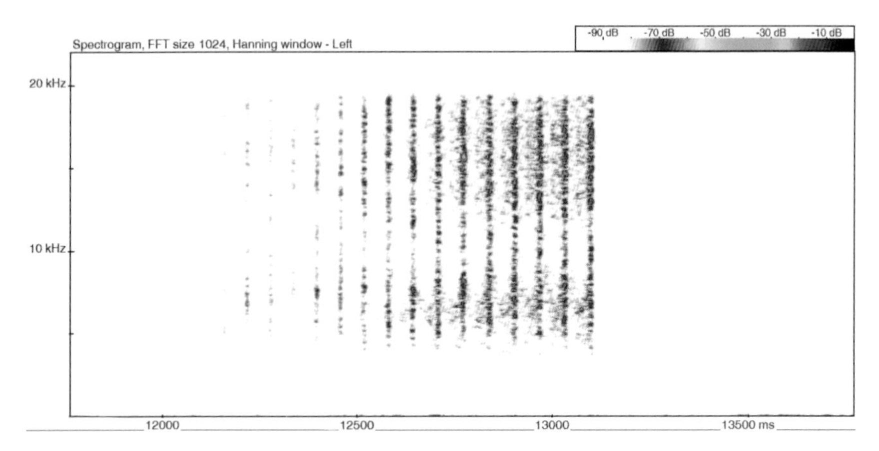

Figure 5.8 Meadow grasshopper call (R Specht, Avisoft Bioacoustics, www.avisoft.com).

Grasshopper or bush-cricket?

Before we go any further let's sort something out. I guess one of the first questions many people have, as I had myself when I first started thinking about this subject, is, *'What's the difference between a grasshopper and a bush-cricket?'* There are in fact many differences, with the things that are easier to see and more relevant from the perspective of acoustic behaviour described in Table 5.3.

Table 5.3 Key differences between grasshoppers and bush-crickets.

Features and behaviour	Grasshoppers	Bush-crickets
Calling (stridulation)[(1)]	Rubs hind legs against wings	Rubs wings together Some species do not stridulate
Frequency of sound[(2)(3)]	Usually well within audible range of humans	Can be audible, but higher in frequency, being ultrasonic at least in part or in entirety
Length of antennae[(1)]	Shorter than body	Longer than body
Positioning of ears[(1)]	At the base of abdomen	On the front legs
Active[(1)]	Usually diurnal (exceptions occur)	Usually nocturnal (exceptions occur)
References [(1)] GRIRS 2019; [(2)] Lee 2004; [(3)] Ragge & Reynolds 1998.		

Tettigoniidae (bush-cricket) species

Within this family of insects sound is used extensively, in particular in connection with mating-related activities, with speculative males singing loudly, continuously and regularly in order to attract females (Robinson & Hall 2002). Therefore the males that sing the loudest and most often are more likely to attract a mate, albeit at the risk of also attracting the attention of a potential predator.

Like grasshoppers, bush-crickets can produce a variety of different calls depending on their behavioural intentions and the scenario in which they find themselves. Table 5.4 provides an overview of many of the different contexts in which bush-crickets produce sound.

Table 5.4 Examples of behavioural contexts in which bush-crickets produce sound (adapted from Robinson & Hall 2002).

Type of call	Behaviour
Advertisement call	Males produce species-specific calls in order to attract females from the same species. In many species these calls are often loud, long in duration and repeated often.
Aggressive call	Males calling to each other as part of an aggressive interaction.
Female response call **Male and female duets**	Receptive female emits a call in response to an advertising male. In response the male may increase his rate of calling. The female will then locate and approach the male. In some species and/or scenarios the female may stay stationary, with the male making the approach. Note that female response calling, thus duetting with the male, does not occur in all species of bush-cricket. In the British Isles, speckled bush-cricket is a good example of a species that shows this behaviour.
Courtship song	Calls produced immediately prior to copulation.

Studying these creatures is in many ways similar to studying bats. Like bats they are more often heard than seen, and generally speaking they are more active during the hours of darkness. Also, they typically produce higher-frequency sound, often within the same range of frequencies produced by bats, and therefore they can easily be picked up on bat detectors.

As mentioned earlier, sound is very important to bush-crickets, especially in relation to their mating behaviour, when their song (usually stridulation) is used to attract a mate. Thankfully, it is often possible with training and experience to identify them to species level based on acoustics alone. However, it has also been shown that each species has a range of acoustic behaviours (i.e. not just one type of call per species), which can make things more complex – or interesting, depending on your perspective (see Table 5.4).

Another thing to bear in mind, from an analysis point of view, relates to the speed at which these insects call (i.e. stridulate). Both the repetition rate and the duration of the calls can be influenced, within degrees of tolerance, by ambient temperature and radiation of heat. The higher the temperature the quicker the call, with call rate slowing down in cooler conditions. This means that while carrying out sound analysis, the speed of the repetition rate is not always the best thing to focus on if you are less experienced, as this can be quite variable for a single species.

Visually, bush-crickets can look very similar to each other, making identification harder in a number of respects. As well as this, they are not easy to see within their environment as they spend most of their time amongst vegetation, where their colouration works well to camouflage their precise location. You would think that it would be quite easy to locate them from the sound being emitted, but quite often as soon as they hear you coming they go silent, especially if like myself you have size 10 flat feet (they call me 'Thumper'). Therefore, a lot of patience and stealth is often required. Either that or a well-positioned static bat detector.

In the British Isles we have 21 species of Tettigoniidae, 10 of which are regarded as native. Bush-crickets, more than any other insect group in our part of the world, have been studied acoustically using bat detectors for many years. Researchers in this area have tended to focus on the oscillogram tool when describing the calls of these insects, with particular attention being paid to call duration, sequence length, variable repetition rates and the relative amplitude between groupings of call components.

In recent years, with the development and wider use of static bat detectors in bat-related projects, it became apparent to some that while these machines were out gathering bat data, they were also, as an indirect consequence of the effort involved, gathering information about other taxa. In particular, adopting this approach, it was considered that bush-crickets would be a useful group to try and gather more data on. I have already mentioned the BTO's contribution to the collection and analysis of bush-cricket data. One of the motivational drivers behind this work was to attempt to get bat workers to participate more in bush-cricket surveys, as well as highlighting these species, in some respects, as being a potential confusion group for those carrying out analysis of bat calls. To this end, on a practical level, in January 2019 one of the interesting outputs from this project was the production of an internet guide to identifying bush-cricket sound for our more commonly occurring species (BTO 2019).

In Table 5.5 I have provided an overview of the more commonly occurring, and therefore more regularly encountered, bush-crickets in the British Isles. Following on from this, Table 5.6 shows the months when you would typically expect to encounter these species acoustically in the field. However, please remember that this behaviour is influenced by local and seasonal weather patterns, as well as temperature at different altitudes.

Following on from the tables, we will consider each of the stridulating bush-cricket species from a sound analysis perspective. When listening to their calls for this purpose I recommend that you do so in time expansion (× 10), as this will allow you to hear differences in call repetition rates and structure more easily.

Table 5.5 Overview of bush-cricket species occurring in the British Isles.

Species	Distribution	Notes on typical habitat and behaviour
Great green bush-cricket *Tettigonia viridissima*	Southern England and south Wales[1]	🔊 Scrubland, hedgerows, bramble, bracken. Produces a continuous series of calls that can be heard audibly (FmaxE *c.* 10 kHz) from late afternoon and well into the night.[2][3]
Roesel's bush-cricket *Metrioptera roeselii*	Southern England and southeast Wales[1]	🔊 Saltmarshes and dunes, expanding into urban wasteland and agricultural set-aside. Also found in areas of longer grass, for example on roadsides and field margins. A long, continuous regular series of calls, which are audible to many of us (FmaxE *c.* 20 kHz).[2][3]
Speckled bush-cricket *Leptophyes punctatissima*	England, Wales, southwest Scotland and Ireland[1]	🔊 Open woodland, scrub, gardens and hedgerows. Produces short ultrasonic call sequences (FmaxE > 30 kHz), which occur at irregular intervals of between 1 and 5 seconds.[2][3]
Dark bush-cricket *Pholidoptera griseoaptera*	Much of England and Wales, and southwest Scotland[1]	🔊 Hedgerows, shrubs, wasteland, scrubland, woodland, saltmarsh, dune systems and clifftops. Usually delivers a call comprising three sequences, the first of which is often weaker. These calls are repeated at regular intervals, and are audible (FmaxE *c.* 12 kHz).[2][3]
Bog bush-cricket *Metrioptera brachyptera*	England, Wales and southwest Scotland	🔊 Nationally scarce, occurring in lowland heaths and clearings in damp heathy woodland. Produces regular ultrasonic calls (FmaxE *c.* 30 kHz) each comprising three to four sequences.[2][3]
Grey bush-cricket *Platycleis albopuntata*	Southern England and Wales[1]	🔊 Nationally scarce, a coastal species, occurring in coarse grass and rough vegetation on sand dunes, shingle banks and south-facing cliffs. Produces a complex call, repeated consistently (Fmin 15–20 kHz, FmaxE *c.* 23 kHz).[2][3]
Long-winged conehead *Conocephalus fuscus*	Southern England and southern Wales[1]	🔊 Coarse vegetation in ungrazed downland, urban wasteland, coastal reedbeds, dry heaths and bogs. Produces ultrasonic calls at an FmaxE of *c.* 26 kHz.[2][3]
Short-winged conehead *Conocephalus discolor*	Northeast and southern England, Wales and southwest Scotland[1]	🔊 Occurs on saltmarshes and sand dunes, especially in areas with rushes and grasses. Inland, may occur on lowland bogs, fens, reedbeds, river floodplains and by fresh water. A regular series of ultrasonic calls is produced, each with equal emphasis (FmaxE *c.* 33 kHz).[2][3]
Wart-biter *Decticus verrucivorus*	Localised to far south of England[1]	Rare, having very specific habitat requirements within chalk grassland. Calls are loud and distinctive, with rapidly repeated clicks that often start slow and get faster.[2][3]

(cont).

Table 5.5 (cont).

Species	Distribution	Notes on typical habitat and behaviour
Oak bush-cricket *Meconema thalassinum*	Southern half of Britain (much of England and Wales)[1]	Arboreal, occurring in hedgerows, woodland and gardens. Tends to be found in tree canopies and unlike many other bush-crickets does not stridulate, but drums a hind leg on a leaf to attract a mate.[2][3]
Southern oak bush-cricket *Meconema meridionale*	Southern England[1]	Non-native species. Arboreal, occurring in hedgerows, woodland and gardens. Tends to be found in tree canopies. Unlike many other bush-crickets does not stridulate, but drums a hind leg on a leaf to attract a mate.[2][3]
References		
[1] GRIRS 2019; [2] BTO 2019; [3] Natural England 2015.		

Table 5.6 Typical months of calling behaviour (shaded in green) for stridulating bush-crickets occurring within the British Isles (GRIRS 2019).

	May	June	July	Aug	Sept	Oct	Nov	Dec
Great green bush-cricket								
Roesel's bush-cricket								
Speckled bush-cricket								
Dark bush-cricket								
Bog bush-cricket								
Grey bush-cricket								
Long-winged conehead								
Short-winged conehead								
Wart-biter								

Great green bush-cricket

Great green bush-cricket (*Tettigonia viridissima*) is one of our louder species, which can be heard audibly. It usually produces very long consistent calls which are repeated in paired sequences (Figures 5.9 to 5.11), each sequence containing multiple components. The FmaxE is usually in the region of 9–11 kHz, with an Fmin of 7–9 kHz. When the call is slowed down (time expansion × 10) the double-pulsing effect of the paired sequences is easily recognised.

This species could perhaps be confused with Roesel's bush-cricket, but note that the frequency tends to be lower for great green bush-cricket and the sequences are in pairs, as opposed to the regular, evenly spaced effect created by Roesel's bush-cricket.

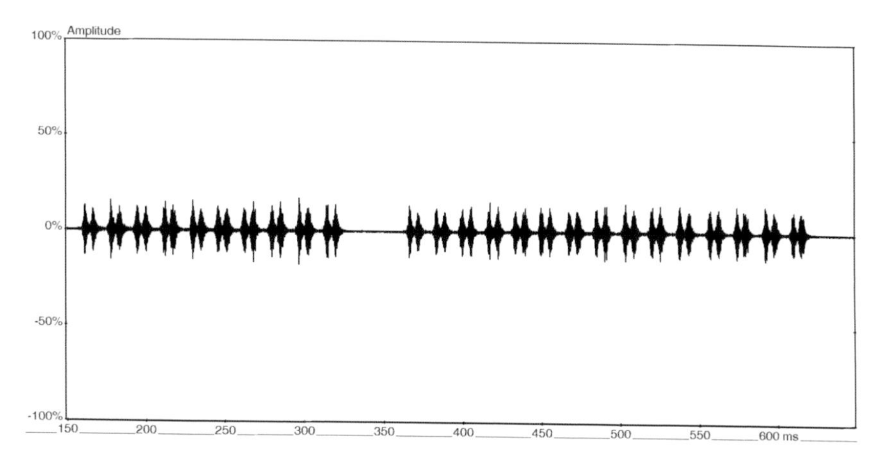

Figure 5.9 🔊 Great green bush-cricket, showing double-pulsing effect on oscillogram (H Lehto, 2015; frame width 500 ms).

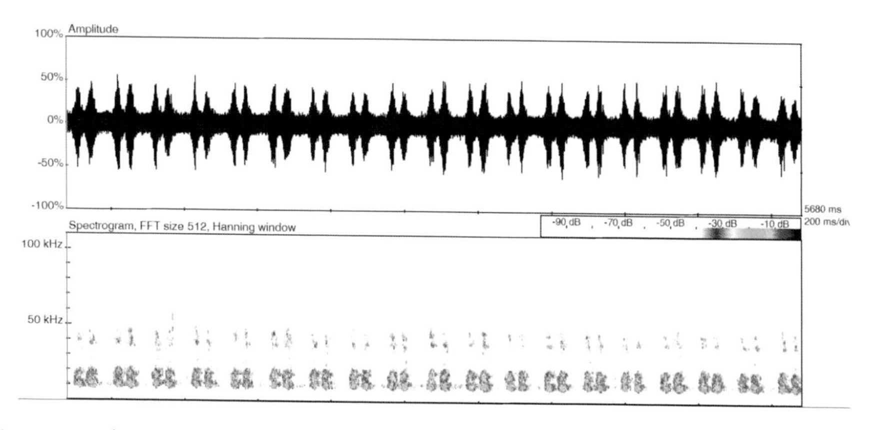

Figure 5.10 🔊 Great green bush-cricket (frame width 2000 ms).

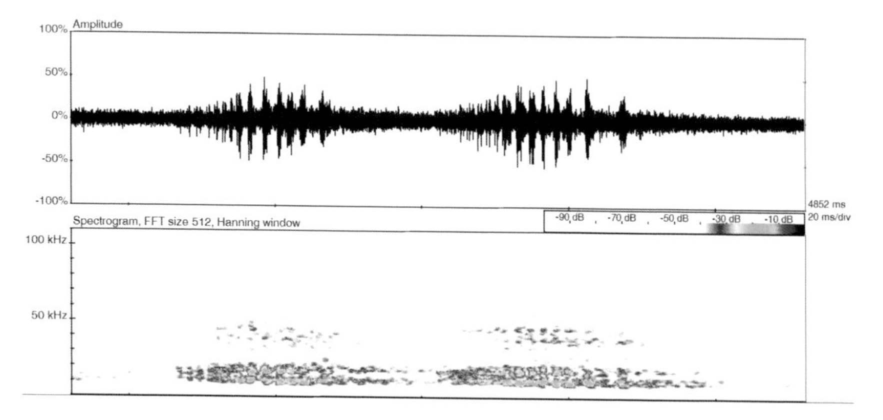

Figure 5.11 🔊 Great Green bush-cricket, showing two groups of components, as in a single paired sequence (frame width 100 ms).

Roesel's bush-cricket

Roesel's bush-cricket (*Metrioptera roeselii*) has been shown to be active not only at night, but from midday onwards (Newson *et al.* 2017). It produces loud and audible long calls. The calls comprise evenly spaced sequences, each containing multiple components (Figures 5.12 to 5.14), typically showing an FmaxE at *c.* 15–24 kHz and an Fmin at *c.* 12 kHz. The audible call has been likened by many ornithologists to Savi's warbler (*Locustella luscinioides*). When it is slowed down (time expansion × 10), the regular, evenly spaced sequences can be easily determined.

This species could perhaps be confused with great green bush-cricket, but note that the frequency tends to be higher (15–24 kHz FmaxE) and the repetition rate is regular, as opposed to the doubling-up effect created by great green bush-cricket. The oscillogram shape usually shows that each sequence gradually increases in amplitude, before stopping and then being repeated. This gives the suggestion of a cone-like shape, with the narrow end of the cone being to the left (Figure 5.2.11). Figure 5.2.12 is a nice example provided by Chris Nason showing not only this species, but also a dark bush-cricket (see later), on top of which is a soprano pipistrelle.

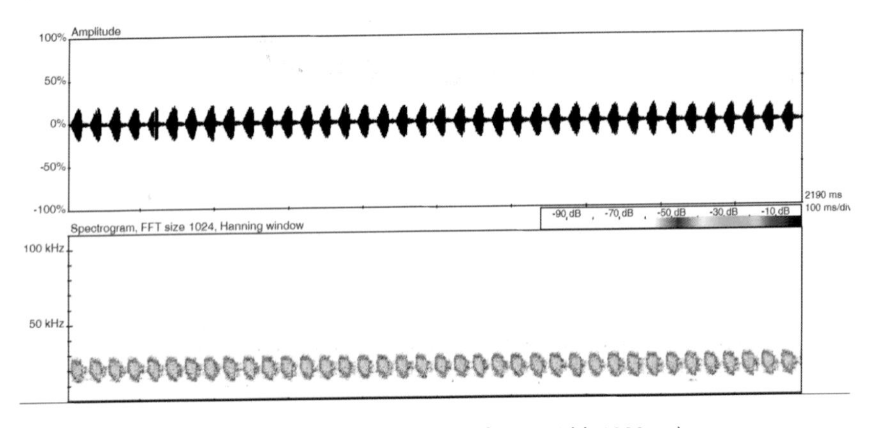

Figure 5.12 Roesel's bush-cricket (C Nason, 2018; frame width 1000 ms).

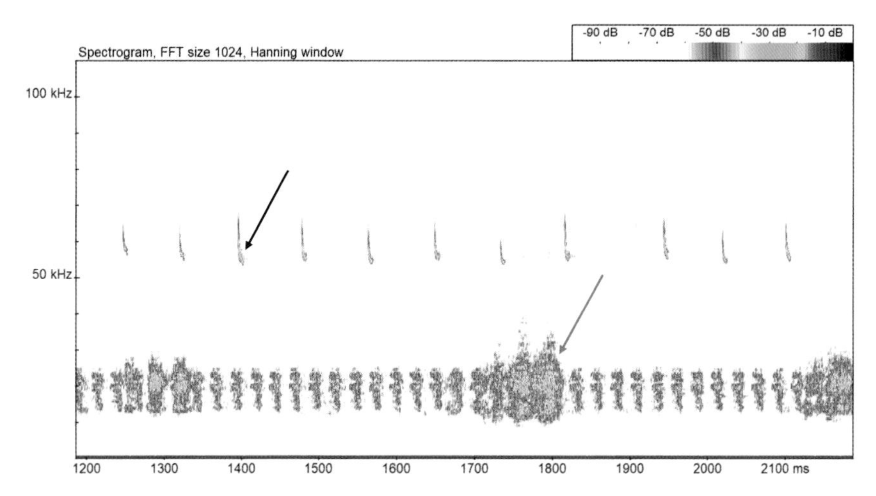

Figure 5.13 Roesel's bush-cricket, with dark bush-cricket (blue arrow) and soprano pipistrelle (black arrow) (C Nason, 2018; frame width 1000 ms).

If we take a closer look at Figure 5.13 and zoom in (Figure 5.14), we can see how the oscillogram tool can make things a little clearer when trying to work out what is going on within a sequence such as this. On the oscillogram in the zoomed-in Figure 5.14 I have shown the differences in amplitude, duration and shape to help clarify that three separate things are occurring.

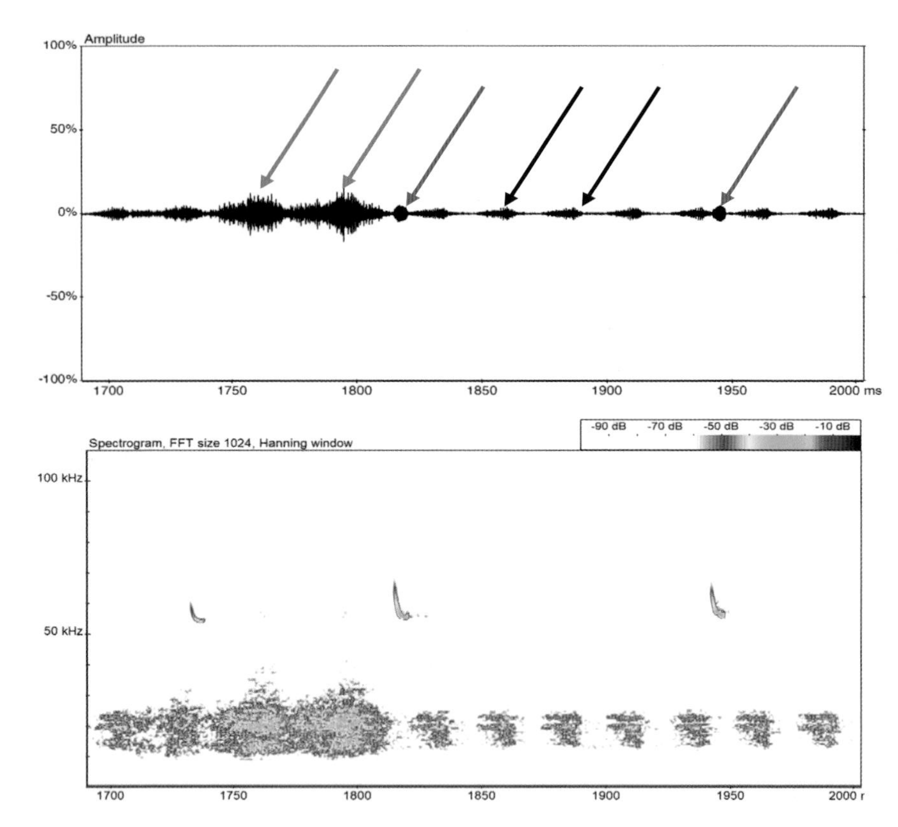

Figure 5.14 Roesel's bush-cricket (black arrows), with dark bush-cricket (blue arrows) and soprano pipistrelle (red arrows). Oscillogram (above), and spectrogram (below) (C Nason, 2018; frame width 300 ms).

Speckled bush-cricket

Speckled bush-cricket (*Leptophyes punctatissima*) is one of the more regularly encountered species, and it has the widest distribution range of all of the bush-crickets occurring within the British Isles. As such, it has the potential to be recorded by many bat workers. The call, comprising a single sequence (Figures 5.15 to 5.17), is totally ultrasonic, repeated at a slow repetition rate, with each call (5–8 ms in duration) being produced intermittently a few seconds apart, or in some cases much further apart and more irregularly spaced. The FmaxE tends to fall within the range 32–40 kHz, with an Fmin range of 24–33 kHz. When listened to in time expansion (× 10) the individual components are heard as a rapid, diminishing clacking sound.

When studied more closely (Figure 5.16) a single sequence can be seen as a group of components emitted extremely quickly. The individual components within the sequence are of short duration, and it is not unusual for the amplitude of each component to reduce as the sequence progresses.

Females produce a shorter version of the male call (1–2 ms in duration) in response to the presence of a male. When a male becomes aware that a female is in the vicinity, the male call rate increases (Robinson & Hall 2002).

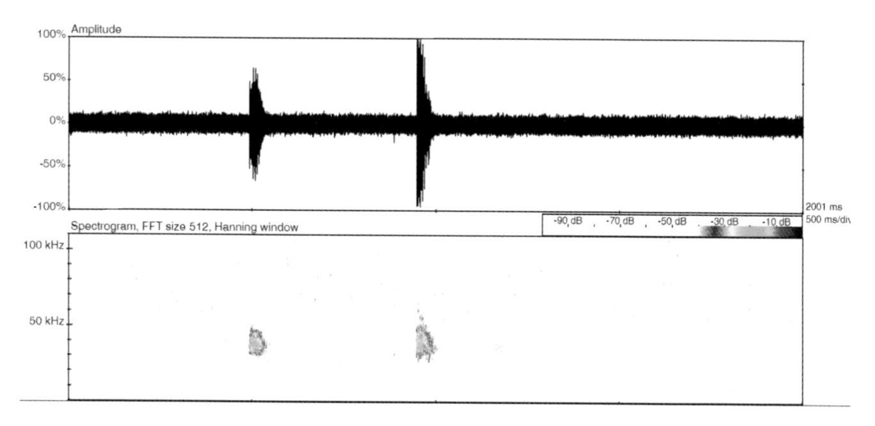

Figure 5.15 Speckled bush-cricket – two calls, each of a single sequence (frame width 2000 ms).

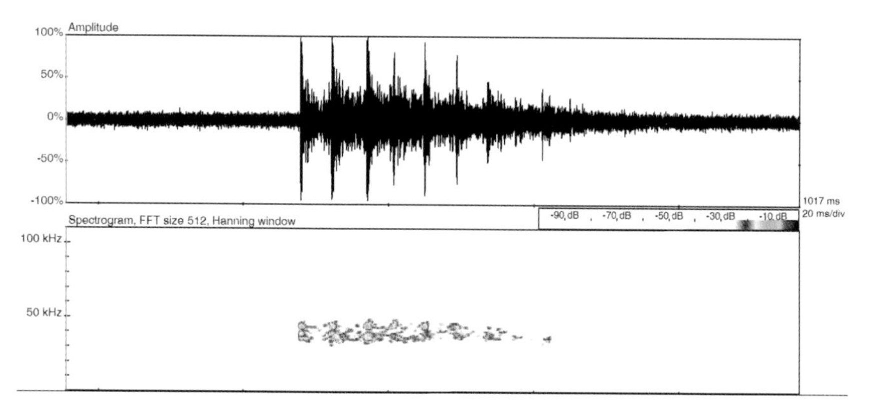

Figure 5.16 Speckled bush-cricket, showing a single multi-component call sequence (frame width 100 ms).

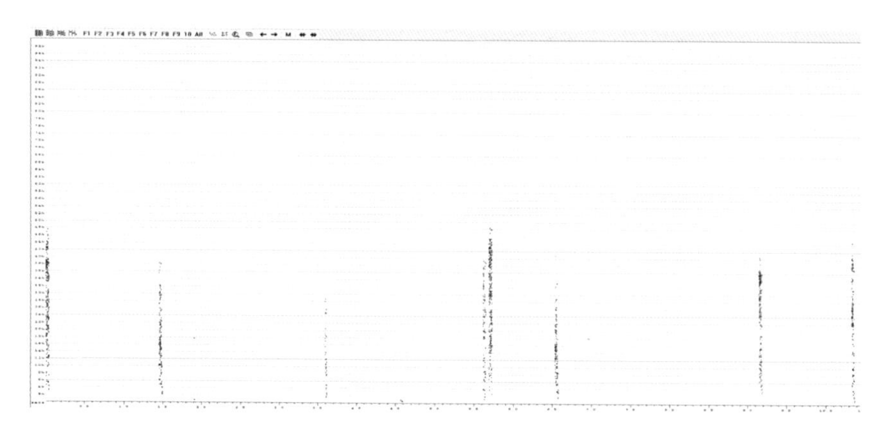

Figure 5.17 Speckled bush-cricket in AnaLookW (S Sowler, 2008; F3, uncompressed, frame width *c.* 10 seconds).

Because the calls of the speckled bush-cricket are entirely ultrasonic, and given the frequency range within which they occur, there is the potential for them to be confused with bat calls. This is especially the case when looked at in detail, as a technician may consider it to be a bat social call (e.g. common pipistrelle), especially if the bush-cricket call is at the lower end of the expected frequency range. This can be even more so when you have a bush-cricket calling along with a social calling bat, as in Figure 5.18. In this example we have a Nathusius' pipistrelle type D call, accompanied by a speckled bush-cricket (see black arrow). Note that there is also a soprano pipistrelle echolocating in the background.

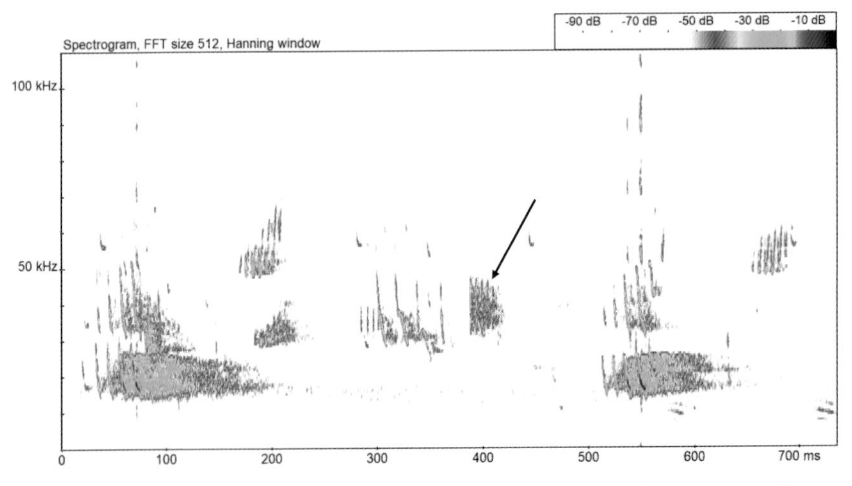

Figure 5.18 Speckled bush-cricket calling (black arrow) within Nathusius' pipistrelle type D social call and soprano pipistrelle (K French, 2018; frame width 700 ms).

For someone familiar with speckled bush-cricket calls, both visually as well as listening in time expansion (× 10), it should become relatively easy to separate the insects from the bats. In the example shown (Figure 5.18) it is far more difficult – and it would be understandable why someone, less experienced or not paying full attention, could attribute all of the sound to the bat involved.

Dark bush-cricket

Dark bush-cricket (*Pholidoptera griseoaptera*) is usually heard audibly as a single sound, repeated at relatively long intervals (e.g. calling once every couple of seconds). On closer inspection you can see that each call is made up from a series of usually three multi-component sequences, the first of which is often much weaker than the remaining two. Calls typically have an FmaxE of *c.* 12 kHz and an Fmin at < 10 kHz. When listening to the call slowed down (time expansion × 10) you will hear the three 'croaking' sequences close together, followed by a long period of silence before the next call (i.e. a series of three sequences) commences.

In terms of the number of calls emitted and the repetition rate, this species could be confused with bog bush-cricket (see later). However, bog bush-cricket can emit more sequences grouped together (up to six) and tends to be at a much higher frequency (e.g. FmaxE at > 25 kHz) and therefore not audible.

For an inexperienced bat worker, or someone not paying attention to frequencies, there is the potential for dark bush-cricket to be misidentified as a type D bat social call, for example the typical three-component call of soprano pipistrelle.

Figures 5.19 and 5.20 show typical full spectrum recordings for dark bush-cricket, and Figure 5.21 gives a ZCA example as shown in AnaLookW. Finally, Figure 5.22 shows an interesting example with both dark bush-cricket and speckled bush-cricket being present.

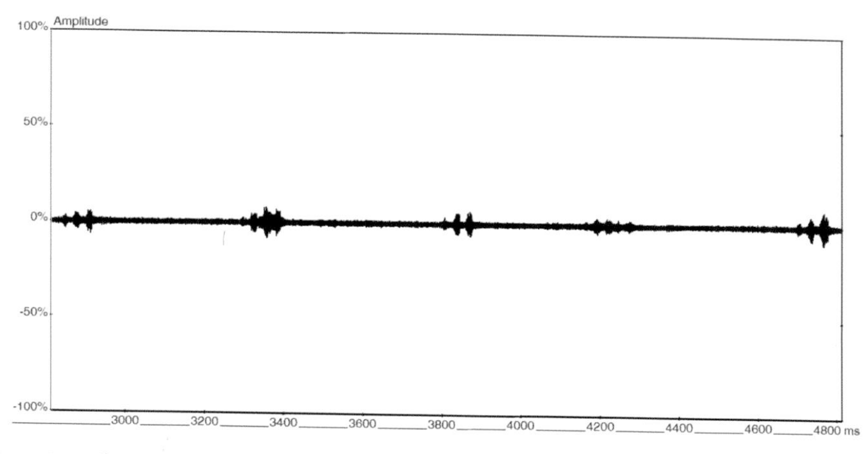

Figure 5.19 🔊 Dark bush-cricket – oscillogram showing weaker first sequence within each series of three sequences (H Lehto, 2015; frame width 2000 ms).

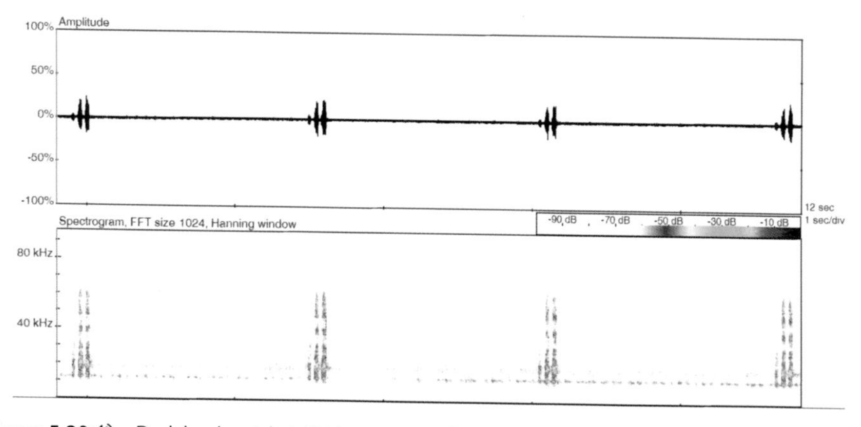

Figure 5.20 🔊 Dark bush-cricket (C Nason, 2018; frame width 5000 ms).

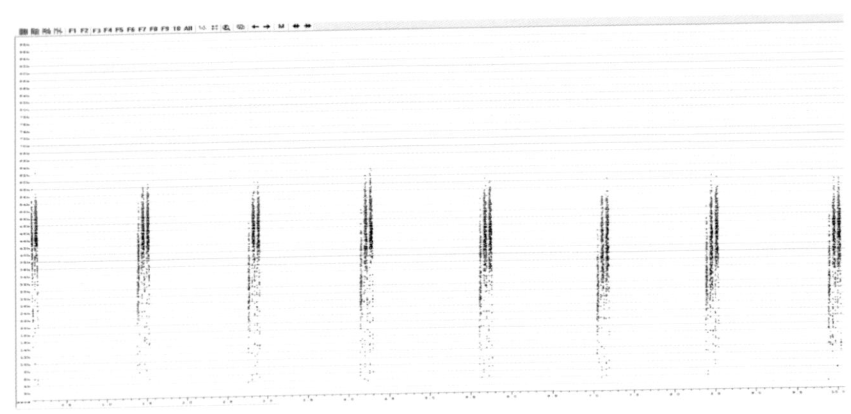

Figure 5.21 Dark bush-cricket (S Sowler, 2008; F3, uncompressed, frame width *c.* 10 seconds)

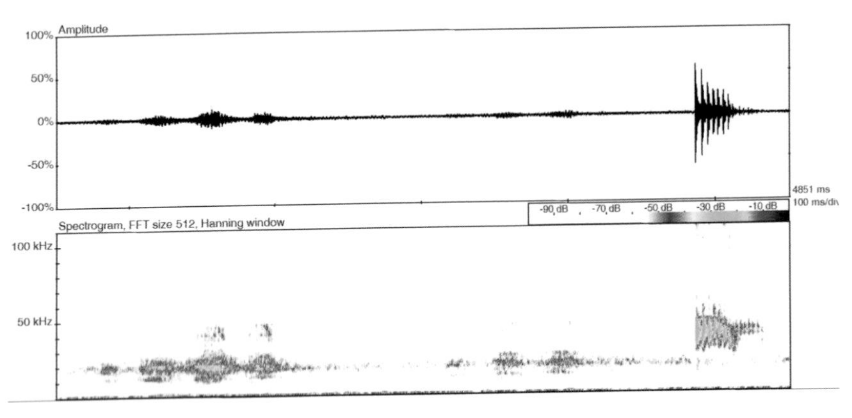

Figure 5.22 Dark bush-cricket with speckled bush-cricket (C Nason, 2018).

Bog bush-cricket

Bog bush-cricket (*Metrioptera brachyptera*) typically emits ultrasonic calls at an FmaxE in the range 28–30 kHz, and an Fmin in the range 20–25 kHz. Often sequences are produced in groups of three to four, but this may increase to as many as six. In terms of the number of grouped sequences emitted, and their repetition rate, bog bush-cricket could be confused with dark bush-cricket, but dark bush-crickets tend to produce calls at a much lower frequency (e.g. FmaxE *c.* 12 kHz) and therefore audible. As in dark bush-cricket, bog bush-cricket tend to show the first sequence within each group as being at a lower amplitude than the others, which can be seen to an extent within Figures 5.23 and 5.24. When listening in time expansion (× 10) the effect is very similar to that of dark bush-cricket, albeit at a higher frequency, although the number of grouped sequences is potentially higher than the typical series of three produced by dark bush-cricket.

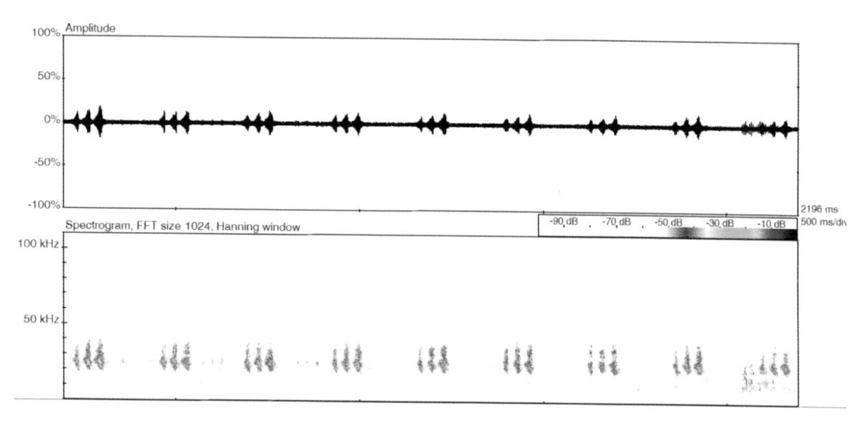

Figure 5.23 Bog bush-cricket – a three-pulse example (S Newson, BTO Thetford, 2015; frame width 2000 ms).

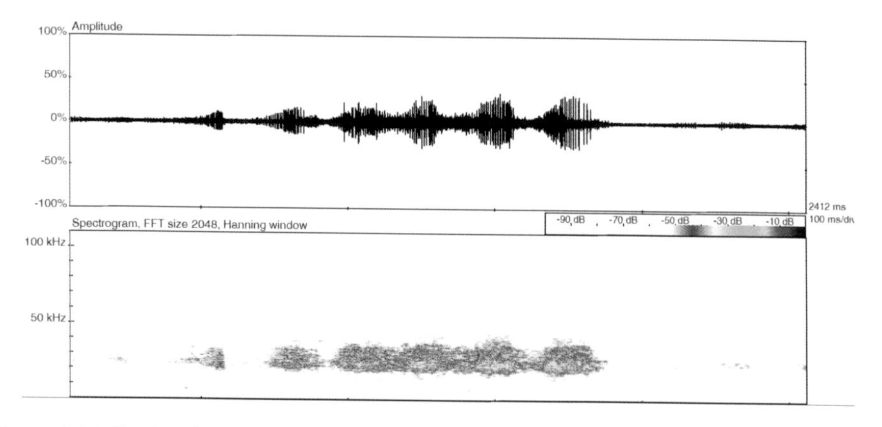

Figure 5.24 Bog bush-cricket – a six-pulse example (S Newson, BTO Thetford, 2015; frame width 500 ms).

Grey bush-cricket

Grey bush-cricket (*Platycleis albopuntata*) produces higher-frequency calls which are hard to hear audibly. Usually they produce four or five sequences per call, repeated at regular intervals, at an FmaxE of *c.* 23 kHz and an Fmin in the range 15–20 kHz. As in some other species, it is usual for the first element of the repeated phraseology to be weaker in amplitude, as easily seen in Figures 5.25 and 5.26. When listening to these calls in time expansion (× 10) the call has a similar feel to that of bog bush-cricket, albeit at a lower frequency and more drawn out.

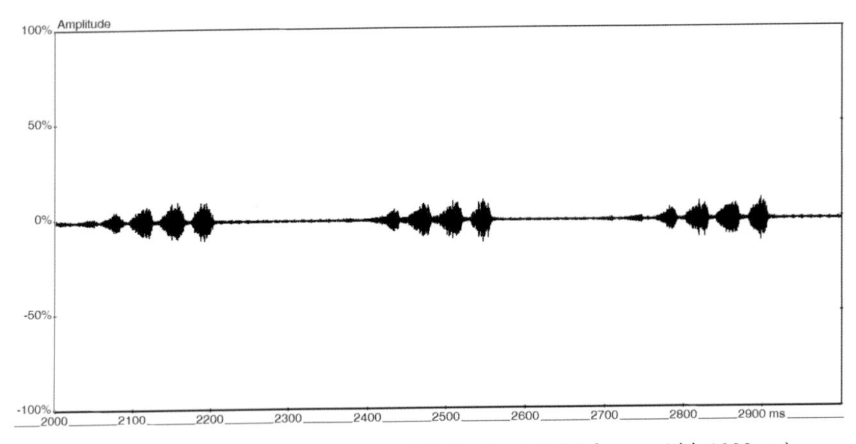

Figure 5.25 Grey bush-cricket (S Newson, BTO Thetford, 2018; frame width 1000 ms).

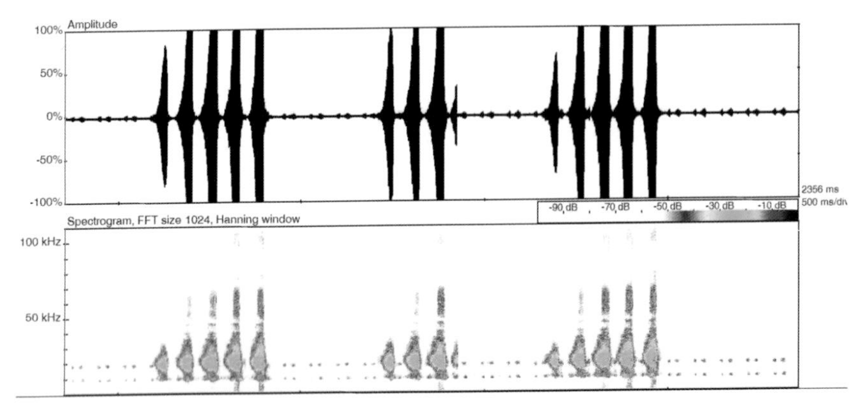

Figure 5.26 Grey bush-cricket, with great green bush-cricket in the background (A Froud, 2013; frame width 2000 ms).

Long-winged conehead

Long-winged conehead (*Conocephalus fuscus*) may be active during daylight hours, to a considerably higher level than would be expected (Newson *et al.* 2017). It can, on occasions, be difficult to separate it with confidence from short-winged conehead. Also, it may be confused with great green bush-cricket (Lee 2004), but listening in time expansion (× 10) should clearly allow you to hear whether the sequences are grouped in pairs or threesomes. Usually the call of long-winged conehead comprises three easily heard sequences grouped together, with the first two sounding short, followed by a third longer sequence. Listening closely, you may be aware of a quick, simple element occurring immediately before the group of three sequences. This can be seen clearly in Figures 5.27 and 5.28. Typically, across the whole call, FmaxE frequencies of *c.* 26–30 kHz would be expected, with Fmin in the region of 25 kHz.

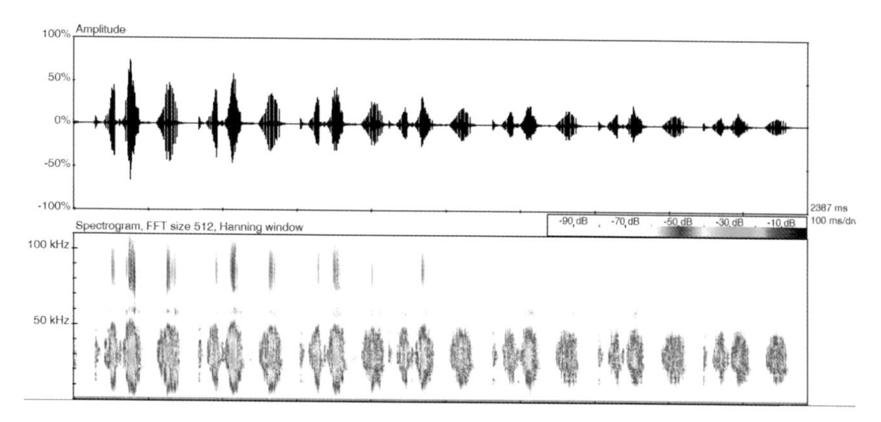

Figure 5.27 🔊 Long-winged conehead – groupings of four sequences. A single click, followed by two short emissions, then a longer gap followed by one longer emission (S Newson, BTO Thetford, 2015; frame width 1000 ms).

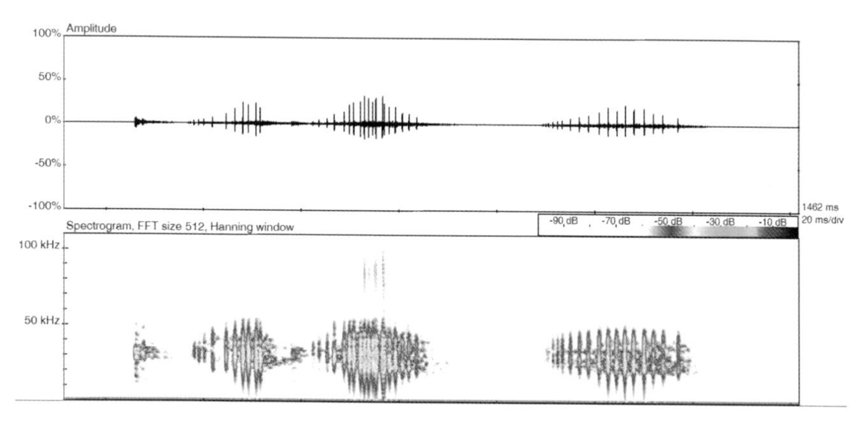

Figure 5.28 🔊 Long-winged conehead – groupings of four sequences. A single click, followed by two short sequences, then a longer gap, followed by one long-sequence emission (S Newson, BTO Thetford, 2015; frame width 150 ms).

Short-winged conehead

Short-winged conehead (*Conocephalus discolor*) has been shown to be heavily active during the day, with 95% of its stridulating time allocated to daylight hours (Newson *et al.* 2017). It may at times be difficult to separate with confidence from long-winged conehead, and there is also potential for confusion with great green bush-cricket (Lee 2004). The calls are repeated in grouped sequences of four, followed by a short space, and repeated consistently over a long period (Figure 5.29). Listening in time expansion (× 10), each sequence sounds of equal length and emphasis. Intermittently during these long calls, single-component sequences may occur, then reverting back to the groups of four. Typically, an FmaxE of *c.* 30–33 kHz would be expected.

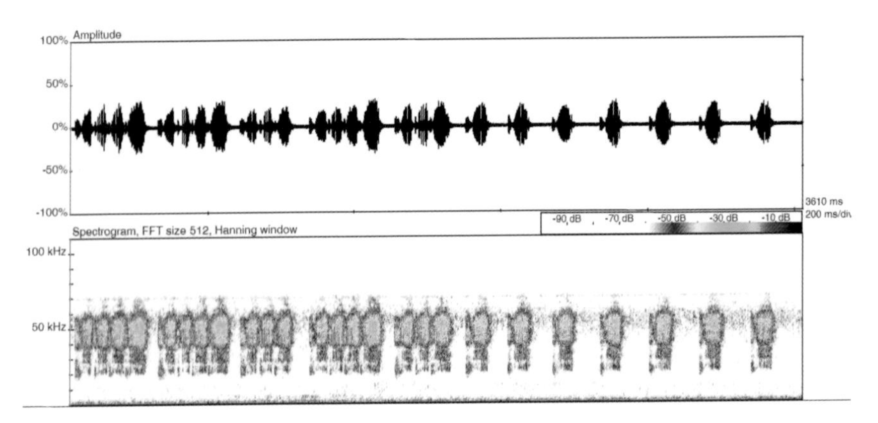

Figure 5.29 Short-winged conehead – groupings of four sequences, followed by single-sequence emissions (S Newson, BTO Thetford, 2015; frame width 1000 ms).

Wart-biter

Wart-biter (*Decticus verrucivorus*) is a rare species for the British Isles, with its range restricted to a small number of sites in the far south of England. The call is audible, being produced on 'hot and sunny days, often lasting for several minutes' (Ragge & Reynolds 1998). Having listened to a number of examples of this call myself, I would describe it as being not dissimilar to long-winged conehead. However, where in that species you would normally expect to see (or hear) two short sequences, followed by a space, and then the final longer sequence, in wart-biter it normally would appear to be the first sequence, followed by a space, and then two sequences close together, each of these three sequences being of similar duration. As always, with this group, it is best to listen in time expansion (× 10). Unfortunately I have not been able to source any recordings myself, but Figure 5.30 demonstrates an artificial example of what I have just described. Please note that Figure 5.30 is artificial, and that the oscillogram produced is not from a real insect.

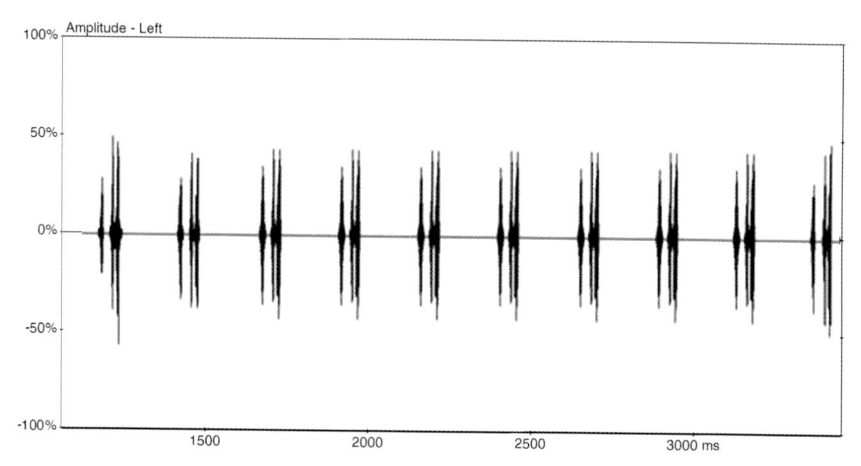

Figure 5.30 Wart-biter – an artificial representation of repetition rate, showing the first sequence, followed by a space and then two sequences close together (frame width 1000 ms).

Oak bush-cricket and southern oak bush-cricket

Compared to the other species of bush-cricket described in this chapter, oak bush-cricket (*Meconema thalassinum*) and southern oak bush-cricket (*Meconema meridionale*) don't quite follow the norm for two reasons. First of all the males don't stridulate when advertising for a mate. Instead they make a drumming noise with a hind leg against a leaf. Secondly, they tend to be arboreal, normally found high up in tree canopies. The chance of this species being recorded during a bat survey is therefore somewhat lower than for the others described in this chapter. Bearing this in mind, despite my disappointment in not being able to show you examples of this species, from a bat worker's perspective it arguably isn't a big omission, unless of course you have placed a static bat detector at height within a tree canopy. OK, possibly it is a big omission after all, giving me something to aim for in the future.

Moths

There are approximately 2500 moth species (67 families) known to occur within the British Isles (UK Moths 2019), and at least some of these are known to emit sound. Before we focus on a small selection of our own species, I feel a wider perspective may be useful. The following paragraph provides more detail relating to moth sounds generally.

Some species of moth are known to make sound for communicative purposes (Conner 1999, Nakano *et al.* 2014), while others (e.g. arctiid species, including tiger moths and hawk-moths) are known to emit 'jamming' (clicking) sounds in response to bat echolocation, with the purpose of confusing predating bats. The subject of predator avoidance is discussed separately, later on. To continue on the theme of sound for communicative purposes, it has been shown that across a range of species, male moths of some species can produce low-intensity, broadband ultrasonic calls when close to females (Nakano *et al.* 2009), and in fact the production of such sound may be far more common in eared moths than was initially perceived to be the case (Nakano *et al.* 2014). The FmaxE ranged from 38 to > 100 kHz across the nine species studied, and it was ascertained that the calls produced were strong enough to be picked up by a nearby potential mate, but too weak to be noticed by 'unintended receivers' (e.g. bat predators). As well as species that produce low-intensity calls, it has also been shown that other species produce louder calls, effective over a longer range (Spangler 1988).

Personally, up until researching this book, I had never considered that moths could be picked up audibly and/or on bat detectors. Keith Cohen corresponded with me to advise of a survey session when he 'spent a while wondering what the interference noise was that I was encountering on a field edge in Northumbria'. He was eventually surprised, and deafened, when a big green moth flew by. It fitted well with the description of the noctuid moth, green silver-lines (*Pseudoips prasinana*). Not only that, but he was able to record the sound, and retained the files – which he has very kindly provided me with, so that I can share them with you here (Figures 5.31 to 5.33).

In order to try and find out more I was fortunate enough to meet, and correspond with, Mark Cubitt, who in turn also consulted with another experienced moth worker, Roy Leverton. Here is Roy's response to my query relating to Keith's observation:

> Male green silver-lines are well known to stridulate in flight. The book Moths and Butterflies of Great Britain and Ireland (Volume 10) (Heath & Emmet 1983) states that it has 'often' been heard to do this, but gives no references. The sound is audible to the human ear, and one of my Banffshire observers, Bill Slater, vividly describes hearing the moth stridulating overhead and netting it during his boyhood collecting days. Because only males have been heard, the sound is believed to be sexual or territorial rather than a bat deterrent as employed by the tiger moths.

Bearing all this in mind, I think this species of moth deserves a bit more attention. Accordingly, Table 5.7 provides more detail about green silver-lines moth, as well as scarce silver-lines (*Bena bicolorana*) and a potentially loud vagrant to the British Isles, death's-head hawk-moth (*Acherontia atropos*).

Table 5.7 Overview of a small selection of moth species that generate acoustic noise in the British Isles.

Species	Distribution	Notes on typical habitat and behaviour
Green silver-lines moth *Pseudoips prasinana*	Fairly common across much of England and Wales; less common, but widely distributed, in Scotland and Ireland	This species is found in woodland habitat as well as gardens, and is one of only a small number of green-coloured moths occurring in the British Isles, including scarce silver-lines (see below), which is larger. It has a wingspan of 30–38 mm, and its flight period is usually May to July. It can be attracted to artificial light.[1] Males of this species are known to make mating calls in flight, these being described as single click noises.[2]
Scarce silver-lines moth *Bena bicolorana*	Distributed widely in England and Wales, as far north as Yorkshire, being locally common in places	This species is found in woodland and parkland, and is one of only a few green moths occurring within the British Isles. It is larger than green silver-lines (see above), with a wingspan of 40–45 mm.[1] Its flight period is June to August, and it can be attracted to artificial light. Males of this species are known to make mating calls in flight, which have been described as comprising numerous clicks.[2]
Death's-head hawk-moth *Acherontia atropos*	A vagrant to the southern UK; has also been recorded in other locations in the British Isles	This species is usually found in gardens and organic potato fields, as well as in the vicinity of bee hives, where it is known to feed on honey. With a wingspan of up to 120 mm, this is the biggest moth known to occur in the British Isles. Vagrants from further south in Europe are recorded annually, later in the summer, through to autumn.[1] It is known to give off an alarm call when distressed.[1] This call is a combination of multiple and very short-duration FM sweeps, followed by a much longer, near-constant-frequency 'whistle' type call (*c.* 30–100 ms), which has an FmaxE in the region of 15 kHz.[3]
The examples given above are merely examples, and are not considered to be the full extent of the moth species occurring within the British Isles which have the ability to make sound.		
References		
[1] Butterfly Conservation 2018; [2] Skals & Surlykke 1999; [3] Sales & Pye 1974.		

Green silver-lines moth

The good thing about many moth species is that their common English names quite often describe what they look like. So guess what? The green silver-lines moth is green, with silver lines. Its flight period occurs 'smack bang' in the middle of our bat survey season, and it can be attracted to artificial light. So watch out if you have your head torch on!

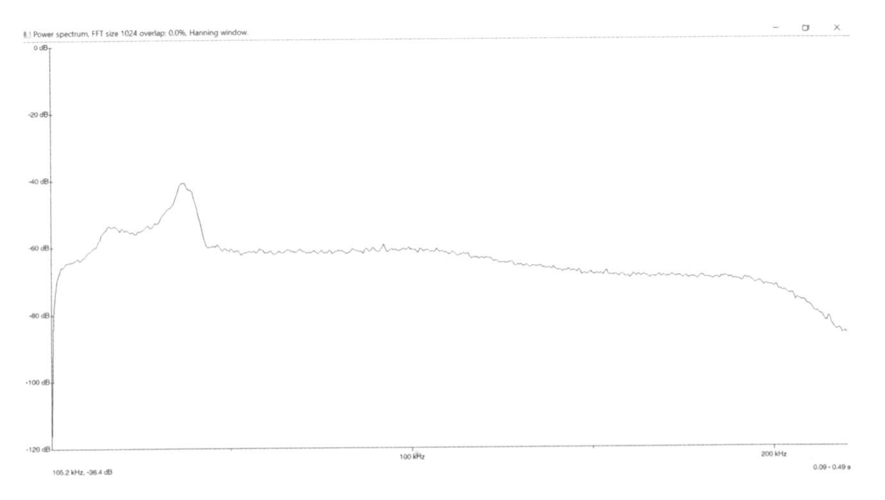

Figure 5.31 Green silver-lines moth – power spectrum showing typical FmaxE (K Cohen, undated).

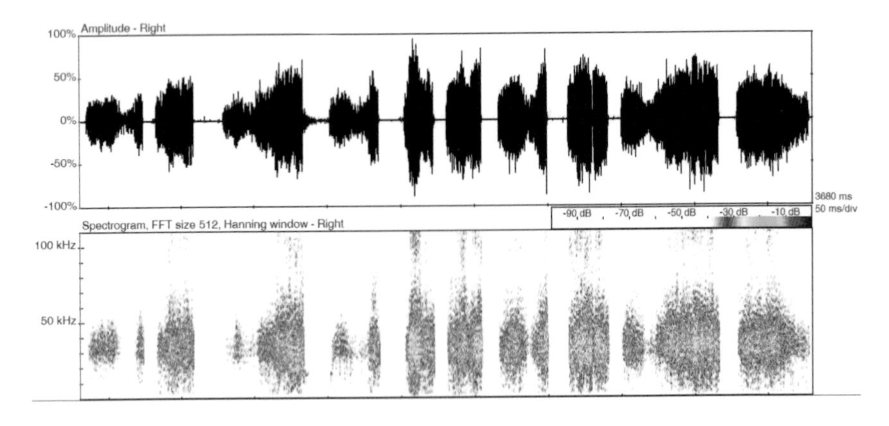

Figure 5.32 Green silver-lines moth – heterodyne recording (K Cohen, undated).

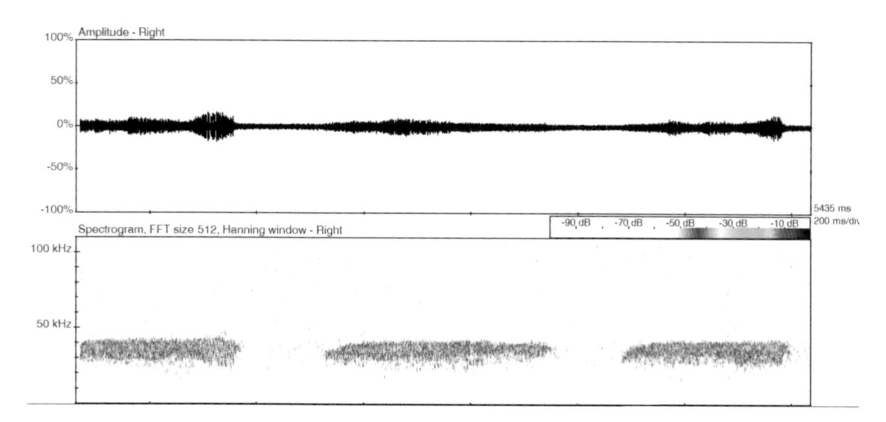

Figure 5.33 Green silver-lines moth – time expansion (× 10) (K Cohen, undated).

Predator versus prey

I feel it is only fair, since we are discussing insects and their relevance to bat work, that I devote at least some space to discussing the relationship between bat predators and their insect prey – especially bearing in mind that sound plays a big part in how the two sides of this conflict carry out their manoeuvres. Also, as bat workers I know many of us regularly make reference to this relationship, for example when giving bat talks to members of the public.

Successful predation has been described as a process involving the following stages: encounter, detection, identification, approach, subjugation and then, finally, consumption (Endler 1991). All bats in the British Isles detect and capture their insect prey using echolocation, although some (e.g. brown long-eared) also use their listening ability to great effect.

From the perspective of the prey, in order to avoid being consumed, there are a number of strategies that can be employed, either in isolation or in combination. A number of insects have developed an ability to hear/sense and react to bat ultrasound, and many of these species (though not all) belong to the following orders:

- Coleoptera (beetles)
- Neuroptera (lacewings)
- Lepidoptera (moths and butterflies)
- Orthoptera (true crickets, bush-crickets and grasshoppers)

In particular, the relationship between insectivorous bats and insects is an interesting one in which to understand more about how bat prey has evolved, in some cases to reduce the risk of capture (i.e. increase the chance of survival). Across a range of insect species a number of strategies have evolved, and many of these are summarised in Table 5.8, before being discussed in more depth, appropriately referenced, in the following paragraphs.

Table 5.8 A summary of many of the tactics adopted by insects for avoiding predators.

Call-related strategy	Description
Predator detection distance	The earlier the prey notices the predator, the easier it is to avoid being predated upon (e.g. the prey flies off in a different direction). Better still if the predator is detected by the prey sooner than the prey is detected by the predator. Upon hearing bats in the vicinity, certain insect species fly away in the opposite direction.
Behavioural cycles	Prey not being active at the same time of day or the same time of year as a potential predator (e.g. night insects avoiding day birds).
Make less noise or create noise at a frequency not detectable	Call less often, and/or call more quietly, and/or produce sound at a frequency that is less noticeable to a predator, taking advantage either of the predator's hearing range or of attenuation, meaning that the sound produced by an insect doesn't carry a great distance.
Switch off sound when predator is heard nearby	For insects that make noise (e.g. while trying to attract a mate), when a bat is sensed to be nearby the insect may stop calling (i.e. go acoustically 'dark') until it perceives that the danger has passed.
Drop to ground and remain motionless until bat has passed by	Certain species of moth are known to drop to the ground in order to escape being detected by a bat. Depending upon where the insect is, it may remain motionless (and therefore audibly invisible to the bat) until the danger passes.
Produce a disturbance sound or jamming sound to confuse a predator at point of attack	Certain species of moth (e.g. tiger moths) emit high-frequency (ultrasonic) signals as a bat is about to attack. This is in order to startle/confuse the predator. It has also been suggested that this also serves as a warning to the pursuing bat that the prey is unpalatable/toxic, as some species of tiger moth undoubtedly are (also see *Mimicry*). In this case, there would be a mutual benefit to both predator and prey.
Mimicry	The prey pretends to be something that may cause the predator harm or may taste unpalatable. It has been suggested that some species of hawk-moth may mimic other toxic species in order to take advantage of a bat's eagerness to avoid toxic prey.
Call within a group	The group, with many eyes and ears, may pick up on the presence of a predator more quickly. Strength in numbers could also mean that each individual is less likely to be attacked, as well as a predator being confused in the dark as to the direction of individual calls.
Call from a protected area	Make calls either from the vicinity of something perceived to be dangerous by your predator, or within a feature that your predator is unable, or would not choose, to enter.
Threat calling	Aggressive threat calls may dissuade a predator from attacking.
Distress calling SOS	At the point of capture the prey item produces distress calls, either to confuse the predator and/or encourage others to come to the assistance of the prey.

The relationship between some species of moths and bats, in the battle between predator and prey, has been the subject of many academic studies. It is widely accepted that ultrasonic hearing/sensing within the relevant moth species has evolved over many millions of years as a defence against being preyed upon by bats. Certain species of bat feed on moths, and this in itself provides a fascinating insight into the world of predator and prey, as the two parties to this relationship have been described as a 'clear-cut example of co-evolution' (Nakano *et al.* 2014).

In some moth species (e.g. Noctuoidea, Geometroidea and Pyraloidae species) it has been shown that they have simple ears that are known to be sensitive to the ultrasonic pulses created by bats during echolocation. These moths are silent, but upon sensing a bat in their vicinity they will take avoiding action (Sales & Pye 1974, Fullard 1988). Relationships between the size of some moth species and their ability to pick up echolocation have also been demonstrated, whereby larger moths, which are more easily picked up by bat echolocation from further away, are more sensitive to lower frequencies, and smaller moths are more sensitive to higher frequencies (Surlykke *et al.* 1999). In addition, some moth species in different parts of the world, where bat echolocation peak frequencies differ, appear to have evolved, to an extent, to demonstrate sensitivity in hearing relative to sympatric predators (ter Hofstede *et al.* 2013). It is also interesting to note that some moths can detect bats from greater distances (20–100 m) than bats can detect moths (1–10 m). This is due to the moths detecting the stronger output created by a bat, as opposed to the bat searching for the weaker returning echoes (Surlykke *et al.* 1999, Surlykke & Kalko 2008, Nakano *et al.* 2014).

Other species of moth (e.g. Grote's tiger moth, *Bertholdia trigona*, which occurs in North America) are known to emit sound in reaction to the presence of an echolocating bat in order to confuse (i.e. jam) the bat's echolocation when it is about to capture the moth (Corcoran *et al.* 2009, 2011). As well as this, some tiger moth species have been shown to adopt a mimicry tactic, whereby the sounds they make, which are similar to those emitted by toxic prey, suggest to a predator that they themselves may also be unpalatable (Corcoran *et al.* 2010). Some 'deaf' species of moth from the genus *Yponomeuta* have been shown to perpetually produce mimicking clicks 'using wingbeat-powered tymbals' to potentially warn bats of their toxicity, these wing-produced sounds being very similar to the sounds produced by other toxic moths (O'Reilly *et al.* 2019).

Another interesting adaptation that appears to have evolved in some nocturnal moths is that their wings absorb ultrasound better than the wings of diurnal moth species (Ntelezos *et al.* 2017). This is advantageous to these nocturnal moths in that both the intensity of any reflective echoes and the detection distance of a predating bat are reduced. In a similar vein, it has also been demonstrated that the thorax fur on some moth species 'acted as an acoustic camouflage absorbing up to 85% of bat echolocation' (Neil 2018).

Moving on, it is not only moth species that have evolved to be able to deploy predator avoidance tactics. Other insect species have been shown to behave in a similar manner to what we see in moths. Mole-crickets and bush-crickets are good examples. In some species it has been shown that they have the ability to detect sound up to as high as 100 kHz (Rheinlaender & Römer 1980, Hutchings & Lewis 1983).

An excellent example, one of many relating to Orthoptera, concerns a species of bush-cricket that occurs within the British Isles. The great green bush-cricket has been shown to react to bat echolocation by adopting up to three tactics depending upon the closeness of the potential predator (Schulze & Schul 2001). Initially it might steer away from the direction of the bat. Secondly, if the risk is perceived to be closer, it ceases to beat its hind wings. The final tactic would be to stop beating its wings altogether and dive quickly to the ground to avoid detection. It was calculated that the steering behaviour could be adopted by the insect when a bat was *c.* 18 m away, the change in wingbeat behaviour at *c.* 10 m, and the dive occurring when the bat was *c.* 5 m away from the insect.

So, as discussed, in an ideal situation it would of course be beneficial to the insect involved, if it was able to detect the presence of a bat before the bat detected the presence of its prey (e.g. through returning echoes). In one particular study it was demonstrated, using a species of bush-cricket (long-winged, *Phaneroptera falcata*) and the echolocation of a potential predator, greater mouse-eared bat (*Myotis myotis*), that the bush-cricket had the ability to pick up the bat echolocation from a distance of 13–30 m away (Schul *et al.* 2000). It was calculated that the insect had more than a one-second advantage over the bat (i.e. before the bat would pick up a returning echo) in which to adopt a predator-avoiding tactic.

This brings us nicely on to a tactic adopted by some bat species whereby they deploy behaviours to overcome the reactions of their prey. This is evident when bats switch off their echolocation when hunting in order to sneak up, unnoticed, on a potential prey item that could hear them if they weren't in stealth mode. Brown long-eared bats, for example, appear to have evolved to counter some of these prey strategies, firstly with a very quiet echolocation signal. This allows them to get much closer to their prey before being detected. In addition to this, they quite often detect prey purely by listening (as opposed to echolocating), which allows them to get close to food without it being aware of their presence. The especially large ears of the brown long-eared bat are ideal for listening for prey and/or receiving the echoes from their quiet ultrasound.

Long-eared bats are not the only species in the British Isles that are thought to have adapted their echolocation in order to behave stealthily when approaching insects that may be able to hear them. Barbastelle bats are thought to produce lower-volume echolocation pulses when hunting, with these pulses becoming even lower in volume as the bat gets closer to its prey, meaning that the bat detects its prey before the moth is aware of the bat's presence (Goerlitz *et al.* 2010), or at the very least, the moth gets less warning notice.

Finally, a third possible strategy is thought to be used by some species of bat to gain an advantage over prey with hearing abilities, for example by a North American species, the spotted bat (*Euderma maculatum*), and by the short-eared trident bat (*Cloeotis percivali*), which occurs in Africa. It appears that evolution may have enabled them to operate at echolocation frequencies outwith the hearing range of their moth prey. It is suggested that horseshoe bat species, such as those occurring in the British Isles, may also be at an advantage while hunting moths, owing to the higher and narrowband nature of the frequency of their echolocation pulses (Jacobs 2015).

Hints and tips

Risks to bat survey methods and analysis

It is quite likely, particularly in the southern half of the British Isles, that calls made by insects could confuse a surveyor in the field, or interfere with the analysis or results of a bat survey. Especially so, bearing in mind that many bush-crickets produce higher-frequency calls within the same range of frequencies used by our bat species. Consequently, it would also be anticipated that there is, at least for some species, the potential for bat-related classifiers to allocate some types of insect noise to a bat species. In addition, a technician may, upon visual inspection, mistakenly feel that the occasional bush-cricket call, of some species, is, for example, a social call from a bat.

For other insect species (e.g. grasshoppers) which tend to produce lower-frequency sound, the bat detector microphone's frequency responsiveness would normally avoid picking up most of these calls, albeit harmonics may still be registered with the potential to cause confusion. During the analysis phase of project work, noise filters could be set appropriately in order to remove such calls, at least where the lower-frequency fundamental elements have been initially recorded. Table 5.9 covers the potential challenges when encountering insects during bat surveys.

Table 5.9 Challenges regarding encountering insects during acoustic bat surveys.

Subject	Problem	Potential solutions
Inadvertently recording	Insects recorded inadvertently and potentially impacting upon sound analysis process	True cricket, mole-cricket and grasshopper calls are often at low frequencies where a bat detector would not pick them up. Bush-crickets produce calls closer to or within the frequency range of a bat detector. As such, knowledge of bush-cricket sound is advisable for anyone doing analysis, when these may be present.
Noise filtering process	Applying a noise filter doesn't remove insect noise	A noise filter applied for sound below a frequency of 10 kHz would remove most sound, other than bush-crickets, without having an impact upon bat recordings. For higher-frequency calls not filtered out, manual analysis may be necessary. Also, carry out manual audit of filtered noise files, to ensure nothing of interest has been misallocated.
Bat classifier error	Insects being misidentified as bats	A manual audit of calls is advisable when bush-cricket species are known to be present, thus avoiding any insect calls being wrongly allocated to a bat species.
Human error	Field surveyor unsure whether call being picked up on detector is from a bat (e.g. social call) or an insect	Surveyors should be taught about insect sound. Remove headphones and listen with ears. Can you hear anything? If the sound isn't audible, then use your bat detector, and point it in different directions. Where is the noise coming from? If from ground level, it's unlikely to be a bat. If the sound is moving around, above your head, then it is almost certainly a flying bat. Bear in mind, however, that bats can make social calls from stationary positions, and some insects may call in flight.
	Person doing bat call analysis mistaking insects for bats	Most bush-cricket calls are unlikely to be confused with bats. However, in some cases (e.g. speckled and dark bush-crickets) there is potential for confusion with bat social calls. Visually and audibly, with experience, these insect calls are separable from social calls.

Author's diary note

Most of my bat work is carried out in the northern half of the UK, and therefore insect noise at night hasn't really been something that I have encountered as often as those conducting bat work further south or in other parts of the world. More often than not, my original irritation caused by the presence of insect noise has occurred during trips overseas. I say *original* irritation, because it is not where I am now on these creatures. I can safely say that I will never hear insect noise again and think bad things. It has been a fascinating 18 months collecting examples and finding out more about this subject. As with some of the other subjects in this book, I can now admit to occasions when I was trying to record insects, that I was actually wishing the bats away from interrupting me and the bush-cricket I was targeting. Those pesky bats! In France I had a spectacular encounter with a non-British species of bush-cricket, when a lesser horseshoe flew by, and this highlighted how torn I had become. 'That was a lesser horseshoe,' I explained to my wife Aileen, '... but ignore it, we need to get the bush-cricket.' Sadly, the bush-cricket in question didn't make the cut, as it was non-British, and it will have to wait for another time and another place, perhaps.

If just a handful of people who have read this chapter become likewise enthused, then that alone would make the time I have spent on it worthwhile. There are far fewer people studying insect acoustics than bat acoustics, so the 'exchange rate' is pretty good at the moment, meaning you could potentially be contributing data that have an even bigger positive contribution to conservation. Don't stop the batting though, that wouldn't be cricket! But it could actually be 'cricket' (lots of them in fact), when you start considering, at a quieter time of the bat worker's year, what can be done with your insect 'bycatch', as Stuart Newson (BTO) describes it.

I owe a huge debt of thanks to Stuart, who not only made his bush-cricket online resource live, coincidentally, a couple of weeks after I started writing this chapter, but also supported me by providing additional information, calls and the confidence that I wasn't alone in the overall concept behind this book.

As I sat down to formally start writing this chapter in January 2019, I admitted to myself that I hadn't been looking forward to it in the same way as the other chapters. It was a 'task' that I knew could potentially be painfully slow. The fact that I was thinking 'task' as opposed to a more positive word says it all. Of all the chapters, it was the subject in which I was weakest. However, like so many tasks that we put off, the reality was nowhere close to what I had feared, and within a matter of days I was completely hooked on this fascinating subject as I began to categorise many of the calls I had gathered myself, and those sent to me by contributors, especially numerous bush-cricket examples received from Harry Lehto, Chris Nason, Andrew Hargreaves, Keith French and Andy Froud. As is always the case, unfortunately, by no means everything that gets sent in can be included – but everything, in its entirety, contributes hugely to gaining an understanding and increased confidence in what is being discussed.

CHAPTER 6

Electrical and mechanical

Don't be overly alarmed, but ...

There is a huge array of electrical and mechanical noise which may occasionally feature during bat surveys. It is useful to know what this looks like on spectrograms as it could, in specific circumstances, cause confusion, not only during manual analysis but also as a result of applying filters and using automated classifiers.

I received numerous thoughts from across the country and much further afield, as to what I could include in this chapter. So much so, I was overwhelmed. It quickly became apparent that I had fallen short in my consideration of the amount of things making a noise that could potentially interfere with bat surveys, or confuse us. I have, unfortunately, needed to be quite ruthless with my selection of what to include and what to leave out. So as valid, and welcome, as suggestions such as combine harvesters and milk floats were, I can only apologise for their omission.

Everything discussed in this chapter would be described as technogenic (Gasc *et al.* 2017), a sub-category within what would, in its widest definition, be termed as anthropogenic. The examples that have been included are merely that: examples. There are so many variations of the sounds that are designed to perform the same function, or associated with the same process, or originating from similar sources. For example, many car models have reverse sensors, and these can sound quite different to each other. I obviously couldn't include them all, and to include more than one or two would give the impression that I had tried. Trust me, I didn't try. Well to be fair, I did try a little bit. One sunny morning I parked up at the local Morrisons supermarket, armed with my bat detector (in broad daylight), and walked rapidly towards any vehicle with reverse sensors in order to position myself to its rear, as the driver manoeuvred backwards. After about 45 minutes I ended up with nothing that useful, other than a near miss with a dodgy Skoda, much to the confusion of its elderly driver. It was round about this point that the store manager came out to ask me what I was up to. I explained that I was undertaking scientific research into engine noises. She told me, 'We have had a complaint from a customer.' I was astonished, as I thought I had been relatively surreptitious in my positioning. Some people are just a bit odd, I guess! My bat detector and I sheepishly departed, never to return.

While I am at it, I should perhaps mention that the local supermarket car park is not the only place I have been asked to leave as a result of my batty endeavours. There was one occasion when I was removed from the bat enclosure at Chester Zoo because I had been hanging around in there too long. But back to the reverse sensors. Other than my own top-of-the-range Jaguar and one the wife's vintage Porsches, I have not included any others. 'One Jag Nelly' – yeah that would be right, but not very practical for harp traps!

Seriously though, to include every variant of every noise made by various car engines, reverse sensors, security systems, trains, aircraft etc. just wouldn't have been feasible. Who knows, maybe one day I'll write a book dedicated to that subject. It would take a whole book to achieve such a feat. Perhaps entitled *The Variability of the Ultrasonic and Audible Noise Emissions Produced by my Nissan X-Trail and Other Friends*. It would be a top seller in its category, I'm sure. On second thoughts I am not that geeky – but Aileen disagrees, and is already beginning to worry that it might really happen.

Overall, the pages that follow include examples of the more common things that you would expect to occur across a range of surveys. I have split these into two broad groups, namely electrical and mechanical. In this chapter (and in Chapter 7) there is relatively little text, since most of the examples are self-explanatory. The figures have been ordered in such a way that, as far as possible, sounds that look similar on a spectrogram are next to each other. This should make it easier for the reader to narrow down where a potential noise has come from, in that comparable alternatives are side by side.

Spectrograms in Chapter 6

Spectrograms in this chapter use the following scales, unless otherwise indicated in the figure legend:

Time (x-axis): 2 seconds (2000 ms)
Frequency (y-axis): 0–110 kHz

When a figure legend includes the 🔊 symbol, this means that the figure has been created from a file in the Sound Library. The figure number matches the file number there. For more information about how to access and download files from the Sound Library please see pages xv–xvi.

Electrical noise

As previously mentioned there is a wide range of noise sources that would fall into this category. The following figures (Figures 6.1 to 6.10) are typical examples relating to the subjects they describe. On a point of caution, many of the examples do have the potential to be accidentally filtered into bat-related groups, as well as providing some interesting results when automated classifiers are applied. When undertaking manual audits of any analysed data, be that from an automated process or from manually identified calls, it is always worth considering the possible presence of electrical noise and its resemblance, in some respects, to genuine bat calls.

Security alarms and lighting

Figure 6.1 relates to a client setting a security alarm system while locking up the premises.

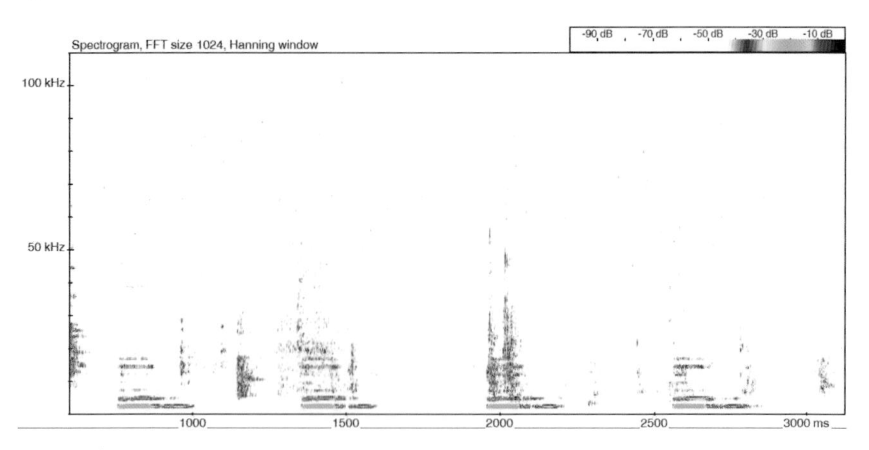

Figure 6.1 🔊 Building alarm system being activated (frame width 2500 ms).

Continuing the theme of security systems, Figure 6.2 is an example of ultrasonic noise produced by a high-powered security light. It was continuous and loud, being picked up from many metres away. It would be obvious to someone using a detector that there was an artificial noise present, and hopefully the surveyor would step away to a distance where the noise was no longer present, or would point the detector in a different direction.

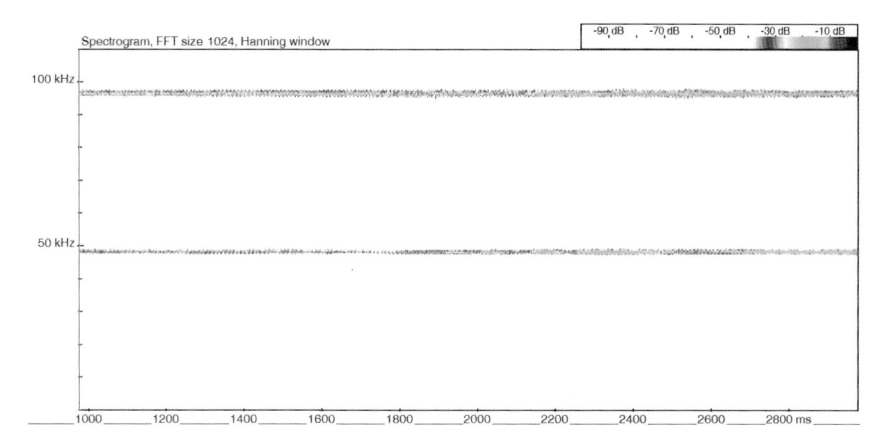

Figure 6.2 🔊 Security light.

Electronic devices

There is a wide range of items that could fall into this category, as well as all of the variations applicable to each. Two examples, from entirely different sources, are given in Figures 6.3 and 6.4. The digital watch is my own Garmin watch, and Figure 6.4 shows the sound it makes every time I switch on the face light in order to check the time in the dark. This is a regular feature during my bat survey season.

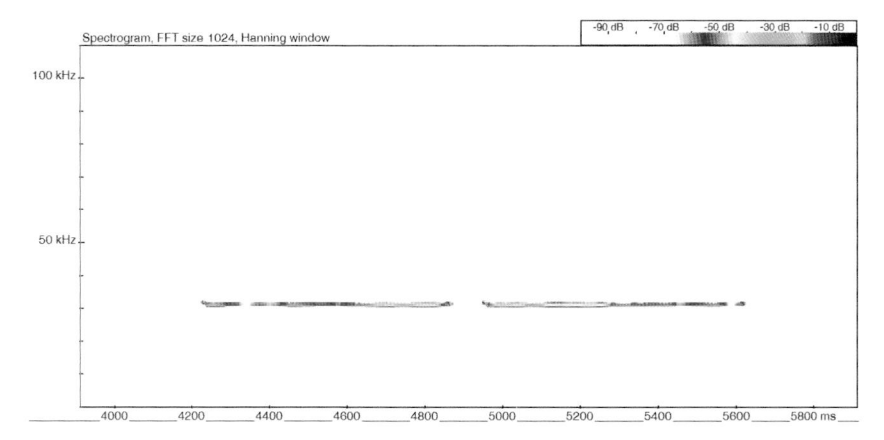

Figure 6.3 🔊 Camera autofocus (H Lehto, 2018).

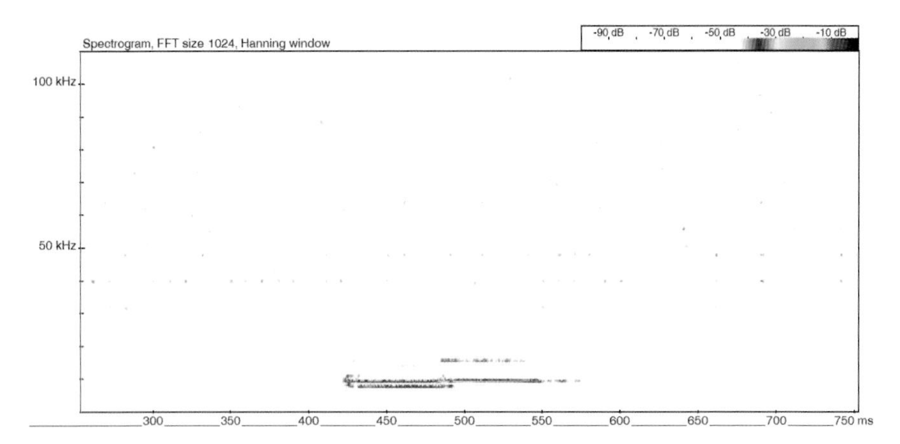

Figure 6.4 🔊 Digital watch prompt (frame width 500 ms).

Vehicle reverse warning beeper

Figure 6.5 gives an example of the audible reverse warning beep made by a heavy goods vehicle (HGV) while manoeuvring. You can see background noise throughout, but at the lowest frequencies you can see the slightly louder pulsed signals.

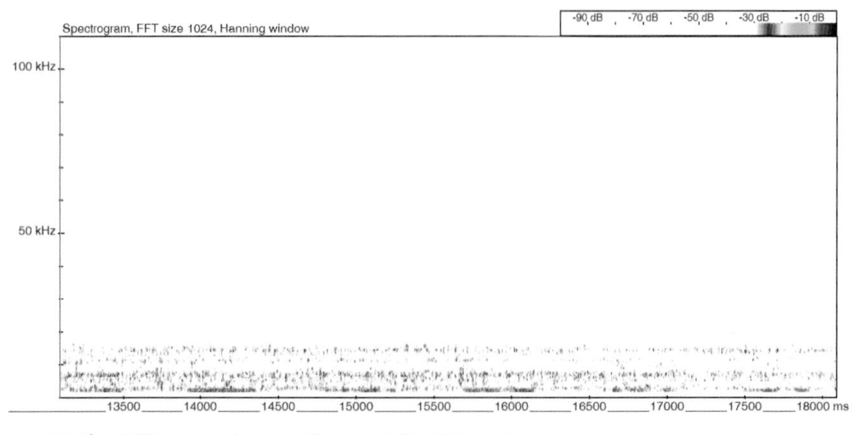

Figure 6.5 🔊 HGV reverse beeper (frame width 5000 ms).

Vehicle reverse sensors

Figures 6.6 and 6.7 show the spectrogram and the oscillogram for the reverse sensors picked up from a vehicle. Note that different models may produce different sounds. There are a couple of things to worry about here. First, in relation to bush-crickets, the double pulsing on the oscillogram (Figure 6.7) looks like a good fit for the great green bush-cricket (see Chapter 5, Figure 5.9), albeit when you look at the frequency (Figure 6.6), it's obviously wrong. Another worry relates to bat classifiers. For the example shown in these two figures, a selection of classifiers gave a high probability (ranging from 55% to *c.* 70%) for common pipistrelle. However, not only this, but the brake squealing received > 80% confidence scoring for greater noctule (*Nyctalus lasiopterus*), which produces one of the lowest echolocation calls in Europe. Once again, and as stated many times now, the importance of auditing is demonstrated. When such a sequence is looked at and listened to by an experienced technician, these are clearly not natural sounds.

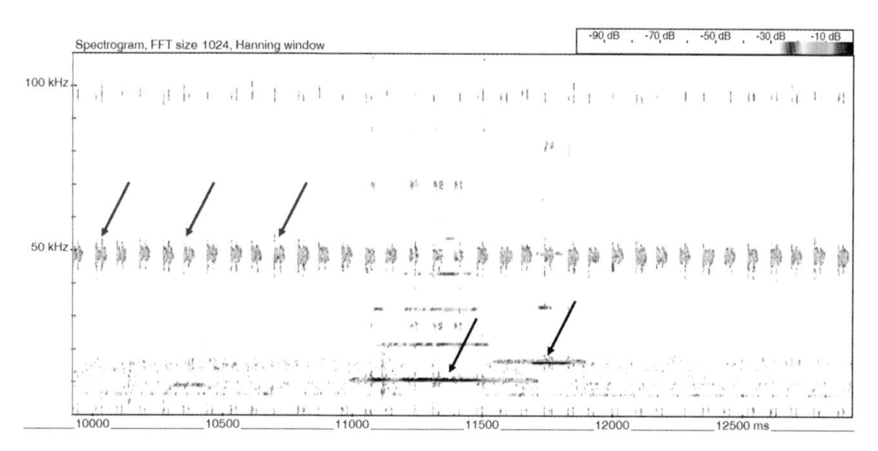

Figure 6.6 Hyundai car reverse sensors (red arrows), with brake squealing (black arrows) (frame width 3000 ms).

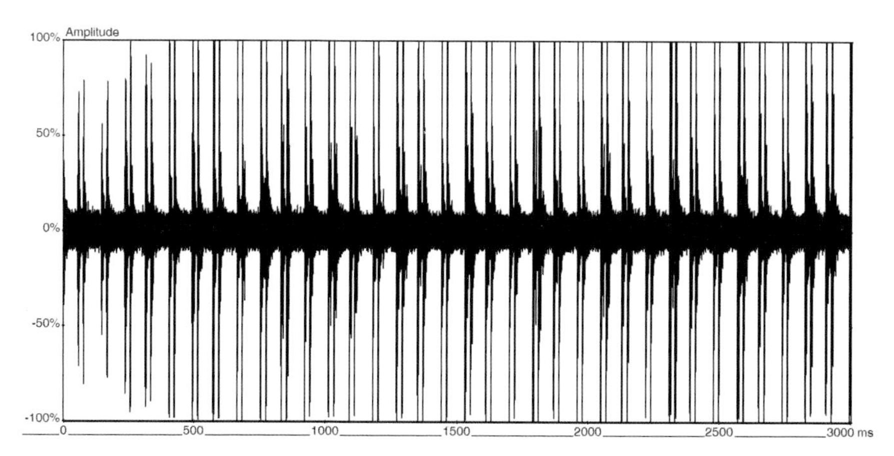

Figure 6.7 Hyundai car reverse sensors – oscillogram showing the double-pulsing effect similar to great green bush-cricket (frame width 3000 ms).

Vehicle braking sensors

In a similar manner to reverse sensors, a number of vehicles are now fitted with obstruction sensors, causing the car to brake if a driver fails to notice an obstruction on the road ahead. Figure 6.8 shows the noise made by a Ford Kuga as it passed a bat detector at about 15 mph (24 km/h). As with reverse sensors, some interesting results were obtained when showing this to bat workers accustomed to manual analysis, as well as when it was processed through an automated classifier, with pipistrelle species featuring heavily.

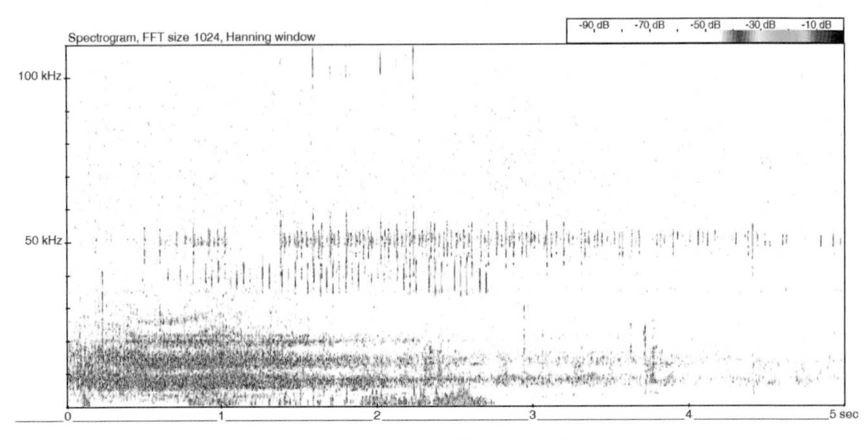

Figure 6.8 Ford Kuga vehicle pass with sensors (frame width 5000 ms).

Deterrents

A wide range of noise-emitting, high-frequency animal and human deterrents are available on the market. The example shown in Figure 6.9 was picked up by Chris Nason during a bat survey. He later established that the noise related to a dog/cat deterrent which was positioned in a garden close to where he had walked past during the transect.

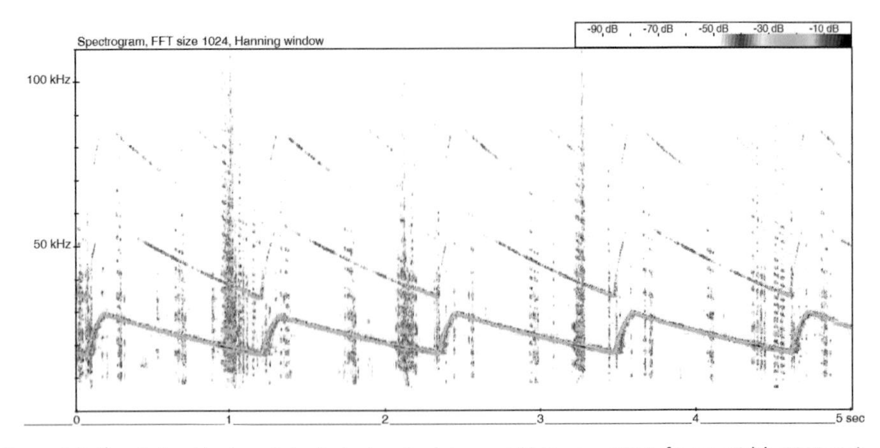

Figure 6.9 Animal (cat and dog) electronic deterrent (C Nason, 2019; frame width 5000 ms).

Mobile phones

With some bat detectors and some mobile phone devices there is certainly a noise that you may have become accustomed to listening out for, if it applies to your kit (Figure 6.10). As soon as you hear it you know that it isn't a bat, but at first you may not realise that it's your phone creating the effect. Having your phone tucked away as far as possible from where you would normally hold the detector, but still on your person, will often help reduce these annoying 'dithering' incidents. Have a listen to the track and see if it is a noise you have encountered in the past.

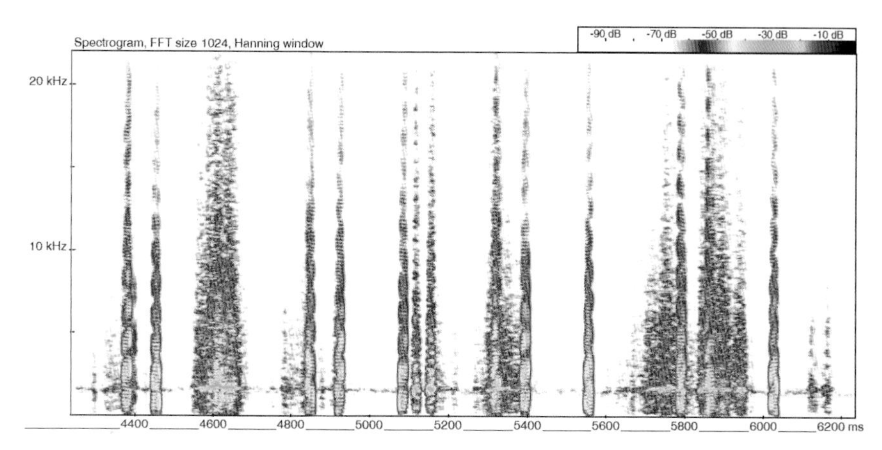

Figure 6.10 🔊 Mobile-phone-related noise (frequency division) as picked up on a Batbox Duet detector

Calibrating bat detectors in a non-digital era

It is sad to say that I have now reached an age when many sentences start off, '*In the old days*...' This has been a gradual process, but exacerbated by the fact that as I have gotten older, bat workers in my presence have gotten younger. As much as it is amusing to explain to youngsters about typewriters, eight-track cassettes, the launch of BBC2 and *Space: 1999*, it is equally refreshing to occasionally have people in the room who can verify that these things did actually exist and the trainer isn't pulling your left tibia.

Along with many of these people of a similar age to myself, I have been fortunate to have lived during a time when some fairly significant technological hurdles were jumped, before which we didn't have computers, the internet, mobile phones or even fax machines. As all of this was occurring, the equipment being used in the bat world was also developing, albeit relatively slowly. My first bat detector was the excellent Batbox II heterodyne from Stag Electronics. I guess there are lots of people out there who would proudly say the same. It was affordable, simple to use and didn't have much that could go wrong with it. In fact, I still have this machine some 25 years later, and it works! One thing that could go wrong, however, was that the tuning dial could be knocked out of sync against the frequency display light. This meant that every now and then we had to calibrate the dial to ensure the detector was giving us the right information. So how did we do that?

Well before the world of digital television we all had cathode-ray-tube TVs in our homes. These TVs (as well as early PC screens) all gave off an ultrasonic, constant-frequency sound at *c.* 31 kHz (thanks to David King at Bat Box Ltd for reminding me of the exact frequency). We switched on the telly and tuned our bat detector, listening out for the point when all of the sound disappeared and we had established what, in the world of heterodyne, is called the 'zero point'. This meant that the frequency that the bat detector was tuned to matched precisely the frequency of the sound source, such that no (i.e. zero) sound was being emitted by the detector. Having found this, we then checked the detector's reading to confirm that it was accurate at 31 kHz, and if not, we then forcibly adjusted our dial in order to have it correctly positioned. *'Switch the telly on, I'm about to head off and do a bat survey.'*

Mechanical noise

Mechanical noise, for the purposes of this book, relates to anything that moves in friction against another object, or collides against another object, be that as a deliberate part of a process, or accidentally. There are some noises that are a combination of mechanical and electrical, and in such cases I have tried to think about what aspect of the sound created is more likely to be picked up on a bat detector.

From the moment we arrive on site and switch on our detectors we have the potential to record mechanical sounds. Just think. You drive up in your vehicle in a hurry, you switch on the detector, turn off the engine, grab your keys, undo your seat belt, open and close the car door, and then gather items from the boot, such as spare batteries and your water bottle, and then you head off to do your survey. Everything I have described there could potentially be recorded, and that's without considering other sounds as covered in Chapter 7 (e.g. footsteps as you scurry across the car park). The following figures (Figures 6.11 to 6.32) have been ordered in such a manner as to keep similar noise sources together, as well as trying to keep similar sounds close to each other.

Keys

The racket that keys make is quite substantial, bearing in mind their close proximity to the detector as the surveyor puts them away in a pocket or a bag. Figures 6.11 and 6.12 provide some examples.

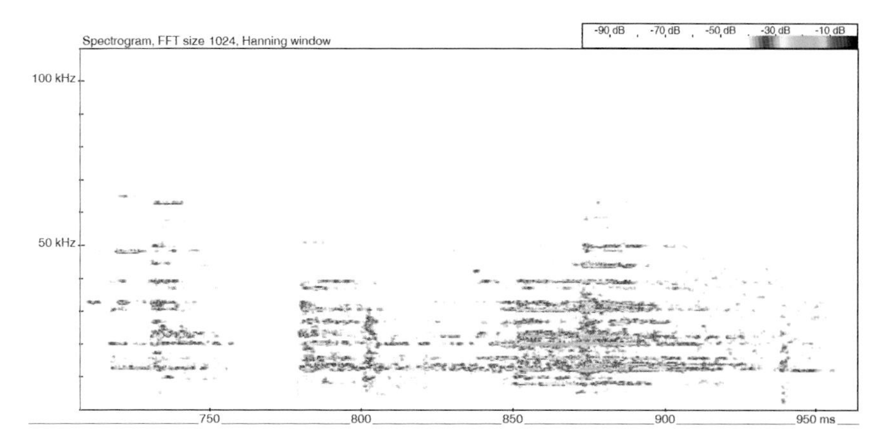

Figure 6.11 Keys (frame width 300 ms).

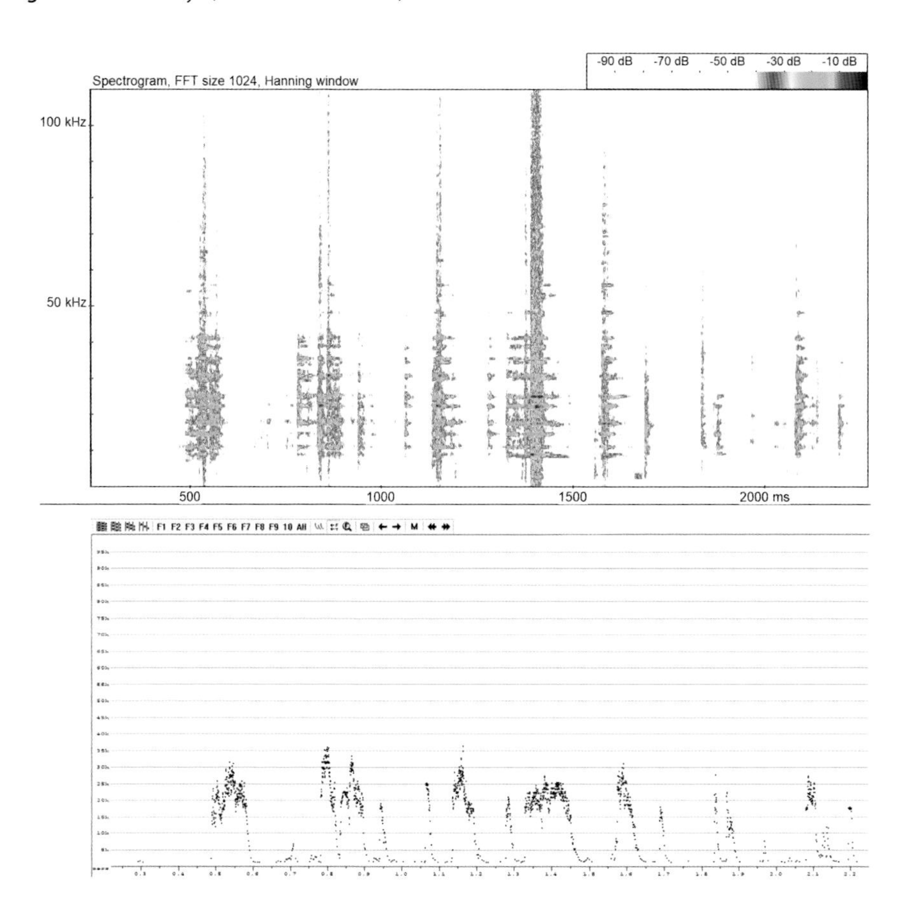

Figure 6.12 Keys – quieter, with AnaLookW example below.

With keys in particular, I find that it is always useful to have them in a pocket that you are unlikely to use while doing the survey. This has two benefits. First, if they are tucked away in a pocket that you won't be accessing for your phone, spare batteries etc. you reduce the risk of irritating key jingling being picked up by the detector. Secondly, if that same pocket is closed and not opened again until you have arrived back at your vehicle or your home, there should be no risk that you have dropped the keys, unnoticed, somewhere remote, with the likelihood of you ever finding them diminishing as quickly as your colleague's vehicle has departed leaving you stranded (though hopefully with your mobile in another pocket and not locked in your own vehicle).

I learned the whole key thing myself, quite painfully, early one morning on a hillside in Aberdeenshire. Six of us out doing a transect. Neil the keyholder. Neil, who lost the keys for the hire van. They were in a pocket with a Mars Bar, no doubt. I knew they had to be somewhere on that heavily tussocked landscape that we looked back towards. It was now 5 o'clock in the morning and we were 150 miles from the hire company's base in Grangemouth. Not that it mattered, as they wouldn't be open until 8 o'clock in any case. Ahead of us we had three hours of doing nothing, other than 'the rest of us' staring daggers at the 'boss of no competence'. But not that team. They got off their tired backsides and we started the pointless search (despite having just completed a dusk to dawn). We were all near to tears with frustration, when suddenly Hannah Paxton announced from afar, 'Found them!' You only ever do that once, and even better if you can learn from the idiot who has already done it, and never do it at all.

Car doors

Figure 6.13 gives an example of a car door being closed, just before a survey is due to commence.

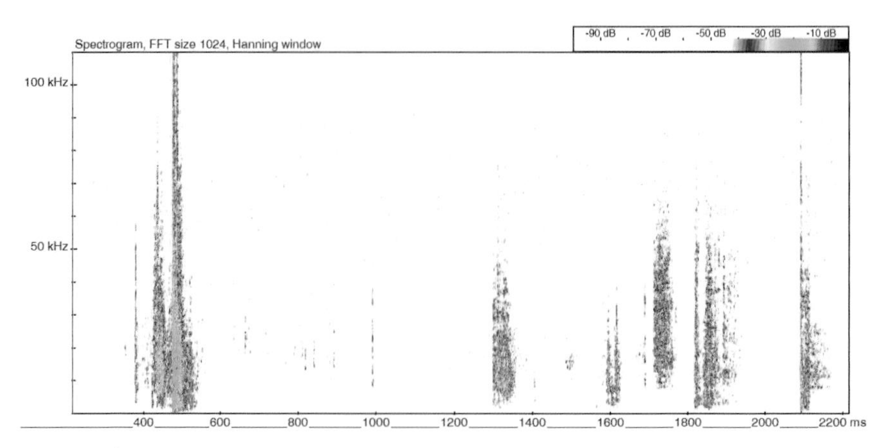

Figure 6.13 Car door closing.

Car engines

I would imagine that car engine noise can feature quite a lot during certain surveys. A colleague's car engine being on at the very start of a survey session perhaps, passing traffic, or the vehicle being used during a driven transect.

Figures 6.14 to 6.21 provide some examples of vehicle-related noise. Note that noise relating to vehicle electronics is covered earlier, under *Electrical noise*. Looking at Figures 6.14 and 6.15, on the assumption that you have read Chapter 5 (*Insects*), don't they look a bit like bush-crickets? Not the double pulsing of great green bush-cricket, as discussed earlier in this chapter (see *Vehicle reverse sensors*), but more like the regular stridulation of Roesel's bush-cricket (Chapter 5, Figure 5.12). So a word to the wise. If you are doing analysis and you think you might have a bush-cricket, listen to the sound before jumping the gun. It might be a car engine.

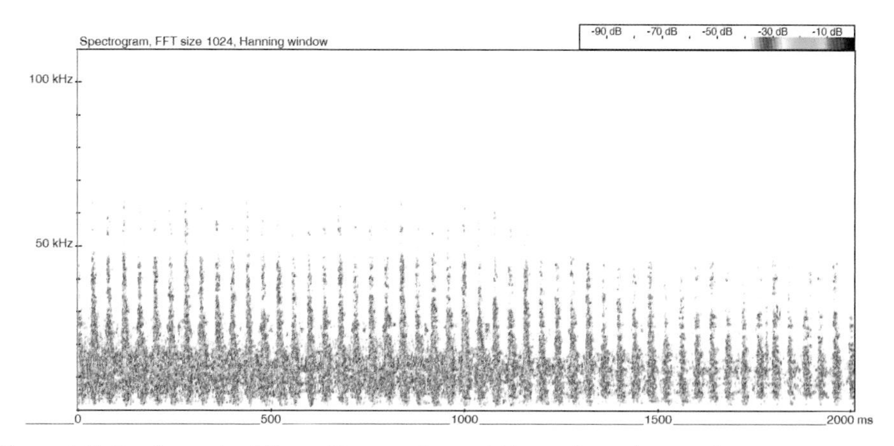

Figure 6.14 🔊 Car engine idling with microphone picking up lower frequencies.

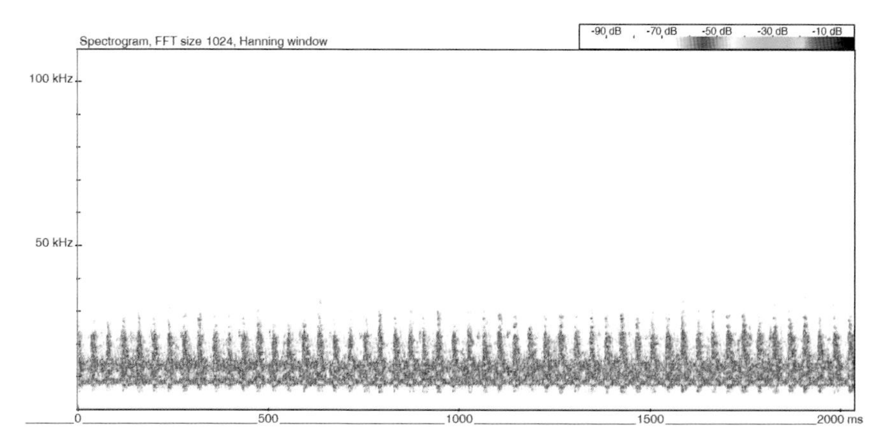

Figure 6.15 🔊 Car engine idling with microphone not picking up lower frequencies.

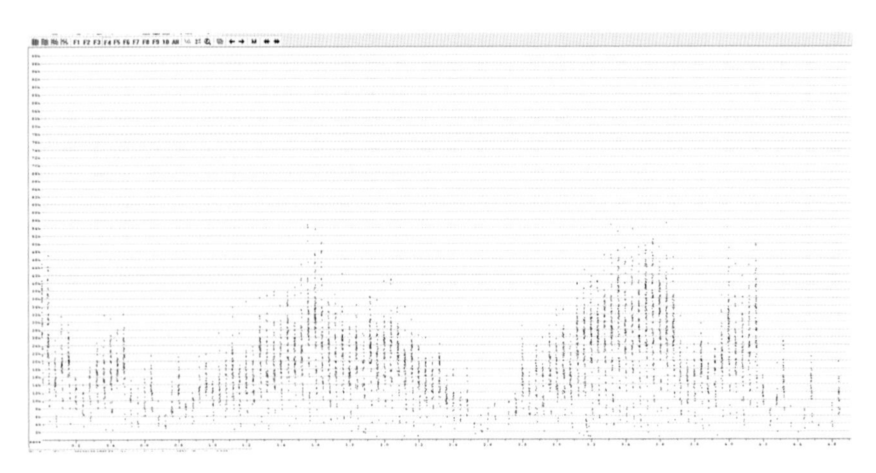

Figure 6.16 🔊 Car engine idling, ZCA example in AnaLookW.

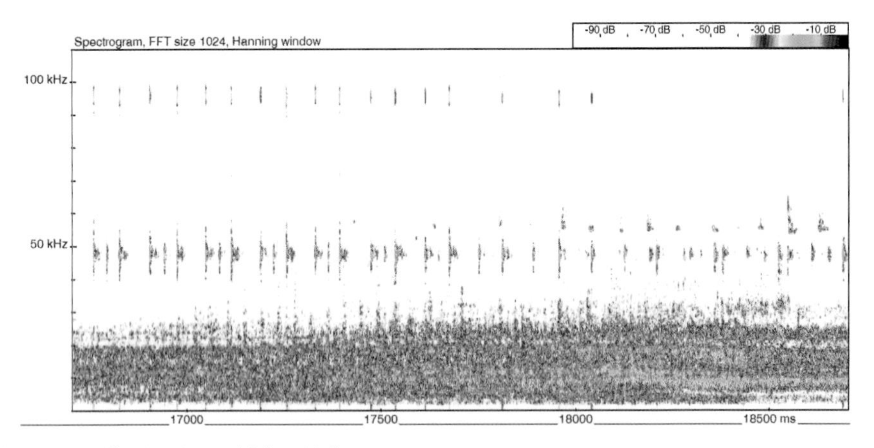

Figure 6.17 🔊 Passing vehicle with bat passes.

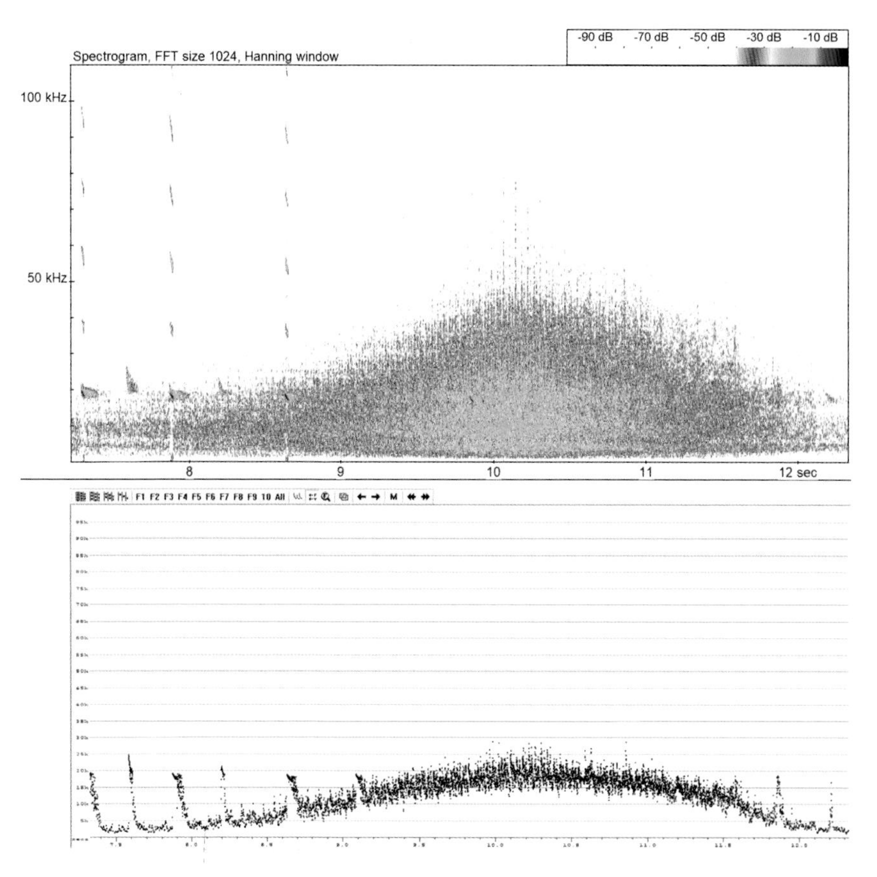

Figure 6.18 🔊 Passing vehicle with noctule bat passes subsequently being drowned out. Full spectrum above and ZCA in AnaLookW below (frame width 4000 ms).

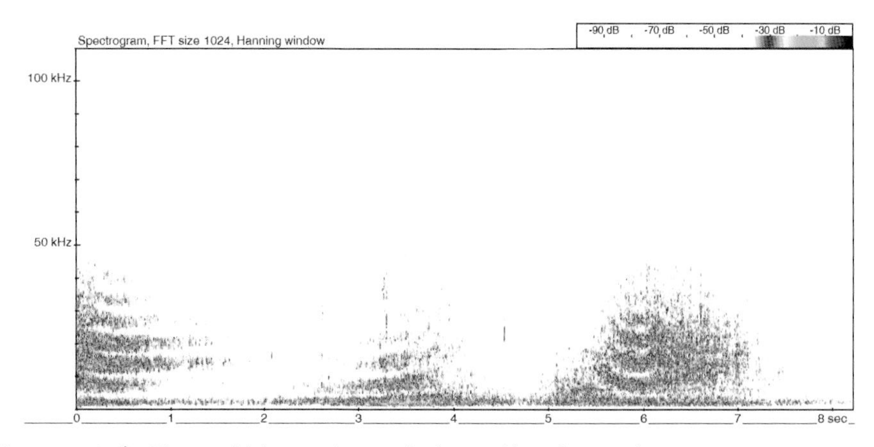

Figure 6.19 🔊 Three vehicles passing, each denoted by a humped appearance representing increases and decreases in noise as they pass by in quick succession (frame width 8000 ms).

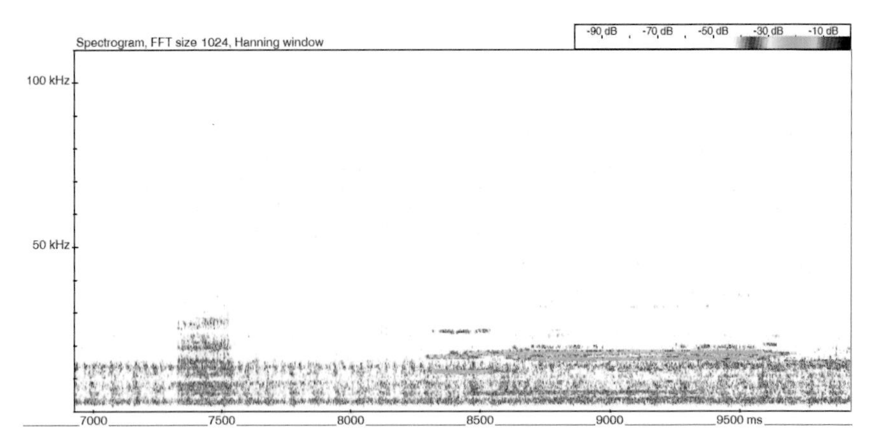

Figure 6.20 HGV with brake squeal (frame width 3000 ms).

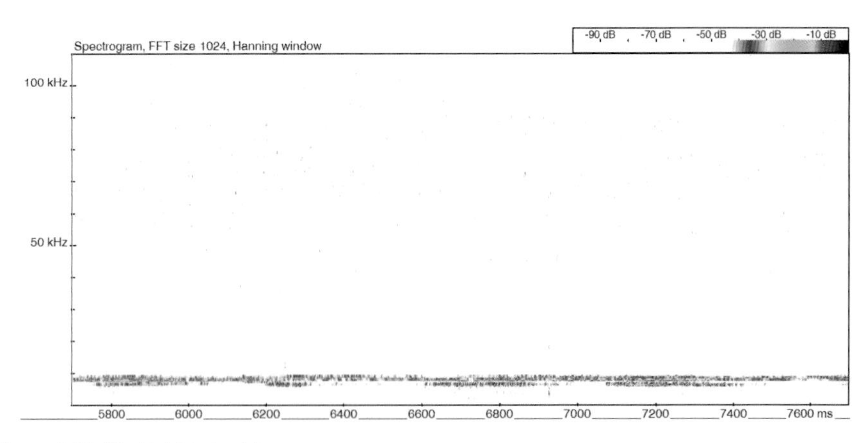

Figure 6.21 Vehicle braking.

Trains

I remember the first time I recorded a train on a bat detector. I didn't have a clue what was going on. I was standing on a footpath close to a railway line and I very quickly became aware of a loud whining noise coming through my headphones. I immediately thought, train. But when looking up and down the track I saw nothing. The noise just kept getting louder, and me more bemused. Then, all of a sudden the train appeared, whizzed by, and the noise began to diminish. Figure 6.22 gives an example of the noise I experienced. So, as well as noise from the train itself, you also can pick up the noise emanating from the tracks, well ahead of the train passing in front of a surveyor.

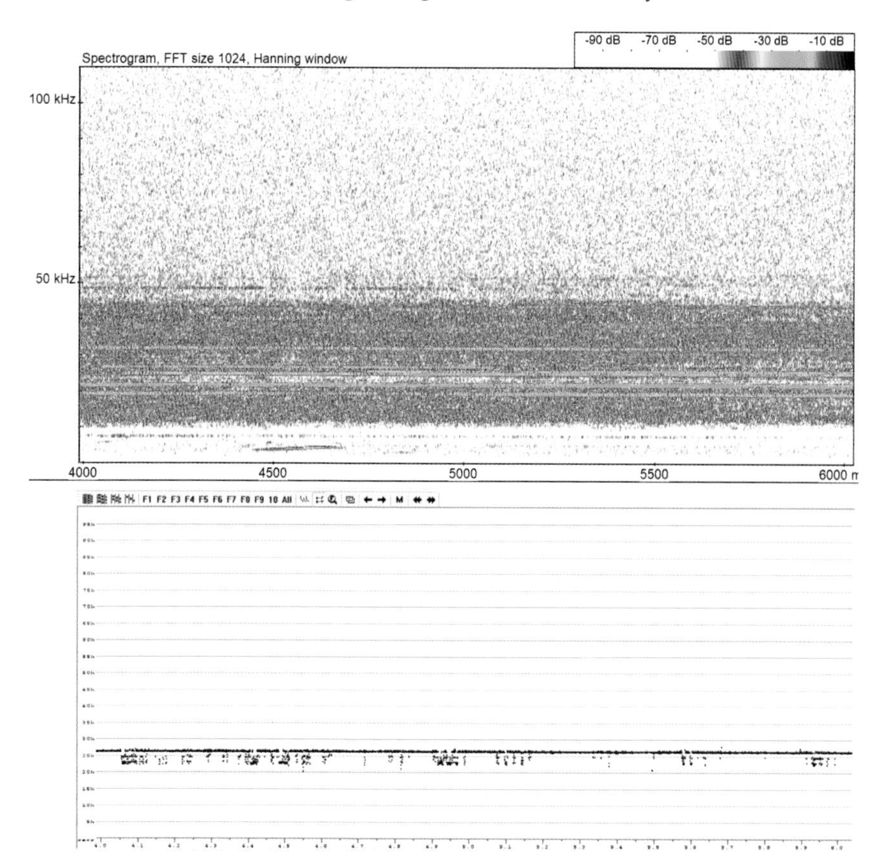

Figure 6.22 🔊 Train passing by – full spectrum above, and ZCA in AnaLookW below.

The examples given in Figure 6.22 show very clearly a major difference between full spectrum and ZCA recording systems. In full spectrum, all noise (including harmonics) is picked up at any given moment in time, whereas in ZCA only the loudest noise is picked up. This means that any other noise, of a lower amplitude, be it at a higher or lower frequency than the loudest sound to occur, is 'cloaked', which means that it doesn't appear on a spectrogram.

Aircraft

To be fair, I was underwhelmed by my attempts to pick up aircraft at a local airport. I was directly under the flight path less than a kilometre from the runway, and after a couple of hours the example shown in Figure 6.23 was the best I could manage. From memory, I think it was an Airbus A320 operated by EasyJet – going somewhere warm, with nice bats and cool beer, no doubt.

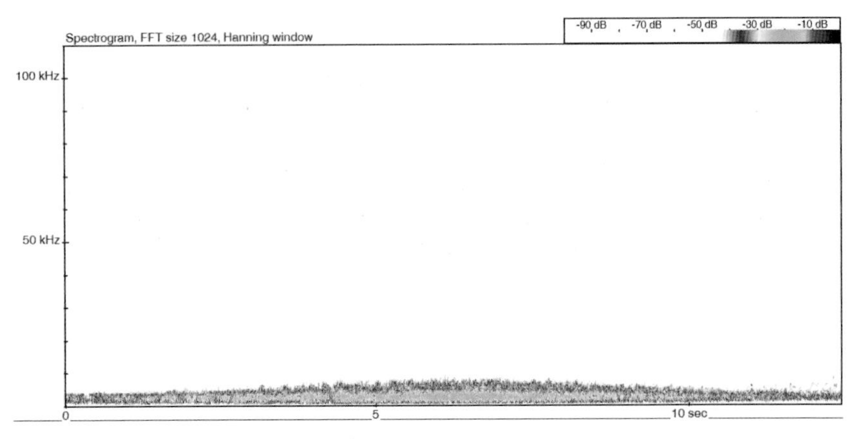

Figure 6.23 Jet engine noise, as plane flew overhead (frame width 12 seconds).

Pedal cycles

Far more interesting than the low-flying jets, the example shown in Figure 6.24 is a passing cyclist on a canal towpath.

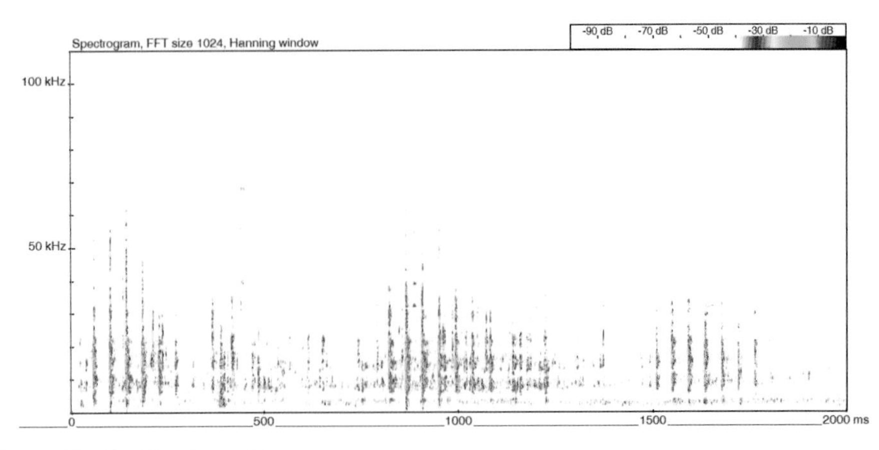

Figure 6.24 Bicycle passing.

Squeaky gates

Quite often we need to go through gates while carrying out a transect. Figures 6.25 and 6.26 are examples of the resulting squeaking sounds that may occur.

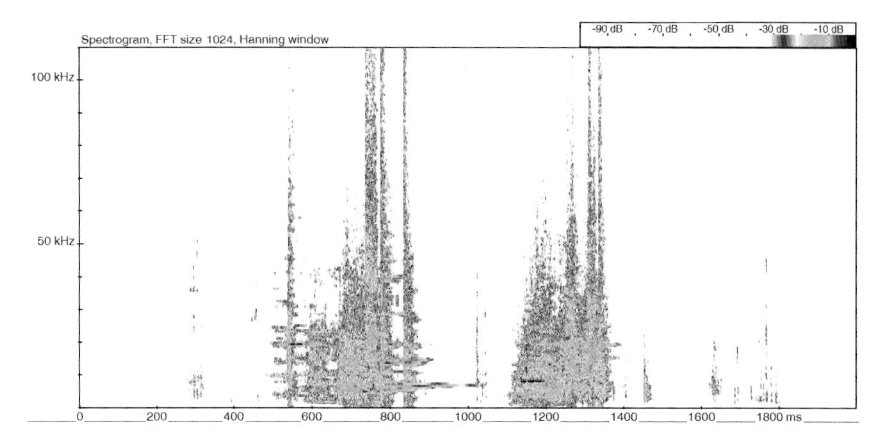

Figure 6.25 🔊 Gate squeaking.

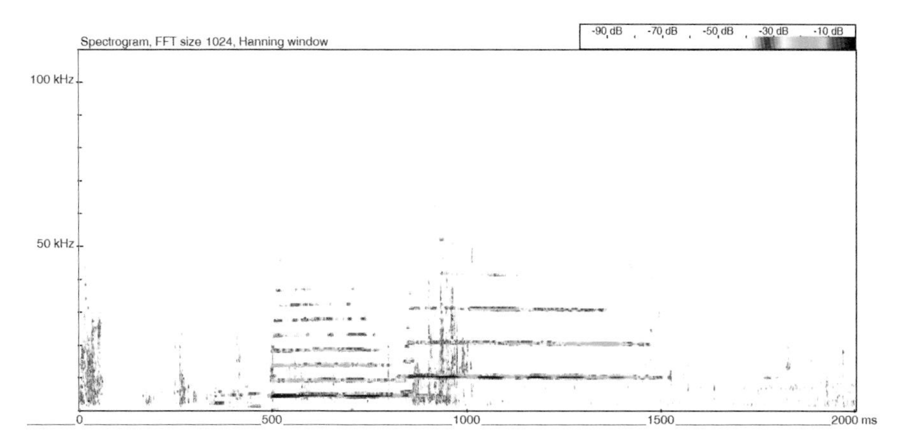

Figure 6.26 🔊 Gate squeaking.

Time out!

It would be fair to say that all of the mechanical noise we have been looking at so far is not very likely to be confused with bats, at least not by an experienced person. The examples that follow, however, are slightly more disturbing.

Chinking jewellery

About ten years ago we had surveyors carrying out transects on a site in the Central Belt. We were using frequency division detectors, and the noise recorded was quieter than the examples shown here (Figures 6.27 to 6.29). The FmaxE was coming in at *c.* 25 kHz, with a repetition rate at what we would expect from a *Nyctalus* in open habitat. It sounded 'sort of rightish', as well. We would get a pass, and then nothing for many minutes, then another pass. It was happening regularly. We were fairly certain after the first survey that we had an unusual occurrence for that area of Scotland. But it was very unusual. A few weeks later the same again, but at a different site. A pattern was developing. It was the same surveyor that was picking up these bats! After a bit of experimenting we established that it was the surveyor's jewellery that was chinking every so often. Problem solved, we all had to stop wearing jewellery while doing surveys (I was well miffed!). OK, so that was frequency division, and some of you might be thinking, well it isn't as good as full spectrum.

Introducing Figure 6.27, a full spectrum example of metallic chinking. Guess what, one of the regularly used classifiers had this labelled as noctule, along with about a dozen other similar examples. Then let's look at Figure 6.28, a ZCA version of the same items. It is easy to see from that example, where only the loudest frequency occurring at any moment in time is registered, why an inexperienced person rushing through analysis would manually do the same as a classifier. Looking closely, however, it's just a bit too perfect. It doesn't look natural, with a very constant CF structure for each pulse.

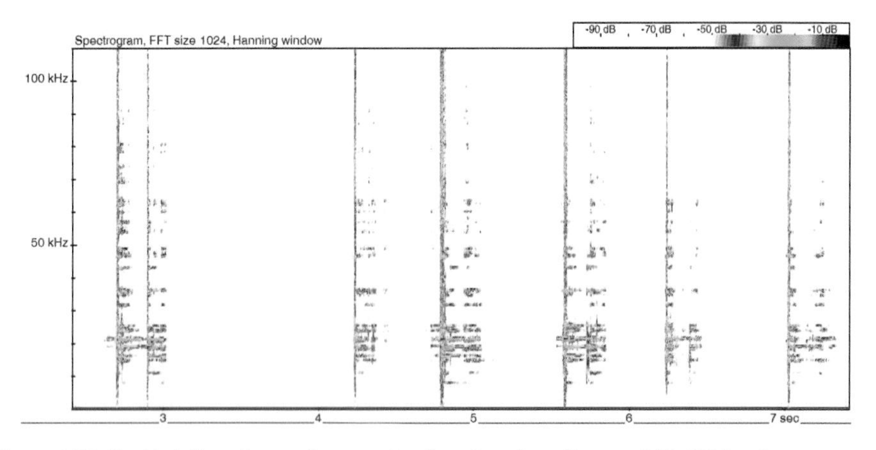

Figure 6.27 🔊 Metallic noise, such as coming from jewellery (frame width 5000 ms).

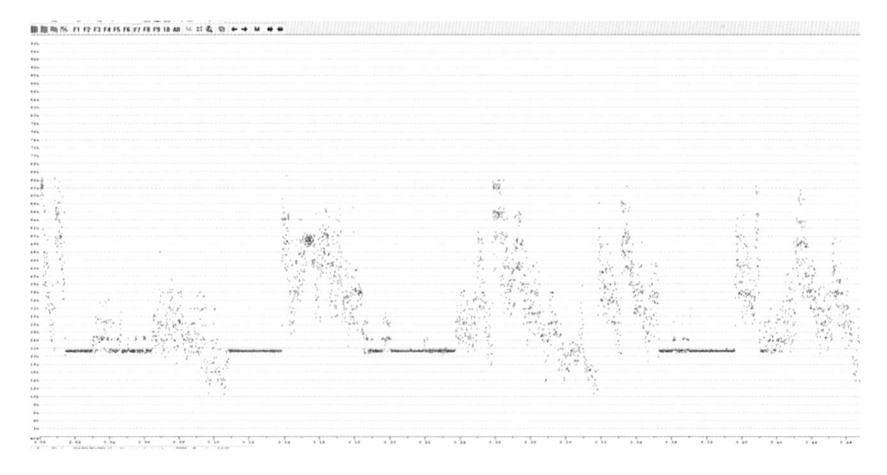

Figure 6.28 🔊 Metallic noise, such as coming from jewellery.

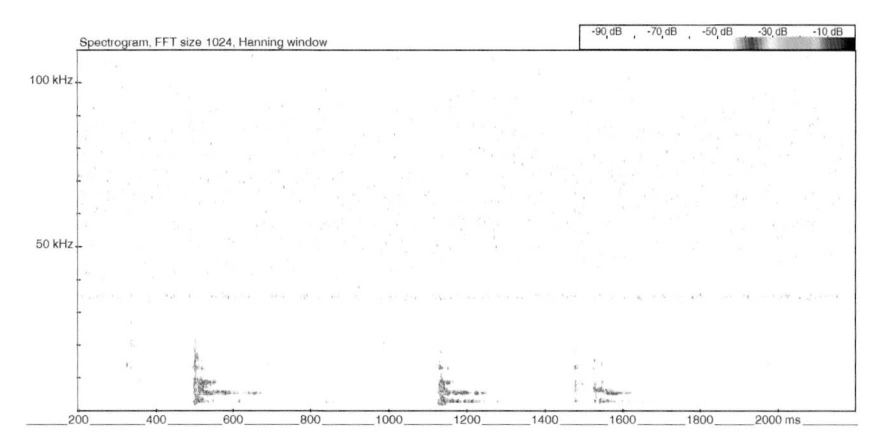

Figure 6.29 🔊 Louder metallic noise – flag pole chinking.

Tally counters

Perhaps the 'so obvious, it isn't obvious' noise sources that create regular potential for being picked up on our bat detecting equipment are the sounds associated with other bat detectors, as well as the associated survey equipment we use. In this respect I would also refer you to Chapter 1, where the issue of speaker sound coming from the surveyor's own detector, or from other bat detectors in the vicinity, is discussed. Aside from that, there is one source of obvious noise that could be present, and even more so as the number of bats being counted emerging from a roost increases. Figure 6.30 relates to a tally counter being used for such a purpose. It is worth bearing in mind that noises such as this may be present during the analysis.

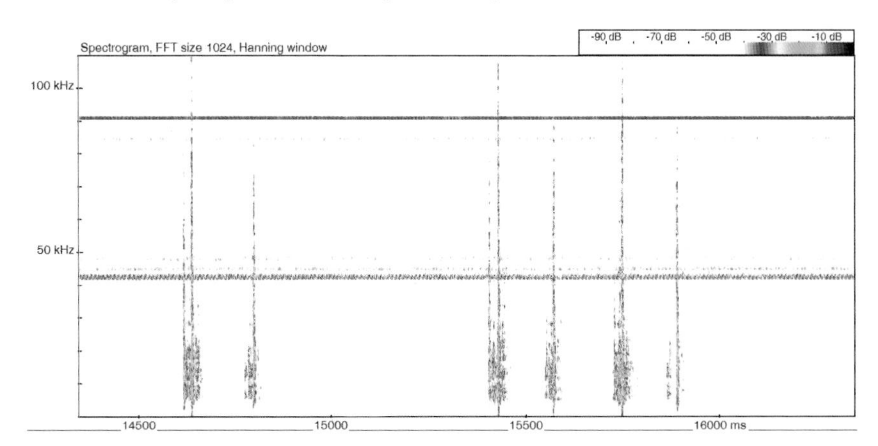

Figure 6.30 🔊 Tally counter.

Wind turbines

As I am sure most readers are aware, quite a lot of survey work centres around bats when it comes to windfarm development. Post-construction monitoring is becoming more of a feature today, albeit not yet commonplace. Imagine my shock when Alexander Hatton sent me some turbine operational noise, maintaining that they could potentially be considered by a species-specific filter to be noctule calls. He very kindly provided some examples, two of which I am sharing with you here. In Figures 6.31 and 6.32 you can see emitted frequencies in the region of 20 kHz. The first example would surely be noticed by an experienced technician as having too long a duration for a bat, but the second example, I'm not so sure. Someone, no matter how experienced, under pressure of time and quickly looking through hundreds of calls, anticipating noctule to be present, could easily misidentify these mechanical noises as the target bat species.

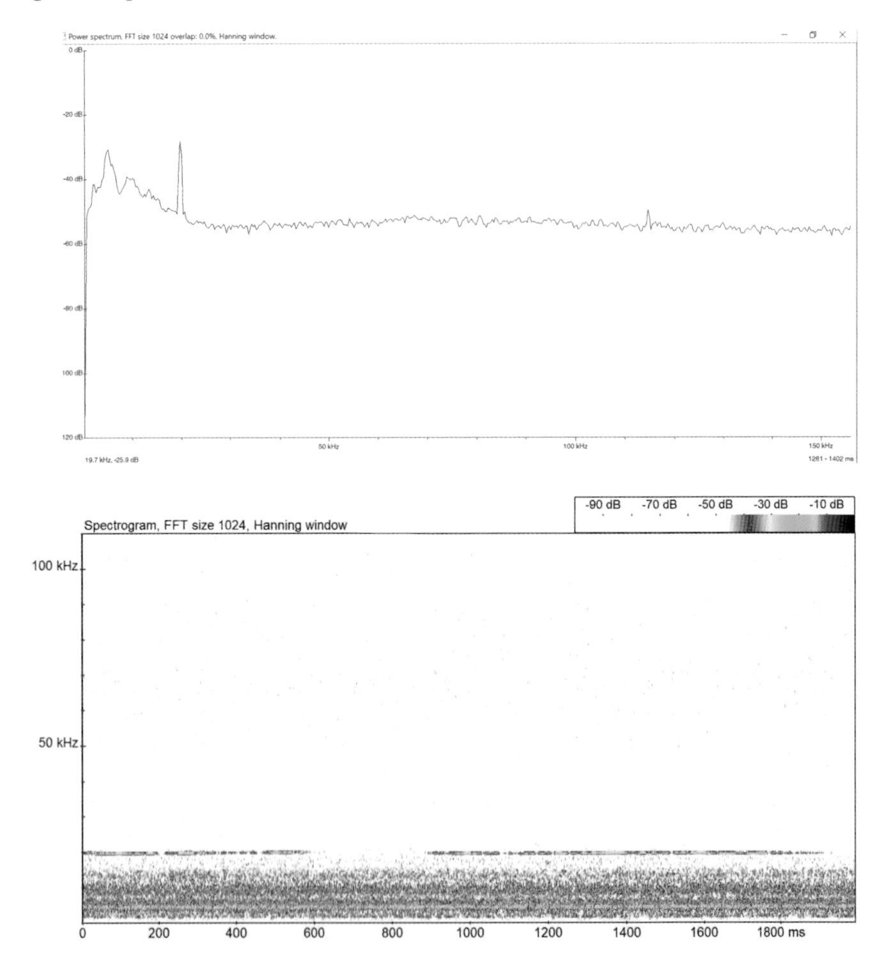

Figure 6.31 🔊 Wind turbine operational noise – with power spectrum (above) and spectrogram (below) (A Hatton, 2018).

Figure 6.32 Wind turbine operational noise – shorter-duration 20 kHz sound (frame width 1000 ms) (A Hatton, 2018).

I already mentioned the contributor's thoughts on a manual filter regarding noise like this, but what about the potential for it being misidentified as noctule by an automated classifier? Surely that couldn't happen. Well, when I put these examples through a couple of classifiers, they both suggested noctule with over 80% confidence. Previous comments regarding caution and auditing apply.

Hints and tips

It is wise, ahead of any survey activity, to anticipate what regular noise episodes are likely to occur at a site. For example, if next to a railway line, the noise caused by trains, junction boxes etc. could all be recorded in advance during a site familiarisation visit and lodged with those doing the analysis for reference purposes. This may also give valuable information needed in order to create appropriate and/or specific filters. In addition to this, there may be noise sources that are not immediately apparent (e.g. ultrasonic sound coming from cabling or light sources). During the daytime visit it could, for some sites, be worthwhile taking a bat detector with you to do a noise scan of the site, as well as your visual assessment. At the very least, switching on a hand-held detector at the location where statics are to be deployed could be beneficial, just to ensure that there isn't something there making a regular noise that you can't hear audibly.

While carrying out hand-held detector surveys, be conscious of and remove (or dampen) the effects of anything that can potentially be picked up by the bat detector (e.g. keys jingling in pocket, loose change, jewellery clinking against other items of jewellery). Doing this will reduce the number of files created, and potentially reduce calls that need to be audited and/or non-bat sound being inadvertently allocated to bat species, either manually or as part of an automated process. Table 6.1 provides a summary of the points already referred to, along with some other considerations.

Table 6.1 Consideration of electrical and mechanical noise during bat surveys and analysis.

Subject	Problem	Potential solutions
Equipment damage	Electrical damage to equipment	Consider using surge protection and/or electrical grounding techniques to protect equipment from power surges and electrostatic discharge.
Inadvertently recording	Inadvertent recording of non-bat sound, thus hindering analysis and causing inefficiencies in the process	Brief the survey team as to actions they can take to avoid recording unwanted artificial noise, including walking away from any obvious sound sources during surveys. Carry out an artificial noise sweep at the site, including the locations where static detectors and surveyors are likely to be present. Take a hand-held detector to static detector locations to establish if any regular noise occurrences are likely. If so, consider moving static to another location. Create a reference library for regular unusual noise sources that are characteristic of the site, if such sources exist.
Noise filtering process	Applying a noise filter removes more than just electrical or mechanical noise	When creating noise filters, consider the sources of likely noise for the site, and at the same time consider an appropriate level of audit for sound being allocated as 'other noise'. Carry out a manual audit of filtered noise files to ensure nothing of interest has been misallocated. Also, consider auditing filtered bat calls for sound that is artificial and incorrectly allocated to a bat species.
Bat classifier error	Electrical and mechanical noise incidents being mistaken for bats	Certain noise events in this category appear, at least occasionally, to be allocated to *Nyctalus* or *Pipistrellus* species. A robust audit of *Nyctalus* and *Pipistrellus* calls is recommended. For any regularly anticipated artificial sound being recorded at the site, test it against the classifier you plan to use, in order to understand better how the classifier may react to such instances. Any information from this process should then be used to inform auditing procedures for the project.
Human error	Person carrying out analysis mistaking artificial noise for bats	For simpler or short-duration sounds this could be an issue, although it becomes less likely with experience and guidance. There is, however, such a vast range of sounds that could potentially occur, to the point that it is conceivable that something electrical or mechanical may be confused for a bat, if looked at quickly and/or not listened to in both real time and time expansion. Those carrying out analysis need to be trained in how to identify non-bat noise in this respect.

Author's diary note

Over the years I have recorded many unusual things on a bat detector, to the point that I thought there wouldn't be that much left to surprise me. Well, a trip to Norway in August 2018 (3rd European Alpine Bat Detector Workshop) was going to throw up a few new encounters, the most surprising being from the workshop organiser himself, Leif Gjerde. Leif was aware that I had been working on this book and he offered me something very special and interesting, which came straight from the heart.

He told me that he had a metallic heart valve fitted, and that it gave off a sound that could be picked up by a bat detector, as well as audibly in quiet conditions. He also explained all about the operation he went through in order to have it fitted, but that is far too much detail, and way too scary for you to read about here. Anyway, back to the metallic valve. There we were, a group of experienced bat detectives, stopped in our tracks by curiosity, holding our respective bat detectors close to Leif's chest.

He very kindly said that I could use it for the book (Figure 6.33), as well as telling me during our experiment that it was dangerous! I immediately panicked and thought, oh my word maybe the bat detectors can give off electronic signals that will impact upon the performance of his heart. Leif then went on to explain that the reason it was dangerous was because if he was looking at an attractive woman, then his increased heart beat would cause the valve to speed up, and immediately give him away. Rest assured he was only looking at me at that point, and I can verify that he was as cool as a cucumber. Have a listen to the track, which includes Leif saying at the end, 'This is kind of dangerous ...'

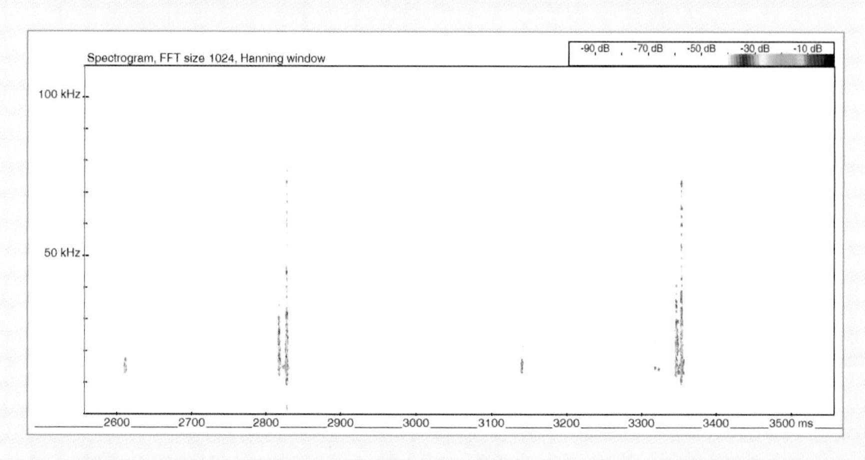

Figure 6.33 Metallic heart valve, courtesy of Leif Gjerde (frame width 1000 ms).

So you now have a reference for Leif's normal 'unexcited' heartbeat. This means that if you are female, and you ever meet him, having a bat detector might be useful in that you will be able to ascertain whether or not he finds you attractive. In fact, on a wider scale, this could develop into a whole new approach to finding a suitable partner in life. The advert reads 'Join match.com today and get your free bat detector!'

CHAPTER 7

Weather, people and other nuisances

Keep your weather ears open

There probably is not a survey occasion that passes, be it hand-held or using static bat detectors, that does not include an element of noise created by weather, water or the presence of a person, moving about (how dare they!), coughing, sniffing or talking. It would be a strange event indeed if none of that was occurring. Maybe a bat detector placed underground, or within an enclosed quiet area such as a loft, could be an exception, but other than these it is hard to imagine where else at least one of these non-bat-related noise events would not potentially be present, at least temporarily.

There is a vast range of sounds that could be recorded, and upon asking for some guidance via social media, once again I was truly overwhelmed, and surprised to be honest, by some of the suggestions. Unfortunately, due to constraints of space, and for the sake of decency (there may be children reading this), I have not included everything that was suggested to me. So hiccups, not to mention all kinds of creatures carrying out all kinds of bodily functions, some of which go well beyond the bounds of printability, are excluded. My aim is to show the more common things that you would expect to occur across a range of surveys, as opposed to more random, and thankfully rare, sources of noise.

As in Chapter 6, I am not going to add much in terms of text, as most of the examples provided are self-explanatory. The figures have been ordered in such a way that sounds that look similar on a spectrogram are next to each other. This means that it should be easier to work out what something is, in that comparable alternatives are side by side.

Spectrograms in Chapter 7

Spectrograms in this chapter use the following scales, unless otherwise indicated in the figure legend:

Time (x-axis): 2 seconds (2000 ms)
Frequency (y-axis): 0–110 kHz

When a figure legend includes the 🔊 symbol, this means that the figure has been created from a file in the Sound Library. The figure number matches the file number there. For more information about how to access and download files from the Sound Library please see pages xv–xvi.

Weather and water-related noise

In this category we are going to look at noise that is created as a result of weather, as well as water being present within the sound environment in which the surveyor or a bat detector is located. These noises would broadly be described as geogenic (Gasc *et al.* 2017). Examples of this would be running water when a detector is placed close to a river, or dripping water from a leaking gutter or tap.

These noises are the very things we would hope that a noise filter would remove most of, prior to us carrying out analysis of bat calls. It is hard to think of any reason why we would want to investigate any such noise, other than perhaps to establish if some sort of weather event was occurring during a static detector survey that could explain an unexpected dip in bat activity over a survey period. To expand upon this, for example, let us say that a good degree of bat activity has been recorded over many nights, but during one night the activity was substantially less. By investigating the filtered noise folder for that night a technician might see or hear sound that suggests heavy rain or strong wind. Of course, this could possibly also be established by checking historical weather information for the site in question, preferably through the use of a weather station located within the site. Irrespective of this, localised weather events can occur, and depending upon where the site is, a local historical weather report may not be available or reliable.

Wind and rain

Excluding extreme weather events (e.g. thunder and lightning), most weather-related noise occurs as a result of two factors, namely wind and rain. In the case of wind, it can be more noticeable as it interacts with habitat (e.g. foliage, trees, structures). Figure 7.1 provides an example of typical wind-related instances as seen on a spectrogram. Generally, the spectrogram view is very busy both in terms of time and frequency, showing different intensities, corresponding with differing volumes of noise. Typically most sound occurs at less than 40 kHz. This means that on any full spectrum type system, bat echolocation occurring above this frequency has a fair chance of being picked up and noticed during analysis, while echolocation below 40 kHz is going to be much tougher to find, and, if it is noticed, the ability to identify calls to species level could very well be inhibited. Using the oscillogram tool is, of course, always an option, but if the amplitude of the wind is similar to or louder than that of any passing bat then again it is unlikely to be noticed. In short, whether or not bats like windy conditions, a bat detector placed in such a way that wind is picked up strongly is not a great idea and should be avoided if at all possible.

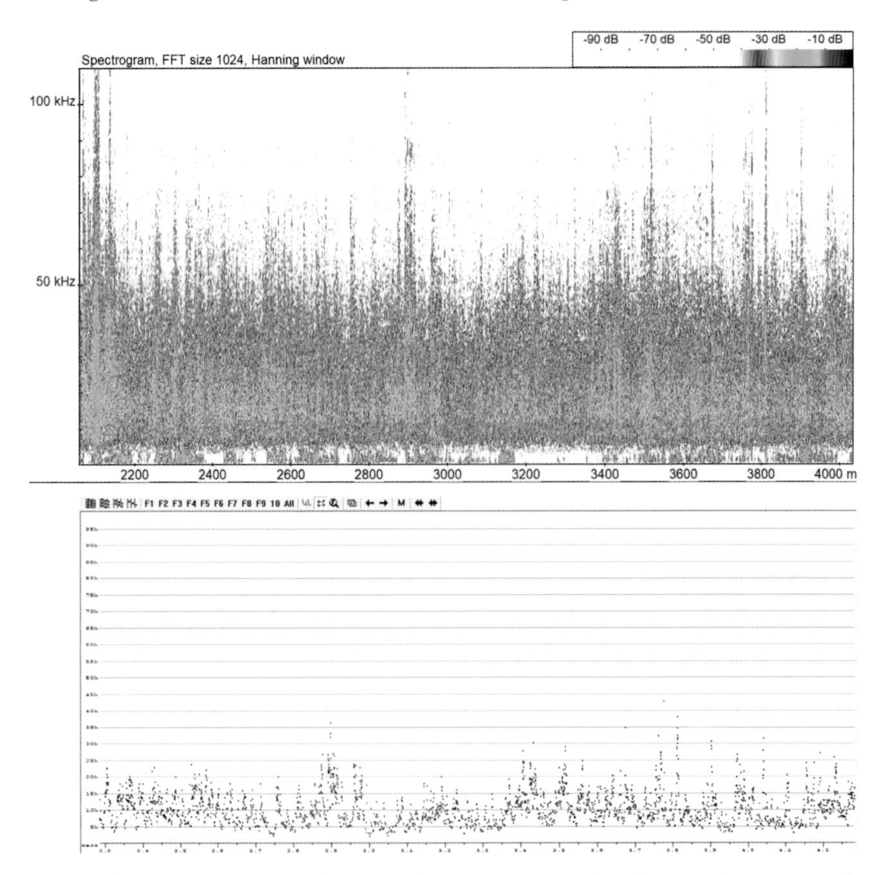

Figure 7.1 Wind with rustling foliage. Full spectrum example (above) with ZCA example in AnaLookW (below).

We all know that bats don't like the rain, but apparently no-one has ever explained this to some of the bats I have encountered over the years. I have certainly witnessed bats flying normally during light showers. On one memorable occasion when Kirsty Morrison and I were carrying out a BCT National Bat Monitoring Programme (NBMP) Waterway Survey for Daubenton's bats, a thunder and lightning storm came over very fast, and we were caught out amongst some trees next to the river we were surveying. All we could do was shelter among the trees (never a good idea!) and hope for the best. About five minutes later, goodness knows why, I switched on our spotlight and shone it over the river, only to see a handful of Daubenton's bats happily foraging (*How do you know they were happy, Neil? Did you see them smiling?*) over the usually calm water above the weir we were next to.

Whether it's a bit of wind or rain, my personal experience is that bats can, on occasions, be relatively tolerant of both, provided there are still insects present to forage on. In much stronger wind, or heavier rain, or indeed a combination of both, then that's when activity will dramatically drop off.

Figure 7.2 provides an example of a recording made during heavy rain. And since we are discussing water, perhaps an example of a dripping tap (Figure 7.3) and a dripping drainpipe (Figures 7.4 and 7.5) might be useful for comparison.

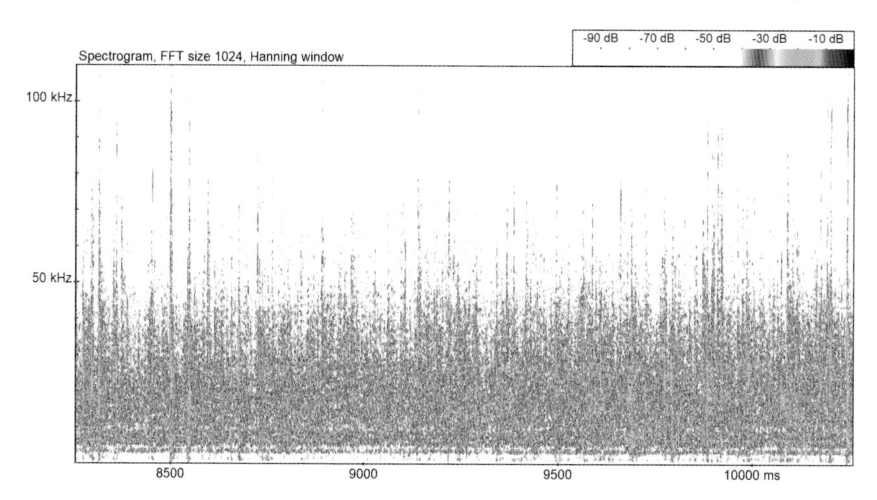

Figure 7.2 🔊 Heavy rain.

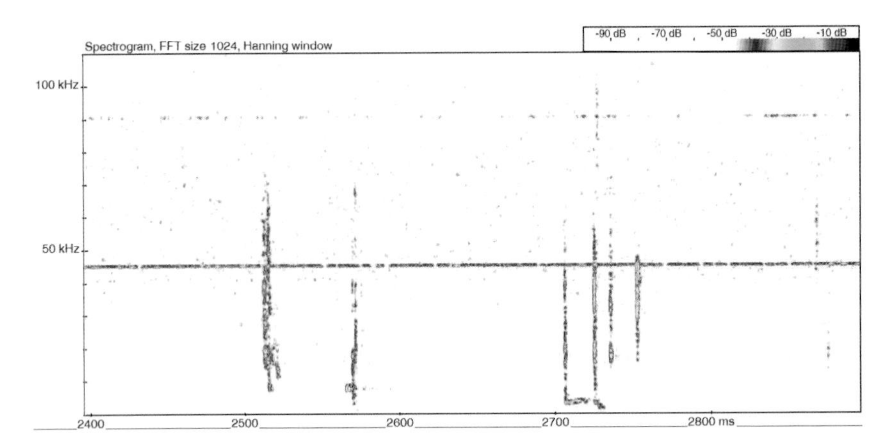

Figure 7.3 Dripping water tap, with electrical noise occurring as constant frequency at *c.* 45 kHz (frame width 500 ms).

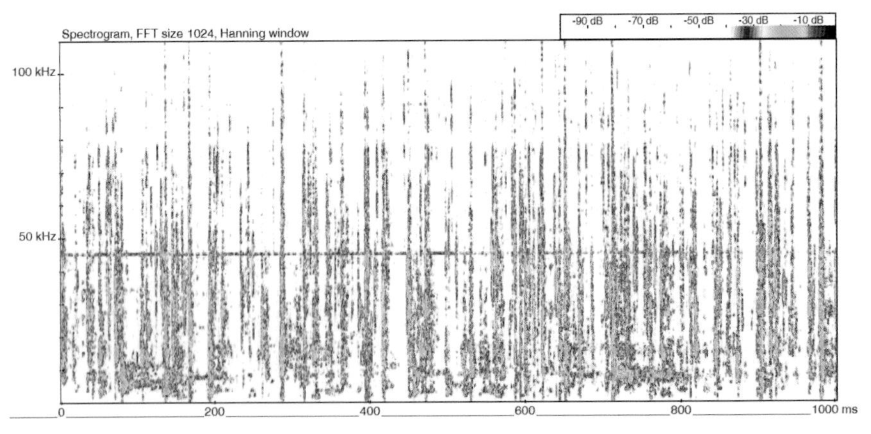

Figure 7.4 Drainpipe dripping after rain event, with electrical noise occurring as constant frequency at *c.* 45 kHz (frame width 1000 ms).

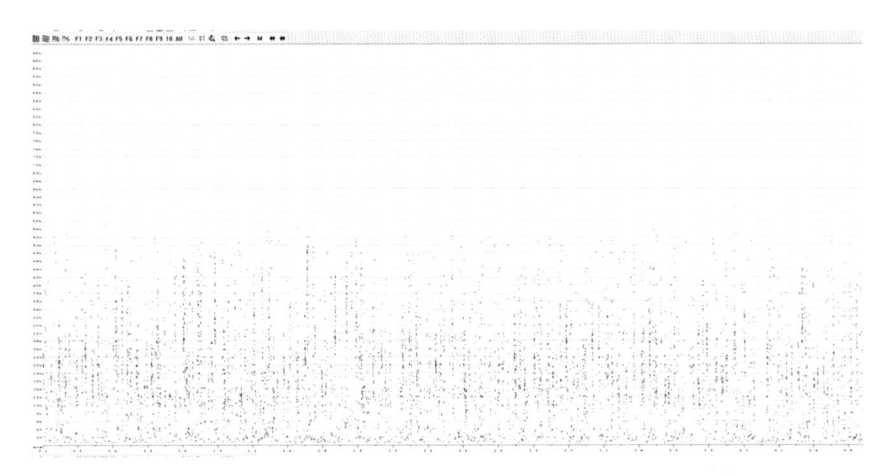

Figure 7.5 Drainpipe dripping after rain event. ZCA in AnaLookW.

Rivers and streams

Continuing with the wet theme, let's now consider a more consistent occurrence of water-related noise, caused by a river or a stream being present. There seems to be a typical occurrence that is noticeable on spectrograms in these instances. Referring to Figure 7.6, look how there appear to be three darker lines, occurring at *c.* 20 kHz and below. This pattern is not as noticeable in Figure 7.7, but if you look at the accompanying power spectrum insert you can see the corresponding amplitude peaks. So this may be a way to differentiate moving surface water from wind, for example, if you didn't know what it was you were looking at. Obviously, if you knew the detector was next to running water at that point (which would be preferable) you wouldn't need a spectrogram to tell you, but it's there as a pointer if you need it.

Depending on how loud the running water noise is, then, as with wind noise, it might be difficult to establish the presence of bats, in that their signals may be hidden amongst this noise. The noise frequency in this case is lower, and therefore it will potentially mask fewer bat species (i.e. species with lower-frequency echolocation). Figure 7.6 shows, very nicely, a soprano pipistrelle surfing well above the waves and clearly visible.

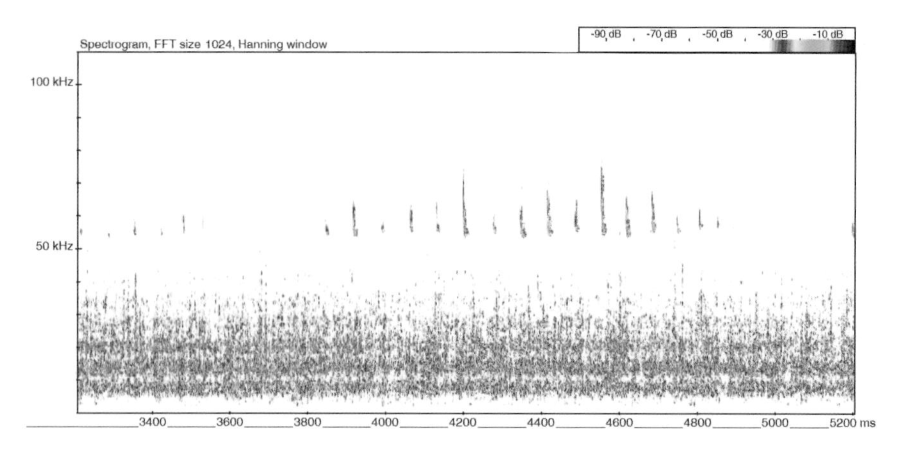

Figure 7.6 Stream with soprano pipistrelle 'wave surfing'.

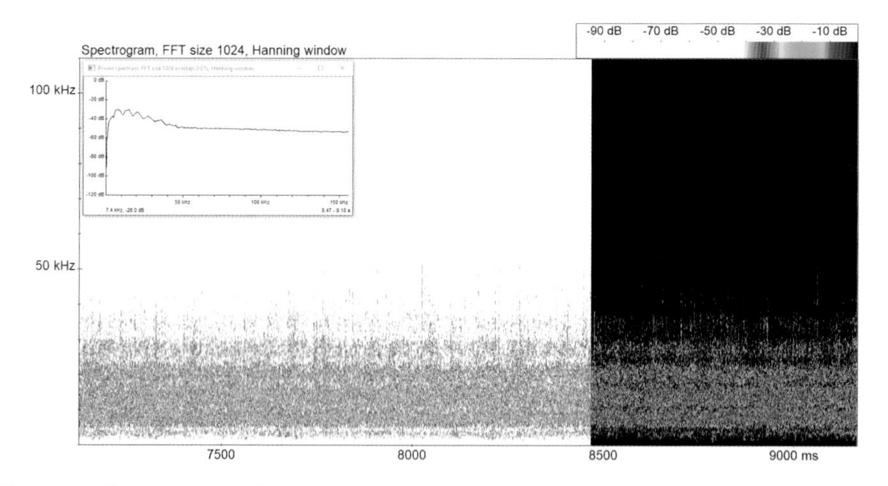

Figure 7.7 Faster water from stream (power spectrum insert).

Human noise

Probably the most regularly occurring non-bat sounds picked up during any hand-held bat detector survey are the various noises that are unavoidably (usually!) created by the surveyors themselves. This is as a result of them being where they are, doing what they are doing, and just being human. The noises covered here are those occurring over and above any electrical or mechanical noise created by the people involved in surveys (see Chapter 6).

With much of the noise in this category, the surveyor should be able to explain what was occurring during the survey (e.g. we were walking on gravel, bashing through vegetation, had a terrible cough throughout). Therefore, a system whereby surveyors documented the potential sources of other unusual noise that they knew was being recorded during the survey could be beneficial in assisting those carrying out the analysis.

Walking

Figures 7.8 to 7.12 provide examples of walking, which obviously is a factor in most surveys, especially transects over variable ground surfaces. The associated noise created is very nondescript and busy. However, unlike rain and wind, there is a pattern to it, that being the speed at which the footsteps were being taken. Listening to the sound should be a good giveaway, or failing that, look at the time scale on the x-axis and imagine how fast footsteps would be, and this may give you a reasonable steer as to the noise source. From all of the examples shown you can clearly see the pattern of steps taken, and when a surveyor is wearing waterproof trousers the 'swish' effect can make such sounds even more obvious.

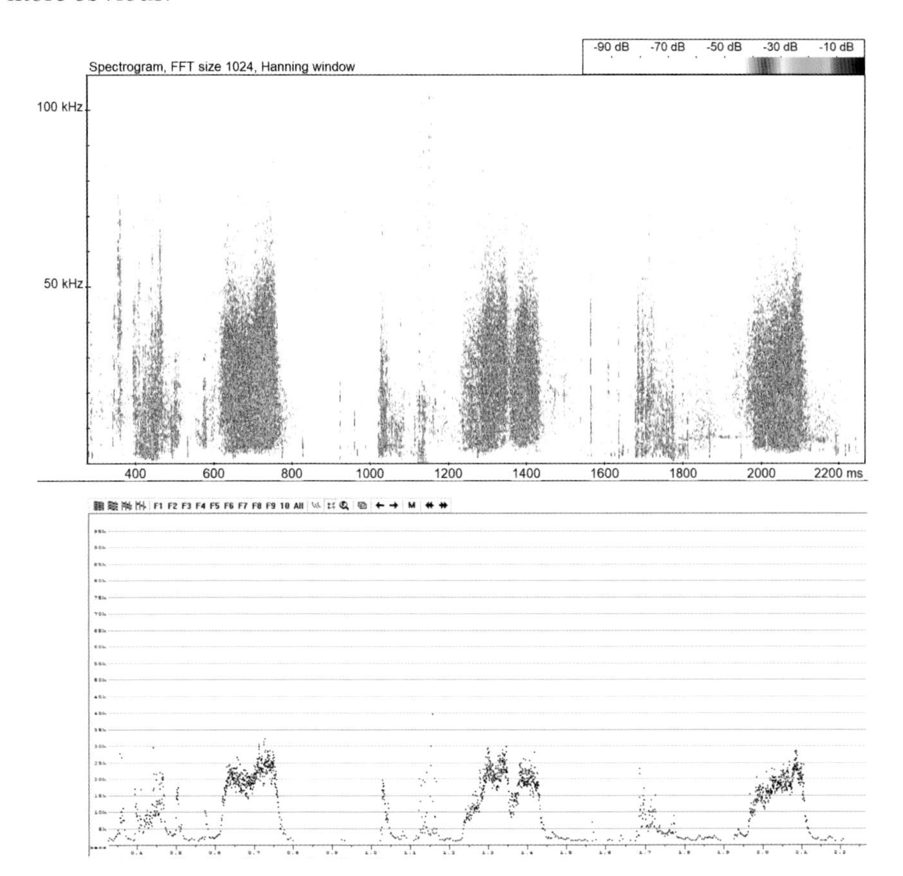

Figure 7.8 Soft footsteps on tarmac. Full spectrum example (above), with ZCA in AnaLookW (below).

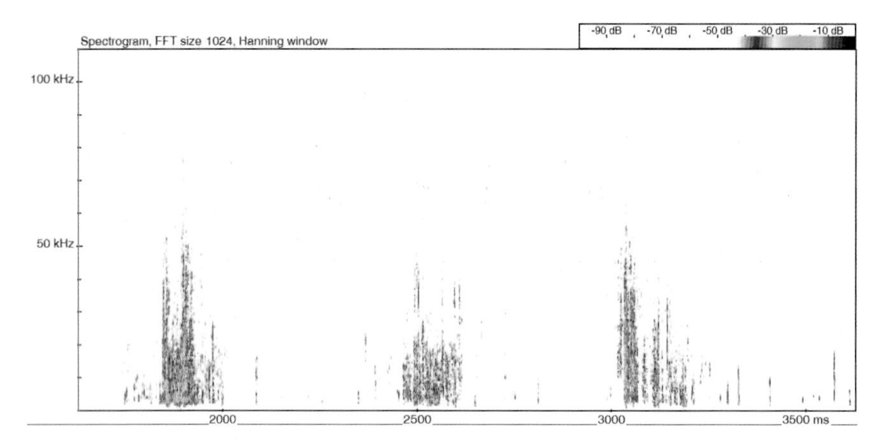

Figure 7.9 🔊 Footsteps on gravel.

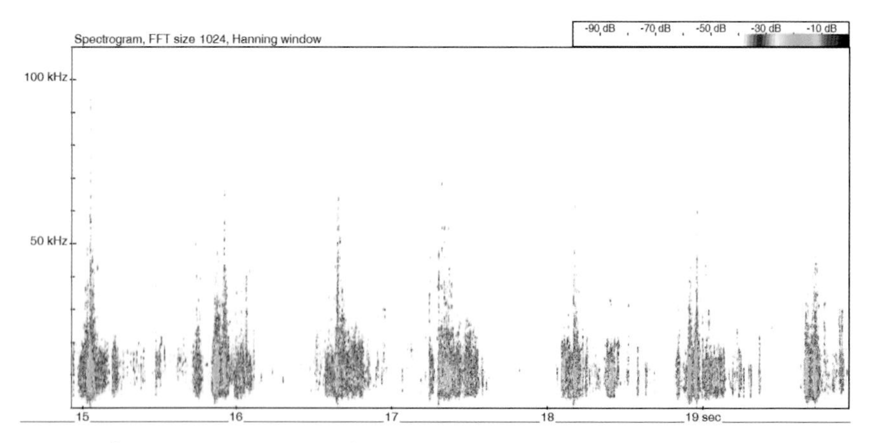

Figure 7.10 🔊 Footsteps on pebbles (frame width 5000 ms).

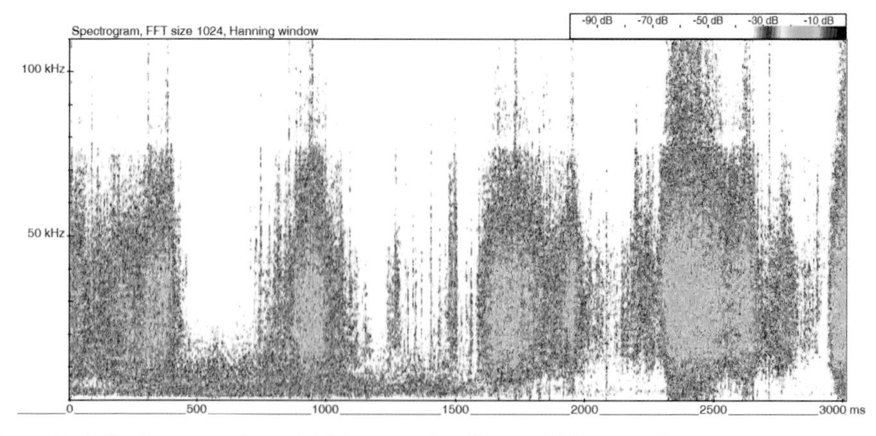

Figure 7.11 🔊 Footsteps through high vegetation (frame width 3000 ms).

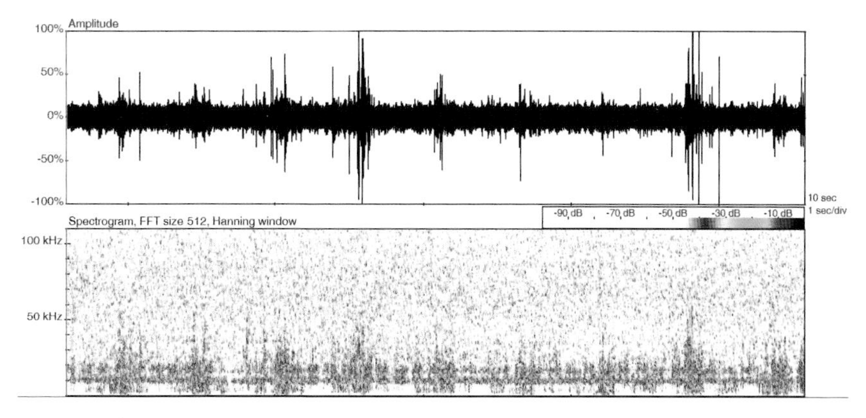

Figure 7.12 🔊 Footsteps with insect noise. Oscillogram (above) with spectrogram (below) (southern France, 2018).

Coughing, sniffing and sneezing

Coughing, sniffing and sneezing can all create sound files, and of course these things can happen in a variety of ways, frequencies and patterns. A single sniff, or a bout of uncontrolled coughing. Again, if you can listen to the sound in its original format that would help greatly to establish its source. Figures 7.13 to 7.17 provide a variety of examples of coughing, sneezing and sniffing. Figure 7.18 introduces some new sounds for you to consider, in the form of a sniff, with tongue clicking, a poor attempt at whistling and a bird calling. Who said this stuff was boring?

We already touched upon sniffing in Chapter 1 (Figure 1.3). Also, however, do bear in mind that there is at least one bird that can look a bit sniff-like (woodcock, Chapter 3, Figure 3.17), as well as at least one small mammal (harvest mouse, Chapter 2, Figures 2.13 and 2.14). The point that I would like to emphasise here is that if you see something that looks like a humped sniff (Figure 7.17), it is always worthwhile listening to it, as it could turn out to be something else altogether.

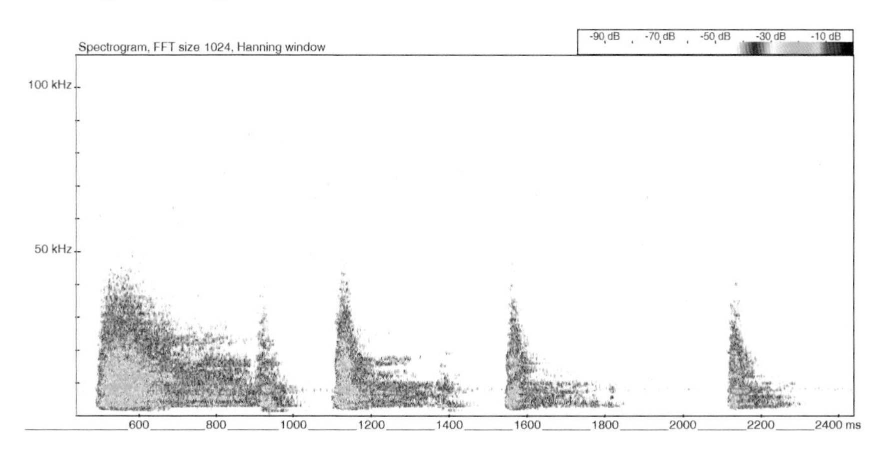

Figure 7.13 🔊 Cough.

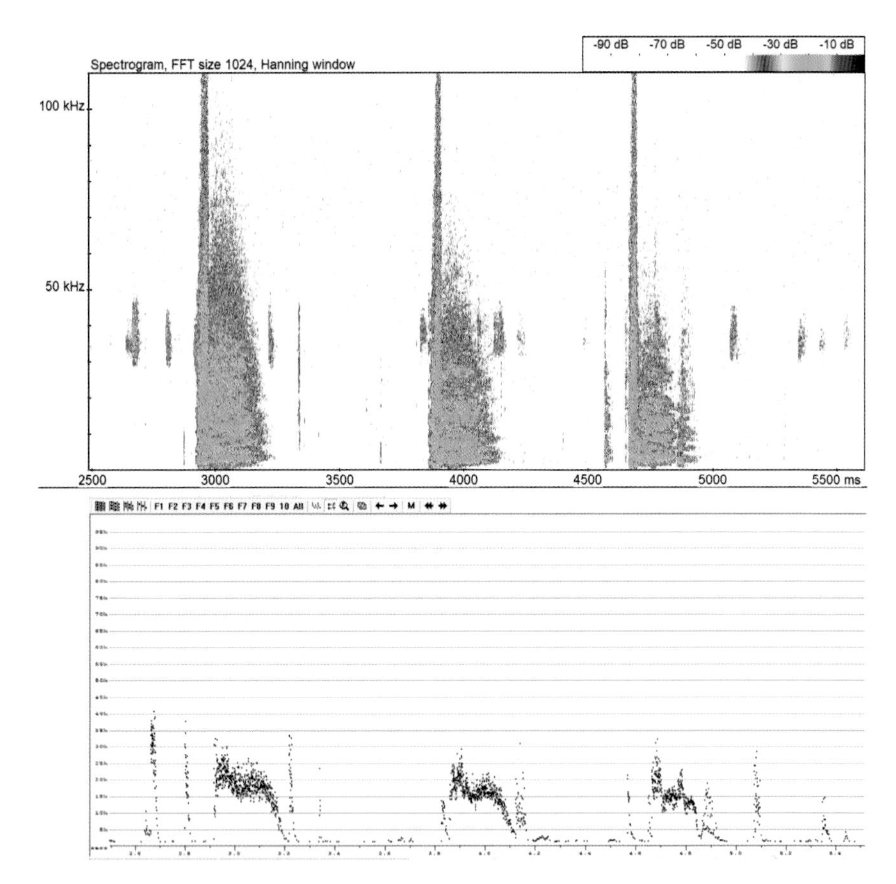

Figure 7.14 🔊 Cough with bush-cricket. Full spectrum (above) with ZCA in AnaLookW (below) (frame width 3000 ms).

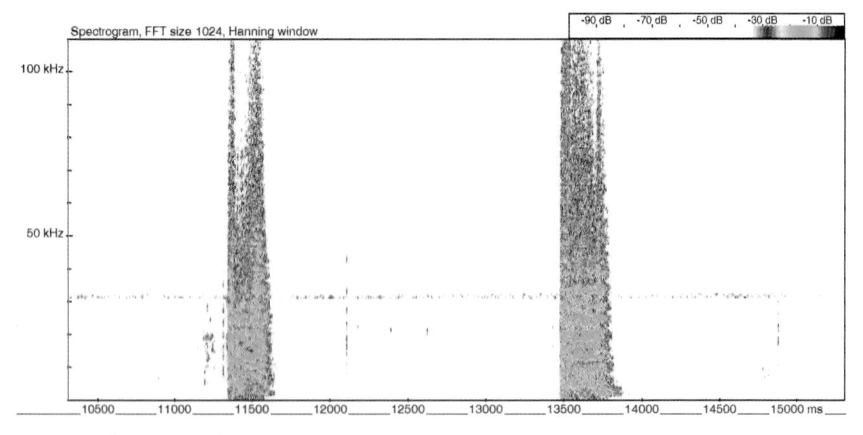

Figure 7.15 🔊 Sneeze (frame width 5000 ms).

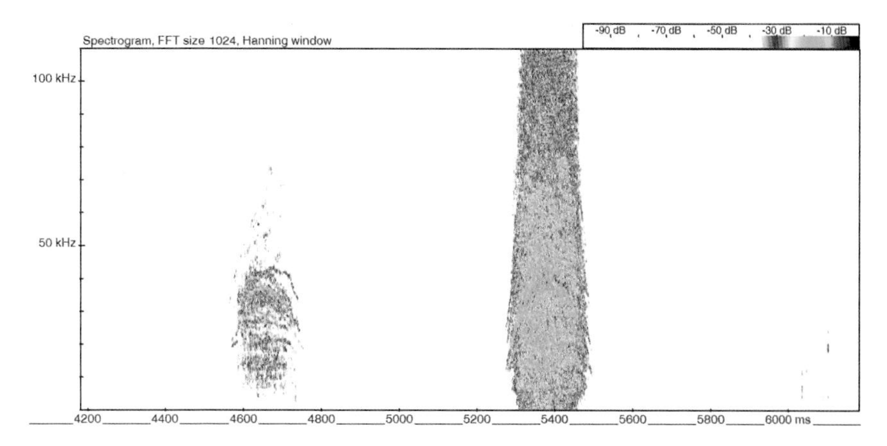

Figure 7.16 🔊 Sniffing – a quiet one, followed by a loud one.

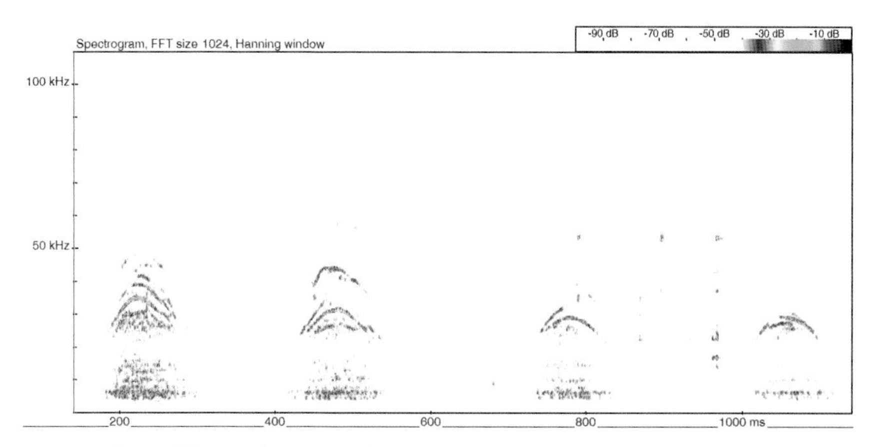

Figure 7.17 🔊 Sniff (K Mandonx, 2018; frame width 1000 ms).

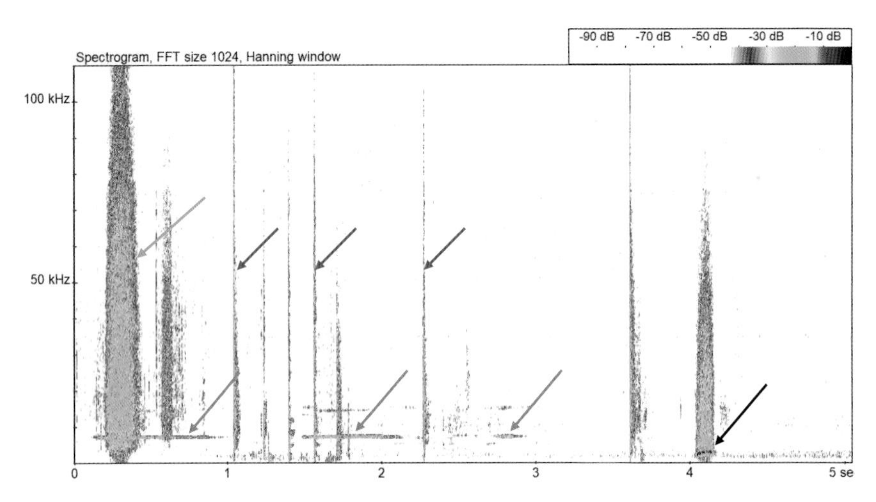

Figure 7.18 Sniff (green arrow), tongue clicking (red arrows), whistle (black arrow) and bird (blue arrows) (frame width 5000 ms).

Speech

The final examples I would like to cover in this chapter relate to voice recordings, made either deliberately or inadvertently by the surveyor. Figure 7.19 shows such a recording as produced using a Batbox Duet's 'Ref' feature, which records voice notes integrated within the sound file being recorded at the time. This is an extremely useful feature, not only on this detector, but also on the Pettersson models D230 and D240X. If the 'Ref' button isn't used and the bat detector inadvertently picks up a voice as a frequency division recording, then it looks like the example shown in Figure 7.20. Figure 7.21 shows a similar voice recording as recorded by a Batlogger M, full spectrum system, which is something a surveyor may choose to do deliberately, or it may be picked up inadvertently if someone just happened to be talking at that point in time. In this final example you can see that a soprano pipistrelle was echolocating immediately prior to the surveyor speaking.

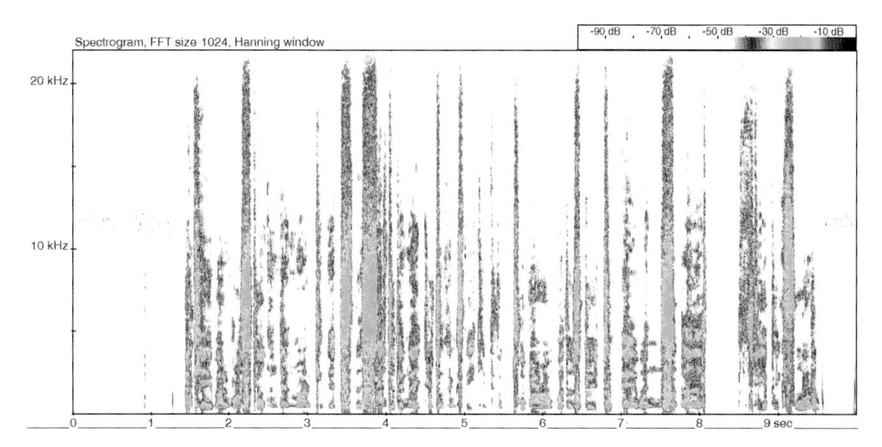

Figure 7.19 Batbox Duet – genuine voice note, deliberately activated (frame width 10 seconds).

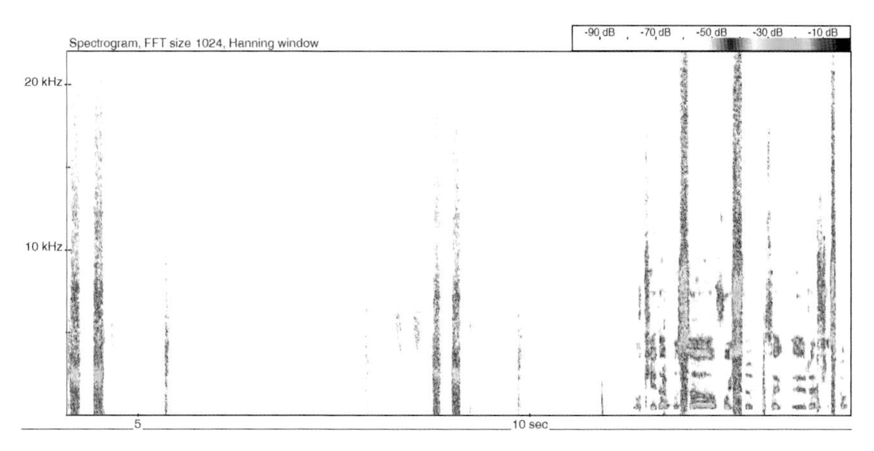

Figure 7.20 Batbox Duet – voice being picked up as a frequency division recording (frame width 10 seconds).

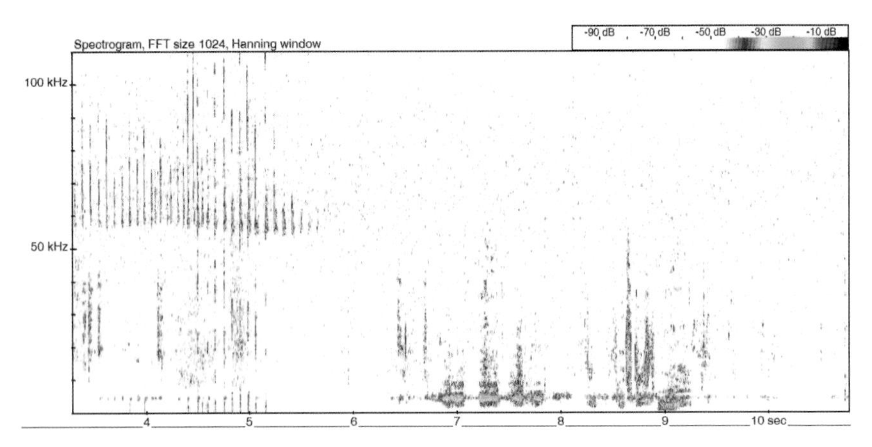

Figure 7.21 Voice note picked up on a full spectrum system, with soprano pipistrelle echolocation immediately before the speech (frame width 7000 ms).

Hints and tips

It is unlikely that noise made by weather, water or humans during a survey would interfere with the analysis or the results. It may therefore appear that there is little need to consider positioning of equipment when it comes to surveys. A bat detector microphone's frequency responsiveness would often reduce the chances of recording the lower frequencies you would expect to occur from such noise, although higher-frequency aspects (e.g. ultrasonic noise created by walking on gravel) could still occur. Also, it would not be anticipated that any classifiers would misidentify these noises as bats, and noise filters set appropriately would more often than not identify such sounds accurately.

One thing that should be considered, however, is the potential for such noise to drown out any bat-related sound (see Chapter 6) in that the noise shown in a spectrogram or heard is so loud that it is difficult to see or hear bat-related sound. Accordingly, the positioning of detectors relative to anticipated noise sources should be a consideration in survey design. Table 7.1 tackles these issues.

Table 7.1 Prevention of weather, water-related and human-generated noise impacting upon bat survey outputs.

Subject	Problem	Potential solutions
Equipment damage	Field equipment being damaged by weather-related incidents	Weather-proofing equipment to deal with normal weather-related incidents would be recommended. Positioning of equipment may need careful consideration (e.g. avoiding river banks or streams that may rise in water level during periods of rain). Consider using surge protection and/or electrical grounding techniques to protect equipment from power surges and electrostatic discharge.
Inadvertently recording	Inadvertent recording of non-bat sound, thus hindering analysis and causing inefficiencies in the process	In many respects the inadvertent recording of these sounds is inherent in the overall survey approach and often difficult to avoid. Loud noise events could impact upon survey results, due to noise drowning out bat calls. Brief the survey team as to actions they can take to avoid recording unwanted or loud noise, including walking away from obvious sound sources, not lingering close to fast-moving water, and keeping speech to a minimum.
Noise filtering process	Applying a noise filter removes more than just weather-related noise etc.	A noise filter removing sound at levels below a frequency of 10 kHz would remove at least some, if not most, of many of these events without having an impact on bat data. Carry out manual audit of filtered noise files to ensure nothing of interest has been misallocated.
Bat classifier error	Weather, water and other noise incidents being misidentified as bats	Unlikely to occur with this group.
Human error	Person analysing bat calls mistaking this noise for bats during sound analysis	Unlikely to occur with this group.
Bear in mind that anything occurring immediately around or beside the microphone could impact upon how it performs or how recorded sound behaves (e.g. sound echoing off a smooth surface close to the microphone). Therefore, any alterations made to a microphone system that could potentially impact upon recorded sound should be tested ahead of being used on case work.		

Author's diary note

There's one final aspect of running water that I feel deserves a mention, and it has occurred numerous times that I can think of, and others have told me stories from their working environment also. Let us be clear about something. If you go for a Jimmy or Jemima Riddle during a survey and you leave your bat detector switched on during the showerette, there is a strong likelihood that the detector will record it. Yes, fast-flowing urine produces ultrasonic noise, just like any other liquid coming into contact with a surface. I really wanted to demonstrate this for you. Or to put it more accurately, I wanted someone else to demonstrate it. I asked many people, on countless occasions, to oblige. They mostly said 'no', with a few saying 'I'll think about it.' But apparently it can be quite off-putting trying to pee on demand. Especially with me hovering at your shoulder with an EM Touch. So what I have done? OK, there wasn't a yellow snowball's chance in an extremely hot place that I was up for it myself. Not even in the name of science. But I did attempt to recreate such an event using some potted plants in my garden and some water in a Fairy Liquid bottle (other brands of washing-up liquid are available). Figure 7.22 shows the result, and no, it isn't included in the Sound Library.

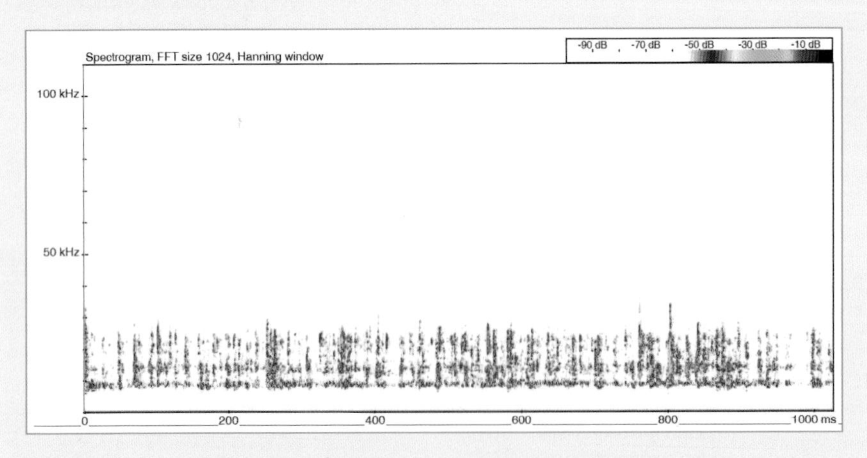

Figure 7.22 Jimmy Riddle does the dishes (that is so, so wrong!).

This is my final diary note. I hope you have enjoyed reading them, as much as I have enjoyed recalling many of the events to which they relate.

Don't hang up

After finishing the writing process, I definitely needed a rest. To announce this, at the time, I posted a picture on social media. A couple of people suggested that it should go into the book, so I have obliged here. Perhaps, having now completed reading the book, you may also be feeling as I did in April 2019. Hopefully, you won't want to hang up your bat detector though. There is so much out there still to learn about, and be fascinated by.

Is That A Bat? Nope, it's Middleton, having finally finished writing the book and now taking a well-earned rest ahead of the forthcoming bat season

APPENDIX 1

Supporting figures and tables

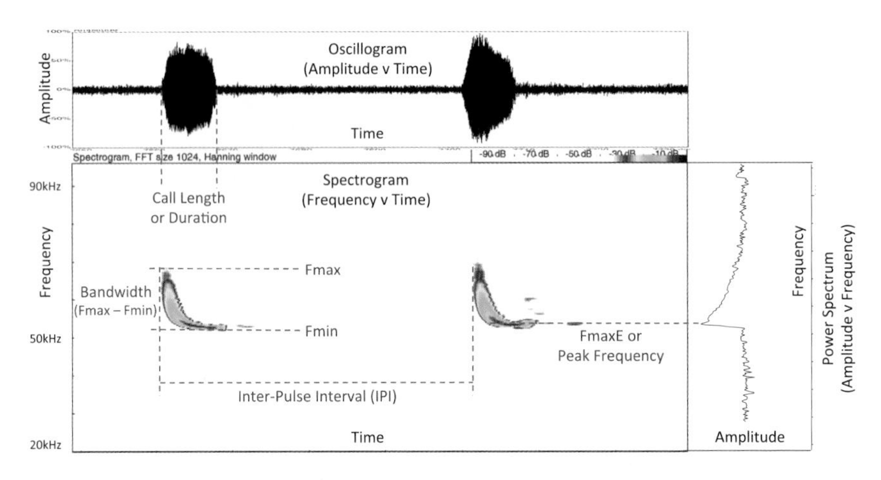

Figure A1.1 Analysis tools and call measurements commonly used in sound analysis.

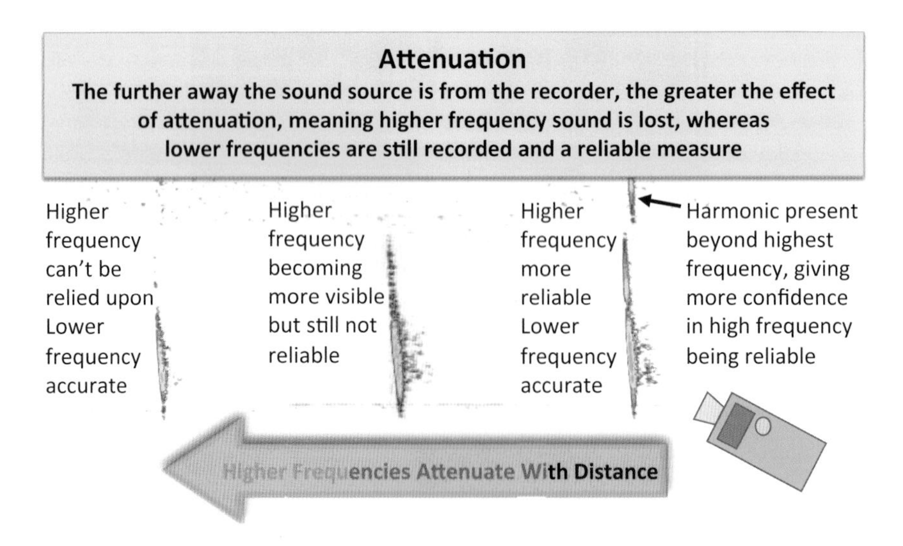

Figure A1.2 The impact of attenuation relative to recording devices.

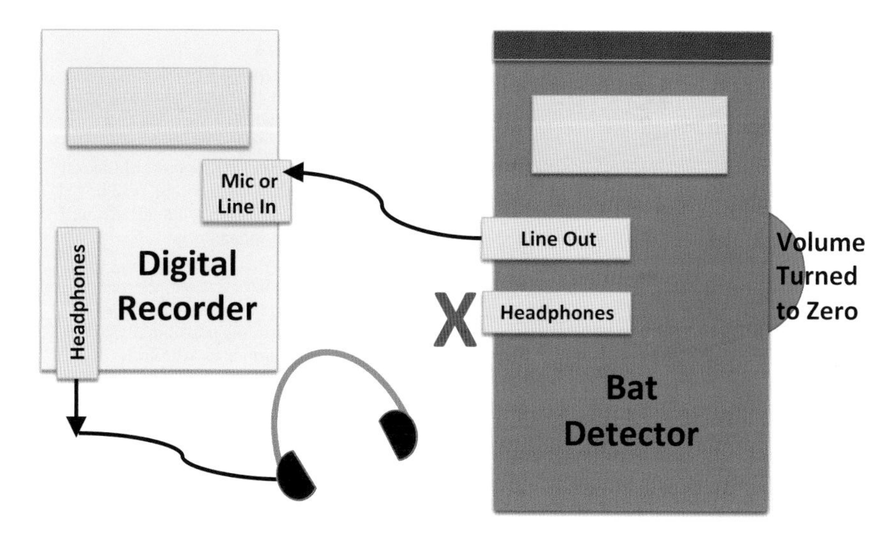

Figure A1.3 Recommended set-up when bat detector and digital recorder are separate units.

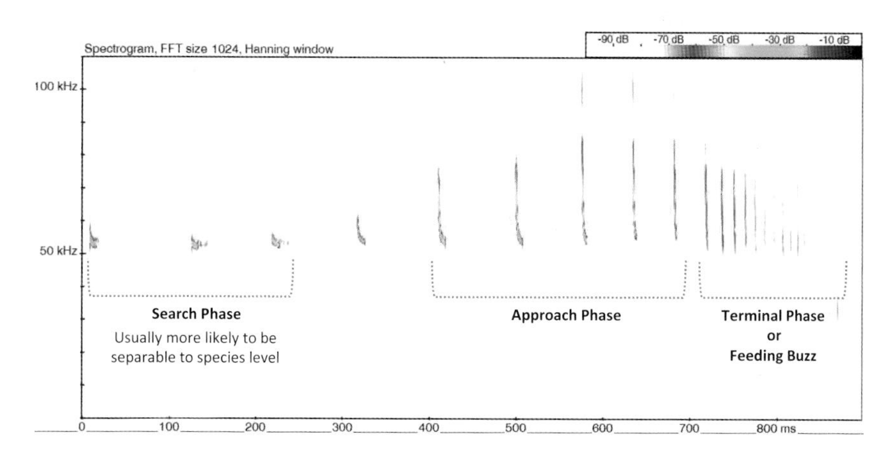

Figure A1.4 Differentiating search, approach and terminal phase echolocation calls in bats.

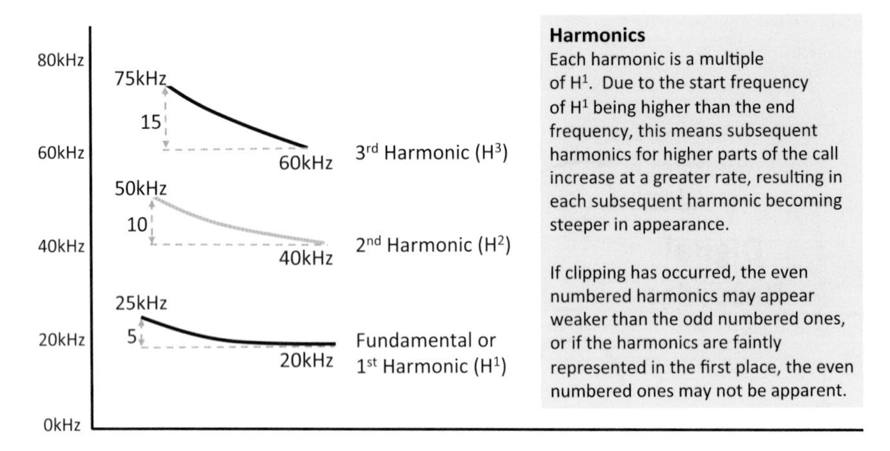

Figure A1.5 Overview of harmonics.

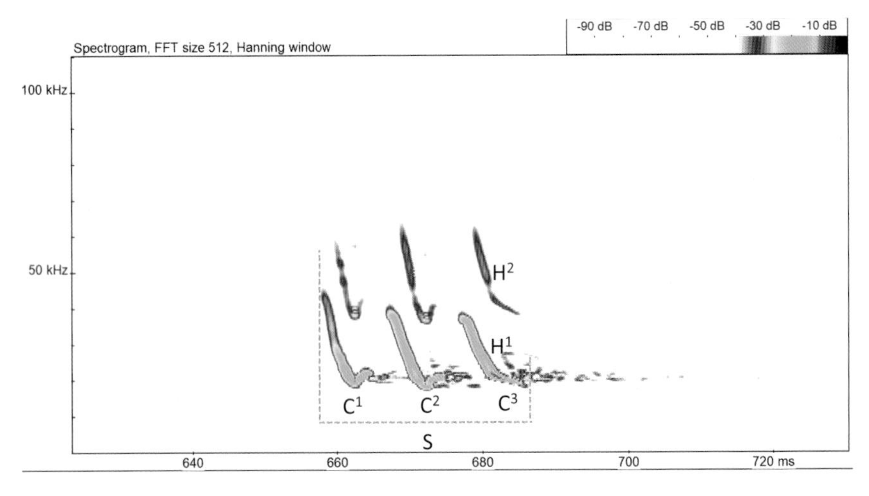

Figure A1.6 Spectrogram of social call showing a call sequence (S), comprising three components (C^1, C^2, C^3), each component comprising two syllables. Also showing position of fundamental (H^1) and 2nd harmonic (H^2) (adapted from Middleton *et al.* 2014).

Table A1.1 Call shape descriptions, as viewed in spectrograms, encountered during sound analysis (adapted from Middleton *et al.* 2014).

Descending FM	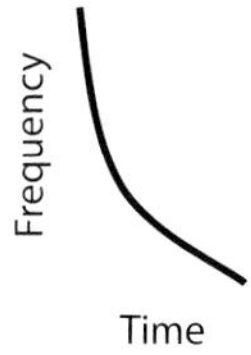 Frequency / Time	**Hockey stick**	
Ascending FM		**Hooked**	
Steep broadband FM		**Steep narrowband FM**	
Shallow narrowband FM		**CF – Constant frequency**	
Ascending QCF		**Descending QCF**	
U		**Arched**	
V		**Modulated oscillated**	
Single component		**Sequence (more than one component)**	
Single syllable (i.e. one sound occurs)		**Multi-syllable (i.e. more than one sound occurs within component)**	

Key

Component – distinct single call separated in time from other calls
Sequence – a group of component calls closely associated to each other
Syllable – each differing aspect in terms of structure, within a single component
FM – frequency modulation (i.e. frequency changes over time)
CF – constant frequency (i.e. frequency remains stable over time)
QCF – quasi-constant frequency (i.e. nearly or almost constant frequency)
Broadband – call travels through a wide frequency range
Narrowband – call travels through a short frequency range

Table A1.2 Time/frequency unit conversion key.

Time	
Seconds (s)	**Milliseconds (ms)**
1.0	1000
0.1	100
0.01	10
0.001	1

Frequency	
Hertz (Hz)	**Kilohertz (kHz)**
100,000	100
10,000	10
1000	1
1	0.001

APPENDIX 2

Problem solving

Here I provide some guidance on a thought process to follow when you are faced with recorded sound and are unsure of the source. It is by no means the only approach, and I can't guarantee that it will work on every occasion, but it is the process that I tend to go through myself in such circumstances.

I also provide two case studies, using a recording sent to me by Jo Richmond (ECOSA Ltd) and another that I have been permitted to use, received as a result of work being carried out by HS2 during 2018. These case studies allow you to see how the problem-solving approach can be followed when faced with a real-life recording. To be clear, both calls were initially sent me for interest only, and on each occasion the contributor had already come to the same conclusion as me, albeit I applied the process shown below.

Step-by-step guide to differentiating non-bat-related sounds from bat sounds

Step 1	Does the time of day or the date of the recording tie in with when you would expect bats to be active?	If timing is close to dusk or dawn this may increase the likelihood that the sound has come from another source (e.g. bird). If the timing is during the day, then it is less likely to be a bat.
Step 2	Double-check the location of the bat detector, as well as surrounding habitat.	This may assist in ruling in or ruling out certain sources of sound.
Step 3	Are you viewing the call in compressed mode?	If yes, then switch to uncompressed mode. Does this help you to determine whether it is a bat or not?
Step 4	Are you able to listen to the recording? What do you hear?	Listen to it as an original 'real-time' audible file (e.g. using Windows Media Player), then, using your analysis software, listen to it as full spectrum, and finally as time expansion (× 10). In original audible does it sound familiar? In full spectrum mode can you hear anything? It is less likely that you would hear bat echolocation and most bat social calls. In time expansion (× 10) does it sound like a bird? If yes, with quick repetitive sounds, it may very well be a bat echolocating. If similar sounding, but more complex, and not repetitive, it may be bat social calls or a small terrestrial mammal. Does it sound like a whale? If yes and you are hearing sounds that are not repetitive, and/or last for a relatively long time/drawn out, it may very well be a bird. Does it sound artificial, for example metallic or electronic?

Step 5	Apart from the confusing sound, is there any definite bat-related sound present in the same sequence, or within a very short time span either side of the file?	The presence of a bat at the same time may indicate that the other sound also belongs to a bat. Be aware, however, that occasionally you may be lucky enough to record two entirely different sound sources simultaneously. The lack of any other bat-related sound may indicate that the sound didn't come from a bat. Also, typically, you would not normally expect sound coming from a single source to interrupt itself. Are you seeing proper and consistent separation of sound events across the sequence? If so, this may mean that all sound is coming from a single source.
Step 6	Using the oscillogram tool, are you able to check amplitude across the length of the entire sequence? Was the detector a static or a hand-held device?	If it was a static device, then any dramatic changes in amplitude may be attributable to an animal moving fast – for example, a quick rise and fall in intensity as a bat flies past the detector. If amplitude is fairly consistent, then this may be due to an animal being stationary or moving much more slowly close to the detector (e.g. small terrestrial mammal). If it was a hand-held device, then changes in amplitude are less reliable, as this could have been due to the surveyor moving the detector at the same time that sound was being recorded.
Step 7	Are there any sound artefacts that may give you more clues?	For example: echoes/comb filtering/clipping/aliasing/speaker noise
Step 8	Double-check the recording parameters/settings of your bat detector, and the frequency responsiveness of the microphone.	Is it possible that the sound you are looking at or listening to has lower-frequency fundamental noise that you are not seeing or hearing because of the recording capabilities of the bat detector? For example, if the microphone is only recording frequencies greater than 10 kHz, is it possible that something audible was occurring that is not present within your spectrogram? One way to check this is if you have recorded a further harmonic (H^{2}?) above what you are currently considering to be the fundamental (H^{1}?). If the fundamental is at 25 kHz, and your next harmonic up is at 37.5 kHz, for example, this would imply that you are in fact looking at a third harmonic (H^{3}) with an invisible fundamental having been emitted at 12.5 kHz.
Step 9	Having carried out all of the above, are you able to rule anything out? By a process of elimination, what are you left with?	Establishing what something isn't is often a useful approach to working out what it is. This could be done at any level of analysis. From the following list, what can you safely dismiss? What's left? electrical/mechanical/weather/human-associated/water/amphibian/insect/small terrestrial mammal/bird/bat echolocation/bat social call

Problem solving: case study 1

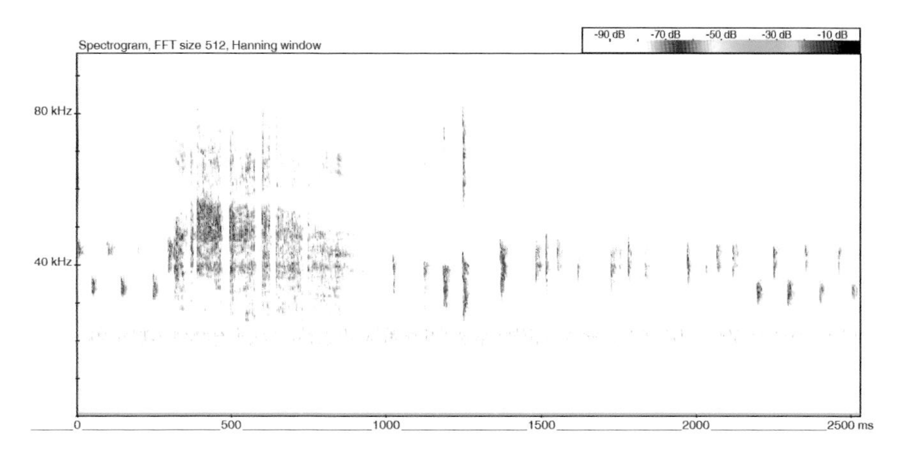

Figure A2.1 Query bat social call (Jo Richmond, ECOSA Ltd, 2018).

Question posed

Is it likely that the noise produced early on in the recording (*c.* 250 ms through to *c.* 900 ms) has been produced by the bat echolocating at the beginning and the end of the event shown?

Thought process

Step 1 It was recorded in June, and the time of the recording is *c.* 01.30 h (i.e. in darkness). Although this doesn't rule out other stuff, the timing does rule in bats.

Step 2 The bat detector location was static and within a woodland setting in southern England.

Step 3 We are viewing the call in true time, as opposed to compressed.

Step 4 There is no audible sound when listening in real time. In time expansion (× 10) it sounds like what you would expect from a bat social call.

Step 5 There is echolocation either side of the query sound, with this echolocation being a good fit for barbastelle both before and after the query noise. The echolocation does not seem to interrupt the query call, which – bearing in mind the speed at which everything has occurred – would make it less likely that there were two separate sound sources.

Step 6 We know that the bat detector was stationary throughout (i.e. a static). The recorded call is of a similar intensity throughout, suggesting all of the sound is coming from the same source, and at a relatively stable distance from the detector (i.e. probably not flying directly towards or away from the machine). The recording is weak in terms of amplitude, but at least this means that there are no issues regarding clipping, albeit higher-frequency sound may have been attenuated and therefore not picked up by the detector.

Step 7 No artefacts are apparent. Also refer to previous thoughts in response to step 6.

Step 8 The frequency relating to the sound in question peaks at *c.* 48 kHz, with an Fmin of *c.* 25 kHz. If this was an H^2, then we would expect to see H^1 (fundamental) at *c.* 24 kHz and 12.5 khz respectively. No sound is occurring at these lower frequencies. The microphone's frequency responsiveness and the bat detector's settings were such that we would expect to pick up sound at those frequencies if it was there.

Step 9 It isn't an insect (as far as I know), mechanical, electrical or weather-related. Given the time of the recording and the frequency, it is unlikely to be a bird. That leaves us with bat or small terrestrial mammal as the only options. The sound coming from another source, separate to the echolocating bat, is less likely as there are no interruptions. As such, by a process of elimination, the bat option is the only one we are left with.

Conclusion

It is very likely a bat, and if so, almost certainly the barbastelle which was echolocating at the start and end of the recording.

Problem solving: case study 2

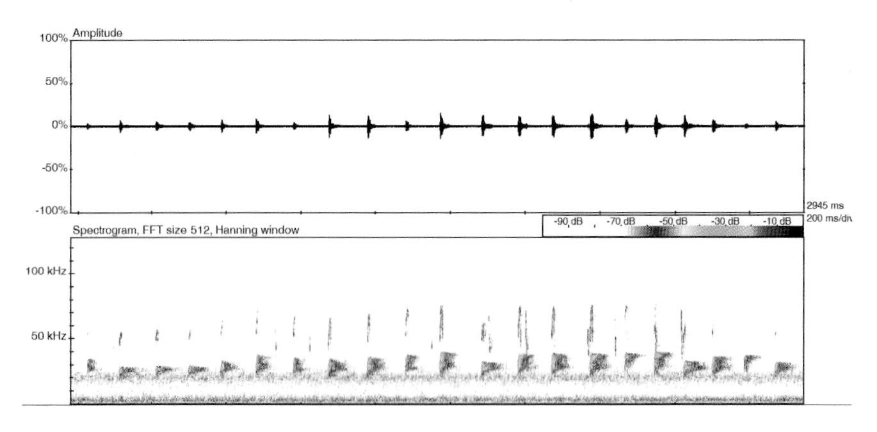

Figure A2.2 Query bat social call (HS2, 2018).

Question posed

A long series of hooked Daubenton's-type social calls has appeared within a sequence along with a small number of *Myotis* echolocation pulses. Is it likely that the social-call-type noises have been made by the same bat that is echolocating, or another bat of the same species? Is it possible that the sound was not produced by a bat, but is from some other source altogether?

Thought process

Step 1 It was recorded in September, a time of year notable for social interaction connected with mating etc. The time of the recording is *c.* 23.00 h (i.e. in darkness). Although this doesn't rule out other stuff, the timing does rule in bats.

Step 2 The bat detector location was static, next to a pond and within a woodland setting in Yorkshire.

Step 3 We are viewing the call in true time, as opposed to compressed.

Step 4 There is no audible sound when listening in real time. In time expansion (× 10) it sounds like what you would expect from bat social calls, and *Myotis* echolocation pulses can also be heard.

Step 5 There is bat echolocation (*Myotis* sp.) occurring during the sequence. The echolocation interrupts the query calls, making it likely that there were two separate sound sources.

Step 6 We know that the bat detector was stationary throughout (i.e. a static). The recorded call is of a different intensity throughout, suggesting the sound was coming from a rapidly moving source, which gets louder and then decreases in amplitude, consistent with a bat in flight. The recording has a reasonable signal-to-noise ratio.

Step 7 There are no issues regarding clipping, but echoes and comb filtering occur within the sequence, suggesting that the bat or the detector was close to a reflective surface. The detector was above vegetation, and therefore it is more likely to be as a result of a bat flying close to the surface of the pond.

Step 8 The frequency relating to the fundamental sound in question has an Fmax at *c.* 35 kHz, with the next harmonic being at *c.* 70 kHz. It therefore looks as if we are seeing the call accurately. The microphone's frequency responsiveness and the bat detector's settings are not thought to be an issue in this instance.

Step 9 It isn't an insect (as far as I know), mechanical, electrical or weather-related. Given the time of the recording and the frequency, it is unlikely to be a bird. That leaves us with bat or small terrestrial mammal as the only options. The sound occurring over ideal Daubenton's habitat, and with such a variable intensity over a very short duration, makes it more likely to be a bat. As well as which, the call structure is good for a Daubenton's hooked social call. The bat option is, by process of elimination, the most likely.

Conclusion

It is very likely a bat, and if so almost certainly a Daubenton's, with one other bat (*Myotis* sp., probably Daubenton's) being present at the same time.

Additional resources

In addition to the resources listed below, there is an organisation based in the UK which you may wish to consider investigating further. The **Wildlife Sound Recording Society (WSRS)** has lots of information that would be of use to anyone interested in recording natural sound. As a member you will also receive their excellent magazine, as well as regularly produced CDs containing examples of recordings made by other members. Formed in 1968, this society has membership from all over the world. It exists in order to encourage *'the enjoyment, recording and understanding of wildlife and other natural sounds'*. More details about WSRS can be found on their website (www.wildlife-sound.org), including how to join, equipment suggestions and planned field visits.

Resources for learning about and recording bat sounds

Acoustic Ecology of European Bats: Species Identification, Study of their Habitats and Foraging Behaviour M Barataud BIOTOPE	This book, which comes with a DVD-ROM, contains a wealth of information about European bat species, echolocation and social calls. The DVD-ROM includes over 300 sound files (both in heterodyne and time expansion), as well as Excel spreadsheets and charts showing call measurement parameters etc.
British Bat Calls: A Guide to Species Identification J Russ Pelagic Publishing	This book is a practical guide to the calls of British bat species, as well as including numerous associated topics. For each of the species included, information is provided on distribution, emergence times, behaviour, habitat and echolocation.
The Bat Detective: A Field Guide to Bat Detection B Briggs and D King Batbox Ltd	This guide is designed to help those using heterodyne bat detectors. It covers both the theoretical and practical aspects of the use of these detectors, as well as many aspects of identification in the field. It comes with a CD, which is referred to throughout the text.
Social Calls of the Bats of Britain and Ireland N Middleton, A Froud and K French Pelagic Publishing	This book brings together the current state of knowledge of bat social calls occurring within Britain and Ireland, with some additional examples from species represented elsewhere. It includes access to a downloadable library of calls to be used in conjunction with the book.
Bat Call Sound Analysis Workshop www.facebook.com/groups/437808079587006	This Facebook site is an excellent resource, with many members contributing examples and offering advice and explanations.

Resources for learning and recording bird sounds

Teach Yourself Bird Sounds CD D Couzens and J Wyatt Subbuteo Natural History Books	A double-CD set covering garden and woodland birds. An excellent resource for those wishing to learn bird calls/songs. The way in which species-specific sound is described is very helpful, and because they are grouped by habitat, the sounds are already 'audited' down to what you would expect to hear in that environment. Unfortunately, it is currently out of print, but if you can get hold of it from somewhere, it is well worth it.
Birds of The Western Palearctic *Interactive* DVD-ROM BirdGuides	Comprehensive multimedia coverage of 970 species from Europe, North Africa and the Middle East. Includes the typical calls/songs for most species. Unfortunately, it is currently not available, but if you can find it somewhere, it is extremely worthwhile. Be careful to check that it's compatible with your PC's operating system.
Raptors: a Field Guide for Surveys and Monitoring J Hardey *et al.* 2009 The Stationery Office, Edinburgh	An excellent book with accompanying CD which provides many examples of different raptor calls. A brilliant reference for owl sounds (as well as all the other raptors), and definitely should be listened to by any ecologist carrying out night-time surveys.
The Sound Approach www.soundapproach.co.uk	This website caters for many aspects of listening to and recording birds. There are links that enable you to buy books on the subject, including the popular *The Sound Approach to Birding*, which aims to 'enhance your field skills and improve your standards of identification'.
Xeno-Canto www.xeno-canto.org	This website is used by birders for sharing recordings from around the world. It aims to 'popularise bird sound recording worldwide, improve accessibility, and increase knowledge'. An excellent site for checking calls/songs you have heard, with European species well represented. It also allows you to contribute to the resource by uploading your own examples if you wish.
Nocmig www.nocmig.com	This website provides 'tips for recording nocturnal migration', with lots of information and advice for those wishing to develop an interest in the subject.

Organisations with resources relating to amphibians

Amphibian and Reptile Conservation (ARC) www.arc-trust.org	A national wildlife charity committed to conserving amphibians and reptiles and saving their habitats.
National Amphibian and Reptile Recording Scheme (NARRS) www.narrs.org.uk www.alienencounters.narrs.org.uk	This scheme is run by ARC (see above) and brings existing surveys for our rarer species together with surveys to collect data on more widespread amphibians and reptiles. Alien Encounters is a website for information on and reporting sightings of non-native amphibians and reptiles in the UK.
Froglife www.froglife.org www.froglife.org/dragonfinder/app/	A national wildlife charity committed to the conservation of amphibians and reptiles – frogs, toads, newts, snakes and lizards – and saving the habitats they depend on. Their Dragon Finder app is an excellent resource that helps you to identify species and report sightings.

Organisations with resources relating to insects

Grasshoppers and Related Insects Recording Scheme of Britain and Ireland www.orthoptera.org.uk	This scheme was launched in 1968 with the support of the Biological Records Centre (BRC), to collect records of grasshoppers and related species, and to map and study their distributions. The information gathered is also used to see how wildlife is responding to changes in land use and climate.
Norfolk Bat Survey (BTO) Bush-cricket Identification www.batsurvey.org/bush-cricket-identification	This resource was launched early in 2019 by the BTO in order to give those carrying out sound-analysis identification (e.g. as generated by bat surveys) guidance on separating the more regularly occurring bush-cricket species. Within the resource there are numerous downloadable sound files that can be used for reference purposes.
Butterfly Conservation www.butterfly-conservation.org	The core aims of Butterfly Conservation are to: • recover threatened butterflies and moths • increase numbers of widespread species • inspire people to deliver species conservation The website is an excellent resource with lots of useful species accounts etc.
Avisoft Bioacoustics www.avisoft.com/sounds.htm	Has a range of insect calls that can be listened to, as well as some spectrogram/oscillogram visuals. Numerous other species are also included within the call library.

Glossary

Algorithm	A set of rules or a procedure that is to be followed in calculations or other problem-solving operations, especially by computer software.
Amplitude	The volume at which a sound is produced.
Anthropogenic	Caused by, relating to, or connected with, human actions.
Appetitive	Characteristic of the natural desire to satisfy needs.
Approach phase	A sequence of echolocation produced by a bat as it approaches a prey target (e.g. an insect) or an obstacle it is seeking to avoid.
Attenuation	The reduction in amplitude of a sound as it travels through air. Higher frequencies usually attenuate more quickly than lower frequencies of the same intensity (see Appendix 1, Figure A1.2).
Audible	A sound that is within the normal hearing range of a human (< 20 kHz).
Bandwidth	A measure of the width of a range of frequencies, or the difference between minimum and maximum frequencies (see Appendix 1, Figure A1.1).
Broadband	A call that travels through a wide frequency range (see Appendix 1, Table A1.1).
Component	A distinct single call separated in time from other calls (see Appendix 1, Figure A1.6 and Table A1.1).
Constant frequency (CF)	Frequency remains stable over time (see Appendix 1, Table A1.1).
Feeding buzz	A rapid series of echolocation pulses produced by a bat as it homes in on a prey target.
FM	*See* Frequency modulation.
Fmax	Maximum or start frequency. The highest frequency emitted during a call (see Appendix 1, Figure A1.1).
FmaxE	Frequency of maximum energy, or peak frequency. The frequency level within a sound containing the most energy (see Appendix 1, Figure A1.1).
Fmin	Minimum or end frequency. The lowest frequency emitted during a call (see Appendix 1, Figure A1.1).
Frequency	The word used to describe the measured sound level in terms of hertz or kilohertz. For example, 'the frequency produced is at 50 kHz.'
Frequency division	A recording format which divides the frequency of the bat call so that it is within the audible range. Rhythm and repetition rate are heard audibly, although, unlike heterodyne systems, this format cannot be tuned in to specific frequencies. Frequency division calls are recorded in real time and can be used for analysis.

Frequency of maximum energy	*See* FmaxE.
Frequency modulation (FM)	Frequency changes over time (see Appendix 1, Table A1.1).
Full spectrum	A bat detecting format used by some bat detectors (e.g. Batlogger M, EM Touch, Anabat Swift) whereby all sound is recorded accurately and in real time. This means that ultrasonic sound cannot be heard by the surveyor, unless the bat detector has a function for creating simultaneous audible noise (e.g. auto-heterodyning). Full spectrum recordings are beneficial for sound analysis, as all sound within the parameters of the detector's capabilities are recorded, including any harmonics produced and/or sound from other sources occurring at the same time.
Fundamental	The first harmonic (H^1) created by any incidence of sound occurring (see Appendix 1, Figure A1.5).
Geogenic	As a result of geological or geographic processes.
Harmonic	Layers of sound occurring on top of each other, in multiples of the fundamental (lowest) noise, which is also called the first harmonic (H^1). Additional harmonics above the first (or fundamental) are described as second (H^2), third (H^3), etc. (see Appendix 1, Figure A1.5).
Heterodyne	A bat detecting format used by some bat detectors (e.g. Batbox Duet) whereby bat ultrasound is converted into an audible representation of the tone, rhythm and repetition rate of the true sound. The audible electronically created calls are broadcast in real time, this being for listening purposes on the part of the user. These electronic noises, however, cannot be used for software analysis.
Hertz (Hz)	A unit of frequency indicating one cycle per second (see Appendix 1, Table A1.2).
Inter-pulse interval	The period occurring from the start of one echolocation pulse to the start of the next echolocation pulse.
Kilohertz (kHz)	A unit of frequency indicating 1000 cycles per second (see Appendix 1, Table A1.2).
Lure	*See* Ultrasonic lure
Narrowband	A call that travels through a narrow frequency range (see Appendix 1, Table A1.1).
Omnidirectional	Receiving sound from all around, as opposed to from one specific direction.
Oscillogram	A graph showing amplitude against time.
Peak frequency	*See* FmaxE.
Power spectrum	A graph showing amplitude against frequency for a recorded sound.
Quasi-constant frequency (QCF)	Nearly constant frequency (see Appendix 1, Table A1.1).

Search phase	A series of echolocation pulses produced by a bat when it is in normal flight and not in the process of locating a precise target or avoiding an obstacle (e.g. when commuting).
Sequence	A group of component calls closely associated with each other (see Appendix 1, Figure A1.6 and Table A1.1).
Signal-to-noise ratio (SNR)	'Signal' is the noise from the target sound source, whereas 'noise' refers to background sound. In general, the louder the signal, relative to the noise, the better the recording.
Spectrogram	A graph showing sound as a visual representation of frequency against time.
Syllable	Each differing aspect, in terms of structure, within a single call component (see Appendix 1, Figure A1.6 and Table A1.1).
Technogenic	Resulting from a process or substance created by human technology.
Terminal phase	*See* Feeding buzz.
Time expansion	A process for recording sound whereby the original noise is slowed down by a factor (usually × 10), meaning that the duration of the event is expanded by the same factor, and the frequency is reduced, in order to become audible. Because all of the original sound is retained (albeit slowed down) time-expanded recordings are beneficial for sound analysis, as all sound within the parameters of the detector's capabilities are recorded, including any harmonics produced and/or sound from other sources occurring at the same time.
Ultrasound	Noise occurring at a frequency beyond what a human would be expected to hear without the aid of a bat detector, for example (> 20 kHz).
Ultrasonic lure	An electronic device emitting high-frequency sound through a speaker in order to attract target species into an area to be trapped and/or for monitoring purposes.
Zero crossing analysis (ZCA)	A recording format used by some bat detectors (e.g. Anabat SD1, SD2, Express) whereby only the strongest frequency in a recording is represented at any moment in time.

References

Ahlén, I. (1990). *Identification of Bats in Flight*. Swedish Society for Conservation of Nature and Swedish Youth Association for Environmental Studies and Conservation.

Alien Encounters (2018). [Online resource providing information and monitoring of non-native amphibians and reptiles.] www.alienencounters.narrs.org.uk.

Altringham, J. D. (1996). *Bats: Biology and Behaviour*. Oxford University Press, Oxford.

Ancillotto, L., Sozio, G., Mortelliti, A. and Russo, D. (2014). Ultrasonic communication in Gliridae (Rodentia): the hazel dormouse (*Muscardinus avellanarius*) as a case study. *Bioacoustics* 23 (2): 129–141.

Ancillotto, L., Mori, E., Sozio, G., Solano, E., Bertolino, S. and Russo, D. (2017). A novel approach to field identification of cryptic *Apodemus* wood mice: calls differ more than morphology. *Mammal Review* 47 (1): 6–10.

Anderson, J. W. (1954). The production of ultrasonic sounds by laboratory rats and other mammals. *Science* 119: 808–809.

Andrews, M. M. and Andrews, P. T. (2003). Ultrasound social calls made by greater horseshoe bats (*Rhinolophus ferrumequinum*) in a nursery roost. *Acta Chiropterologica* 5 (2): 221–234.

Andrews, M. M., Andrews, P. T., Wills, D. F. and Bevis, S. M. (2006). Ultrasound social calls of greater horseshoe bats (*Rhinolophus ferrumequinum*) in a hibernaculum. *Acta Chiropterologica* 8 (1): 197–212.

Andrews, M. M., McOwat, T. P., Andrews, P. T. and Haycock, R. J. (2011). The development of the ultrasound social calls of adult *Rhinolophus ferrumequinum* from infant bat ultrasound calls. *Bioacoustics* 20 (3): 297–316.

Andrews, M. M., Hodnett, A. M. and Andrews, P. T. (2017). Social activity of lesser horseshoe bats (*Rhinolophus hipposideros*) at nursery roosts and a hibernaculum. *Acta Chiropterologica* 19 (1): 161–174.

Baker, P. A. and Harris, S. (2008). Chapter 9 (pp. 407–422), in Harris, S. and Yalden, D. W. (eds.), *Mammals of the British Isles: Handbook*, 4th edition. Mammal Society, London.

Barataud, M. (2015). *Acoustic Ecology of European Bats. Species Identification, Study of their Habitats and Foraging Behaviour*. Biotope, Muséum national d'Histoire naturelle.

Beattie, R. C. (1987). The reproductive biology of common frog (*Rana temporaria*) populations from different altitudes in northern England. *Journal of Zoology* 211: 387–398.

Beebee, T. (2013). *Amphibians and Reptiles*. Naturalists' Handbooks 31. Ecology and identification. Pelagic Publishing, Exeter.

Berry, R. J., Tattersall, F. H. and Hurst, J. (2008). Chapter 5 (pp. 141–149), in Harris, S. and Yalden, D. W. (eds.), *Mammals of the British Isles: Handbook*, 4th edition. Mammal Society, London.

Bogert, C. M. (1960). The influence of sound on the behavior of amphibians and reptiles. In Lanyon, W. E., and Tavolga, W. N. (eds.), *Animal Sounds and Communication*. American Institute of Biological Sciences Publication no. 7. Intelligencer Printing Company, [Washington, D.C.], pp. 137–320.

Brabant, R., Laurent, Y., Dolap, U., Degraer, S. and Poerink, B. J. (2018). Comparing the results of four widely used automated bat identification software programs to identify nine bat species in coastal Western Europe. *Belgian Journal of Zoology* 148 (2): 119–128.

Bright, P. W. and Morris, P. A. (2008). Chapter 5 (p. 80), in Harris, S. and Yalden, D. W. (eds.), *Mammals of the British Isles: Handbook*, 4th edition. Mammal Society, London.

Brudzynski, S. M. (2009). Communication of adult rats by ultrasonic vocalization: biological, sociobiological, and neuroscience approaches. *Institute for Laboratory Animal Research (ILAR) Journal* 50 (1): 43–50.

BTO (2019). Norfolk Bat Survey (BTO): bush-cricket Identification. www.batsurvey.org/bush-cricket-identification.

Buckley, J. and Foster, J. (2005). Reintroduction strategy for the pool frog *Rana lessonae* in England. English Nature Research Reports, number 642.

Buley, K., Gibson, R. and Pinel, J. (2001). The agile frog, *Rana dalmatina*, species action plan. Agile Frog Group, States of Jersey Environmental Services Unit.

Burgdorf, J., Kroes, R. A., Moskal, J. R., Pfaus, J. G., Brudzynski, S. M. and Panksepp, J. (2008). Ultrasonic vocalizations of rats (*Rattus norvegicus*) during mating, play, and aggression: behavioral concomitants, relationship to reward, and self-administration of playback. *Journal of Comparative Psychology* 122 (4): 357–367.

Butterfly Conservation (2018). Green Silver-lines *Pseudoips prasinana*. https://butterfly-conservation.org/moths/green-silver-lines.

BWP*i* (2009). *Birds of the Western Palearctic interactive*, Version 2.0.2. BirdGuides Ltd, www.birdguides.com.

Chapman, N. G. (2008). Chapter 11 (p. 569), in Harris, S. and Yalden, D. W. (eds.), *Mammals of the British Isles: Handbook*, 4th edition. Mammal Society, London.

Churchfield, S. (2008). Chapter 5 (pp. 271–275), in Harris, S. and Yalden, D. W. (eds.), *Mammals of the British Isles: Handbook*, 4th edition. Mammal Society, London.

Churchfield, S. and Searle, J. B. (2008a). Chapter 5 (pp. 257–265), in Harris, S. and Yalden, D. W. (eds.), *Mammals of the British Isles: Handbook*, 4th edition. Mammal Society, London.

Churchfield, S. and Searle, J. B. (2008b). Chapter 5 (pp. 267–271), in Harris, S. and Yalden, D. W. (eds.), *Mammals of the British Isles: Handbook*, 4th edition. Mammal Society, London.

Ciechanowski, M. and Sachanowicz, K. (2014). Fat dormouse *Glis* (Rodentia: Gliridae) in Albania: synopsis of distributional records with notes on habitat use. *Acta Zoologica Bulgarica* 66 (1): 39–42.

Coffey, K. R., Marx, R. G. and Neumaier, J. F. (2019). DeepSqueak: a deep learning-based system for detection and analysis of ultrasonic vocalizations. *Neuropsychopharmacology* 44: 859–868. doi: 10.1038/s41386-018-0303-6.

Conner, W. E. (1999). 'Un Chant D'appel Amoureux': acoustic communication in moths. *Journal of Experimental Biology* 202 (13): 1711–1723.

Constantine, M. and The Sound Approach (2006). *The Sound Approach to Birding: a Guide to Understanding Bird Sound*. The Sound Approach, Dorset.

Corcoran, A. J., Barber, J. R. and Conner, W. E. (2009). Tiger moth jams bat sonar. *Science* 325 (5938): 325–327.

Corcoran, A. J., Conner, W. E. and Barber, J. R. (2010). Anti-bat tiger moth sounds: form and function. *Current Zoology* 56 (3): 358–369.

Corcoran, A. J., Barber, J. R., Hristov, N. I. and Conner, W. E. (2011). How do tiger moths jam bat sonar. *Journal of Experimental Biology* 214: 2416–2425.

Delahay, R., Wilson, G., Harris, S. and Macdonald, D. W. (2008). Chapter 9 (p. 431), in Harris, S. and Yalden, D. W. (eds.), *Mammals of the British Isles: Handbook*, 4th edition. Mammal Society, London.

Endler, J. A. (1991). Interactions between predators and prey. In Krebs, J. R. and Davies, N. B., (eds.), *Behavioural Ecology: an Evolutionary Approach*. Blackwell, Oxford, pp. 169–196.

Flowerdew, J. R. and Tattersall, F. H. (2008). Chapter 5 (pp. 125–137), in Harris, S. and Yalden, D. W. (eds.), *Mammals of the British Isles: Handbook*, 4th edition. Mammal Society, London.

Forsman, K. A. and Malmquist, M. G. (1988). Evidence for echolocation in the common shrew, *Sorex araneus*. *Journal of Zoology* 216 (4): 655–662.

Franke, P. (2018). Recording nocturnal bird migration. *Wildlife Sound* 14 (4): 33–35.

Froglife (2018). [Online resource relating to pool frog (*Pelophylax lessonae*).] www.froglife.org/info-advice/amphibians-and-reptiles/pool-frog.

Fullard, J. H. (1988). The tuning of moth ears. *Experientia* 44 (5): 423–428.

Gasc, A., Francomano, D., Dunning, J. B. and Pijanowski, B. C. (2017). Future directions for soundscape ecology: the importance of ornithological contributions. *The Auk* 134: 215–228.

Gent, T. and Gibson, S. (2012). *Herpetofauna Workers' Manual*. Pelagic Publishing, Exeter.

Gillings, S., Moran, N., Robb, M., van Bruggen, J. and Troost, G. (2018). A protocol for standardised nocturnal flight call monitoring. Version 1. Download available at www.trektellen.nl/static/doc/Protocol_for_standardised_nocturnal_flight_call_monitoring_v01.pdf.

Gleed-Owen, C., Banks, B. and Buckley, J. (2013). *Competencies for Species Survey: Natterjack Toad*. CIEEM Technical Guidance Series. Chartered Institute of Ecology and Environmental Management, Winchester.

Gnoli, C. and Prigioni, C. (1995). Preliminary study on the acoustic communication of captive otters (*Lutra lutra*). *Hystrix* 7 (1–2): 289–296.

Goerlitz, H. R., ter Hofstede, H. M., Zeale, M. R. K., Jones, G., and Holderied, M. W. (2010). An aerial-hawking bat uses stealth echolocation to counter moth hearing. *Current Biology* 20 (17): 1568–1572.

Gould, E., Negus, N. C. and Novick, A. (1964). Evidence for echolocation in shrews. *Journal of Experimental Zoology* 156: 19–38.

Green, J., Green, R. and Jefferies, D. J. (1984). A radio tracking survey of otters *Lutra lutra* on a Perthshire river system. *Lutra* 27: 85–145.

Grenander, U. (1959). *Probability and Statistics: the Harald Cramér Volume*. Wiley, New York.

GRIRS (2019). Grasshoppers and Related Insects Recording Scheme of Britain and Ireland. www.orthoptera.org.uk.

Gyger, M. and Schenk, F. (1983). Semiotical Approach to the ultrasonic vocalization in the woodmouse *Apodemus sylvaticus* L. *Behaviour* 84 (3): 244–257.

Gyger, M. and Schenk, F. (1984). Ultrasonic vocalization in the Woodmouse *Apodemus sylvaticus* L. *Acta Zoologica Fennica* 171: 97–99.

Halls, S. A. (1981). The influence of olfactory stimuli on ultrasonic calling in murid and cricetid rodents. PhD thesis, University of London.

Hammerschmidt, K., Radyushkin, K., Ehrenreich, H. and Fischer, J. (2012). The structure and usage of female and male mouse ultrasonic vocalizations reveal only minor differences. *PLoS One* 7 (7): e41133. doi: 10.1371/journal.pone.0041133.

Hardey, J., Crick, H., Wernham, C., Riley, H., Etheridge, B. and Thompson, D. (2009). *Raptors: a Field Guide for Surveys and Monitoring*. Stationery Office, Edinburgh.

Harris, S. and Lloyd, H. G. (1991). Chapter 10, in Corbet, G. B. and Harris, S. (eds.), *Handbook of British Mammals*, 3rd edition. Blackwell Science, Oxford.

Heath, J. and Emmet, A. M. (1983). *The Moths and Butterflies of Great Britain and Ireland*, Volume 10, Noctuidae (Cuculliinae to Hypeninae, and Agaristidae). Harley Books, Colchester.

Heinzmann, U. (1970). Untersuchungen zur Bio-Akustik und Ökologie der Geburtshelferkröte, *Alytes o. obstetricans* (Laur.) (Bio-acoustic and ecological investigations in the midwife toad, *Alytes o. obstetricans* (Laur.)). *Oecologia* 5 (1): 19–55.

Hewison, A. J. M. and Staines, B. W. (2008). Chapter 11 (pp. 611–612), in Harris, S. and Yalden, D. W. (eds.), *Mammals of the British Isles: Handbook*, 4th edition. Mammal Society, London.

Hoffmeyer, I. and Sales, G. D. (1977). Ultrasonic behaviour of *Apodemus sylvaticus* and *Apodemus flavicollis*. *Oikos* 29 (1): 67–77.

Hutchings, M. and Lewis, B. (1983). Insect sound and vibration receptors. In Lewis, B. (ed.), *Bioacoustics: a Comparative Approach*. Academic Press, London, pp. 181–205.

Hutterer, R. and Peters, G. (2001). The vocal repertoire of *Graphiurus parvus*, and comparisons with other species of dormice. *Trakya University Journal of Scientific Research* Series B, 2 (2): 69–74.

Iesari, V., Catorci, A., Scocco, P., Bieber, C. and Fusani, L. (2017). Vocal behaviour of the edible dormouse (*Glis glis*) during the mating season. Masters degree in applied geobotany, experimental thesis produced by V. Iesari. School of Biosciences and Veterinary Medicine, University of Camerino.

Inns, H. (2009). *Britain's Reptiles and Amphibians: a Guide to the Reptiles and Amphibians of Great Britain, Ireland and the Channel Islands*. Amphibian and Reptile Conservation. WildGuides Ltd, Hampshire.

Jacobs, D. (2015). Explainer: the evolutionary arms race between bats and moths. *The Conversation*. http://theconversation.com/explainer-the-evolutionary-arms-race-between-bats-and-moths-43675.

Jefferies, D. J. and Woodroffe, G. L. (2008). Chapter 9 (p. 441), in Harris, S. and Yalden, D. W. (eds.), *Mammals of the British Isles: Handbook*, 4th edition. Mammal Society, London.

Jürgens, K. D. (2002). Etruscan shrew muscle: the consequences of being small. *Journal of Experimental Biology* 205: 2161–2166.

Juškaitis, R. and Büchner, S. (2013). *The Hazel Dormouse,* Muscardinus avellanarius. Verlags KG Wolf (formerly Westarp Wissenschaften).

Kapusta, J., Sales, G. D. and Czuchnowski, R. (2007). Aggression and vocalization of three sympatric vole species during conspecific and heterospecific same-sex encounters. *Behaviour* 144 (3): 283–305.

Kapusta, J. and Sales, G. D. (2009). Male-female interactions and ultrasonic vocalization in three sympatric species of voles during conspecific and heterospecific encounters. *Behaviour* 146 (7): 939–962.

Kazial, K. A. and Masters, W. M. (2004). Female big brown bats, *Eptesicus fuscus*, recognize sex from a caller's echolocation signals. *Animal Behaviour* 67 (5): 855–863.

Kick, S. A. and Simmons, J. A. (1984). Automatic gain control in the bat's sonar receiver and the neuroethology of echolocation. *Journal of Neuroscience* 4 (11): 2725–2737.

Kim, E. J., Kim, E. S., Covey, E. and Kim, J. J. (2010). Social transmission of fear in rats: the role of 22-kHz ultrasonic distress vocalization. *PLoS One* 5 (12): e15077. doi: 10.1371/journal.pone.0015077.

Lahvis, G. P., Alleva, E. and Scattoni, M. L. (2011). Translating mouse vocalizations: prosody and frequency modulation. *Genes, Brain, and Behavior* 10 (1): 4–16.

Lambin, X. (2008). Chapter 5 (pp. 99–107), in Harris, S. and Yalden, D. W. (eds.), *Mammals of the British Isles: Handbook*, 4th edition. Mammal Society, London.

Langbein, J., Chapman, N. G. and Putman, R. J. (2008). Chapter 11 (p. 601), in Harris, S. and Yalden, D. W. (eds.), *Mammals of the British Isles: Handbook*, 4th edition. Mammal Society, London.

Lee, M. (2004). Bat detectors: a beginner's guide for orthopterists. www.erccis.co.uk.

López-Baucells, A., Torrent, L., Rocha, R., Bobrowiec, P. E. D., Palmeirim, J. M. and Meyer, C. F. J. (2019). Stronger together: combining automated classifiers with manual post-validation optimizes the workload vs reliability trade-off of species identification in bat acoustic surveys. *Ecological Informatics* 49: 45–53.

Lupanova, A. S. and Egorova, M. A. (2015). Vocalization of sex partners in the house mouse (*Mus musculus*). *Journal of Evolutionary Biochemistry and Physiology* 51 (4): 324–331.

Marchlewska-Koj, A. (2000). Olfactory and ultrasonic communication in bank voles. *Polish Journal of Ecology* 48: 11–20.

Marsh, A. C. W. and Montgomery, W. I. (2008). Chapter 5 (pp. 137–141), in Harris, S. and Yalden, D. W. (eds.), *Mammals of the British Isles: Handbook*, 4th edition. Mammal Society, London.

Mathews, F., Coomber, F., Wright, J. and Kendall, T. (2018). *Britain's Mammals 2018: The Mammal Society's Guide to their Population and Conservation Status*. Mammal Society, London.

McInerny C. J. and Minting P. J. (2016). *The Amphibians and Reptiles of Scotland*. Glasgow Natural History Society, Glasgow.

Middleton, N. (2006). A study of the emission of social calls by *Pipistrellus* spp. within central Scotland; including a description of their typical social call structure. *BaTML Publications* 3: 23–28.

Middleton, N., Froud, A. and French, K. (2014). *Social Calls of the Bats of Britain and Ireland*. Pelagic Publishing, Exeter.

Miska-Schramm, A., Kapusta, J. and Kruczek, M. (2018). Copper influence on bank vole's (*Myodes glareolus*) sexual behaviour. *Ecotoxicology* 27 (3): 385–393.

Morris, P. A. (2008). Chapter 5 (pp. 82–83), in Harris, S. and Yalden, D. W. (eds.), *Mammals of the British Isles: Handbook*, 4th edition. Mammal Society, London.

Murray, K. L., Britzke, E. R. and Robbins, L. W. (2001). Variation in search-phase calls of bats. *Journal of Mammalogy* 82 (3): 728–737.

Nakano, R., Takanashi, T., Fujii, T., Skals, N., Surlykke, A. and Ishikawa, Y. (2009). Moths are not silent, but whisper ultrasonic courtship songs. *Journal of Experimental Biology* 212: 4072–4078.

Nakano, R., Takanashi, T. and Surlykke, A. (2014). Moth hearing and sound communication. *Journal of Comparative Physiology* 201: 111–121.

NARRS (2018). National Amphibian and Reptile Recording Scheme. Online resource providing information about surveying of native amphibians and reptiles. www.narrs.org.uk.

Natural England (2015). A review of the Orthoptera (grasshoppers and crickets) and allied species of Great Britain. Orthoptera, Dictyoptera, Dermaptera, Phasmida. Species Status no. 21. Natural England Commissioned Report NECR187.

Neil, T. R. (2018). Moths survive bat predation through acoustic camouflage fur. Presentation delivered to the Acoustical Society of America. Reported in *ScienceDaily*, 6 November 2018.

Newson, S. E., Bas, Y., Murray, A. and Gillings, S. (2017). Potential for coupling the monitoring of bush-crickets with established large-scale acoustic monitoring of bats. *Methods in Ecology and Evolution* 8 (9): 1051–1062.

Newton-Fisher, N., Harris, S., White, P. and Jones, G. (1993). Structure and function of red fox *Vulpes vulpes* vocalisations. *Bioacoustics* 5: 1–31.

Ntelezos, A., Guarato, F. and Windmill, J. F. C. (2017). The anti-bat strategy of ultrasound absorption: the wings of nocturnal moths (Bombycoidea: Saturniidae) absorb more ultrasound than the wings of diurnal moths (Chalcosiinae: Zygaenoidea: Zygaenidae). *Biology Open* 6: 109–117.

Oddie, B. (1995). *Bill Oddie's Little Black Bird Book*. Robson Books, London.

Olshausen, B. A. (2000). Aliasing, PSC 129 – sensory processes. www.rctn.org/bruno/npb261/aliasing.pdf.

O'Reilly, L. J., Agassiz, D. J. L., Neil, T. R. and Holderied, M. W. (2019). Deaf moths employ Müllerian mimicry against bats using wingbeat-powered tymbals. *Scientific Reports* 9: 1444. doi: 10.1038/s41598-018-37812-z.

Osipova, O. and Rutovskaya, M. V. (2000). Information transmission in bank voles by odor and acoustic signals (signalling communication). *Polish Journal of Ecology* 48: 21–36.

Pellet, J. and Schmidt, B. R. (2004). Monitoring distributions using call surveys: estimating site occupancy, detection probabilities and inferring absence. *Biological Conservation* 123: 27–35.

Penone, C., Le Viol, I., Pellissier, V., Julien, J. F., Bas, Y. and Kerbiriou, C. (2013). Use of large-scale acoustic monitoring to assess anthropogenic pressures on Orthoptera communities. *Conservation Biology* 27 (5): 979–987.

Pfalzer, G. and Kusch, J. (2003). Structure and variability of bat social calls: implications for specificity and individual recognition. *Journal of Zoology* 261: 21–33.

Putman, R. J. (2008). Chapter 11 (p. 592), in Harris, S. and Yalden, D. W. (eds.), *Mammals of the British Isles: Handbook*, 4th edition. Mammal Society, London.

Quy, R. J. and Macdonald, D. W. (2008). Chapter 5 (pp. 149–155), in Harris, S. and Yalden, D. W. (eds.), *Mammals of the British Isles: Handbook*, 4th edition. Mammal Society, London.

Ragge, D. R. and Reynolds, W. J. (1998). *The Songs of the Grasshoppers and Crickets of Western Europe*. Harley Books in association with the Natural History Museum, London.

Rheinlaender, J. and Römer, H. (1980). Bilateral coding of sound direction in the CNS of the bushcricket *Tettigonia viridissima* L. (Orthoptera, Tettigoniidae). *Journal of Comparative Physiology* 140: 101–111.

Robinson, D. J. and Hall, M. J. (2002). Sound signalling in Orthoptera. *Advances in Insect Physiology* 29: 151–278.

Rodolfi, G. (1994). Dormice *Glis* activity and hazelnut consumption. *Acta Theriologica* 39 (2): 215–220.

Russ, J. (2012). *British Bat Calls: a Guide to Species Identification*. Pelagic Publishing, Exeter.

Russo, D. and Voigt, C. C. (2016). The use of automated identification of bat echolocation calls in acoustic monitoring: A cautionary note for a sound analysis. *Ecological Indicators* 66: 598–602.

Russo, D., Ancillotto, L. and Jones, G. (2017). Bats are still not birds in the digital era: echolocation call variation and why it matters for bat species identification. *Canadian Journal of Zoology* 96 (2): 63–78. doi: 10.1139/cjz-2017-0089.

Rydell, J., Nyman, S., Eklöf, J., Jones, G. and Russo, D. (2017). Testing the performances of automated identification of bat echolocation calls: a request for prudence. *Ecological Indicators* 78: 416–420.

Sales, G. D. (2010). Ultrasonic calls of wild and wild-type rodents. In Brudzynski, S. M. (ed.), *Handbook of Mammalian Vocalization*, pp. 77–88.

Sales, G. D. and Pye, J. D. (1974). *Ultrasonic Communication by Animals*. Chapman & Hall, London.

Sánchez-Bayo, F. and Wyckhuys, K. A. G. (2019). Worldwide decline of the entomofauna: a review of its drivers. *Biological Conservation* 232: 8–27.

Schneider, H. and Sinsch, U. (2007). *Amphibian Biology*, Volume 7, Chapter 8. Surrey Beatty & Sons, Chipping Norton, NSW.

Schul, J., Matt, F. and von Helverson, O. (2000). Listening for bats: the hearing range of the bushcricket *Phaneroptera falcata* for bat echolocation calls measured in the field. *Proceedings of the Royal Society B, Biological Sciences* 267 (1454): 1711–1715.

Schulze, W. and Schul, J. (2001). Ultrasound avoidance behaviour in the Bushcricket *Tettigonia viridissima* (Orthoptera: Tettigoniidae). *Journal of Experimental Biology* 204: 733–740.

Sewell, D., Griffiths, R. A., Beebee, T. J. C., Foster, J. and Wilkinson, J. W. (2013). Survey protocols for the British Herpetofauna. Version 1.0. Amphibian and Reptile Conservation, DICE (University of Kent) and University of Sussex.

Shore, R. F. and Hare, E. J. (2008). Chapter 5 (pp. 88–99), in Harris, S. and Yalden, D. W. (eds.), *Mammals of the British Isles: Handbook*, 4th edition. Mammal Society, London.

Siemers, B. M., Beedholm, K., Dietz, C., Dietz, I. and Ivanova, T. (2005). Is species identity, sex, age or individual quality conveyed by echolocation call frequency in European horseshoe bats? *Acta Chiropterologica* 7 (2): 259–274.

Siemers, B. M., Schauermann, G., Turni, H. and von Merton, S. (2009). Why do shrews twitter? Communication or simple echo-based orientation. *Biology Letters* 5: 593–596. doi: 10.1098/rsbl.2009.0378.

Simeonovska-Nikolova, D. M. (2004). Vocal communication in the bicoloured white-toothed shrew *Crocidura leucodon*. *Acta Theriologica* 49 (2): 157–165.

Skals, N. and Surlykke, A. (1999). Sound production by abdominal tymbal organs in two moth species: the green silver-line and the scarce silver-line (Noctuoidea: Nolidae: Chloephorinae). *Journal of Experimental Biology* 202: 2937–2949.

Spangler, H. G. (1988). Moth hearing, defense and communication. *Annual Review of Entomology* 33: 59–81.

Speybroeck, J., Beukema, W., Bok, B. and van der Voort, J. (2016). *Field Guide to the Amphibians and Reptiles of Britain and Europe*. Bloomsbury, London.

Staines, B. W., Langbein, J. and Burkitt, T. D. (2008). Chapter 11 (p. 581), in Harris, S. and Yalden, D. W. (eds.), *Mammals of the British Isles: Handbook*, 4th edition. Mammal Society, London.

Stoddart, D. M. and Sales, G. D. (1985). The olfactory and acoustic biology of wood mice, yellow-necked mice and bank voles. *Symposia of the Zoological Society of London* 55: 117–139.

Strachan, R. (1999) *Water Voles*. British Natural History Series, Volume 27. Whittet Books, London.

Surlykke, A., Filskov, M., Fullard, J. H. and Forrest, E. (1999). Auditory relationships to size in noctuid moths: bigger is better. *Naturwissenschaften* 86: 238–241.

Surlykke, A. and Kalko, E. K. V. (2008). Echolocating bats cry out loud to detect their prey. *PLoS One* 3 (4): e002036. doi: 10.1371/journal.pone.0002036.

Svensson, L., Mullarney, K., Zetterström, D. and Grant, P. J. (2010). *Collins Bird Guide*, 2nd edition. Harper Collins, London.

Takahashi, N., Kashino, M. and Hironaka, N. (2010). Structure of rat ultrasonic vocalizations and its relevance to behavior. *PLoS One* 5 (11): e14115. doi: 10.1371/journal.pone.0014115.

ter Hofstede, H. M., Goerlitz, H. R., Ratcliffe, J. M., Holderied, M. W. and Surlykke, A. (2013). The simple ears of noctuid moths are tuned to the calls of their sympatric bat community. *Journal of Experimental Biology* 216: 3954–3962.

Thomas, J. A. and Jalili, M. S. (2004). Review of echolocation in insectivores and rodents. In Thomas, J. A., Moss, J. F. and Vater, M. (eds.), *Echolocation in Bats and Dolphins*. University of Chicago Press, Chicago, IL, pp. 547–563.

Toledo, L. F. and Haddad, C. F. B. (2005). Reproductive biology of *Scinax fusco-marginatus* (Anura, Hylidae) in south-eastern Brazil. *Journal of Natural History* 39 (32): 3029–3037.

Toledo, L. F., Martins, I. A., Bruschi, D. P., Passos, M. A., Alexandre, C. and Haddad, C. F. B. (2014). The anuran calling repertoire in the light of social context. *Acta Ethologica* 18: 87–99.

Trout, R. C. (1978). A review on studies of captive harvest mice (*Micromys minutus* (Pallas)). *Mammal Review* 8: 159–175.

Trout, R. C. and Harris, S. (2008). Chapter 5 (pp. 117–125), in Harris, S. and Yalden, D. W. (eds.), *Mammals of the British Isles: Handbook*, 4th edition. Mammal Society, London.

Twigg, G. I., Buckle, A. P. and Bullock, D. J. (2008). Chapter 5 (pp. 155–158), in Harris, S. and Yalden, D. W. (eds.), *Mammals of the British Isles: Handbook*, 4th edition. Mammal Society, London.

UK Moths (2019). Online guide to the moths of Great Britain and Ireland. https://ukmoths.org.uk.

von Merten, S. (2011). Spatial exploration and acoustic orientation in shrews: a comparative approach on the three sympatric species *Sorex araneus, Sorex minutus* and *Crocidura leucodon*. Dissertation, Tübingen.

von Merten, S., Hoier, S., Pfeifle, C. and Tautz, D. (2014). A role for ultrasonic vocalisation in social communication and divergence of natural populations of the house mouse (*Mus musculus domesticus*). *PLoS One* 9 (5): e97244. doi: 10.1371/journal.pone.0097244.

Walkowiak, W. and Brzoska, J. (1982). Significance of spectral and temporal call parameters in the auditory communication of male grass frogs. *Behavioral Ecology and Sociobiology* 11 (4): 247–252.

Wells, K. D. (1977). The social behaviour of anuran amphibians. *Animal Behaviour* 25: 666–693.

Wöhr, M. (2018). Ultrasonic communication in rats: appetitive 50-kHz ultrasonic vocalizations as social contact calls. *Behavioral Ecology and Sociobiology* 72 (1): 14.

Wöhr, M. and Schwarting, R. K. W. (2013). Affective communication in rodents: ultrasonic vocalizations as a tool for research on emotion and motivation. *Cell and Tissue Research* 354 (1): 81–97.

Wong, J., Stewart, P. D. and MacDonald, D. W. (1999). Vocal repertoire in the European Badger (*Meles meles*): structure, context and function. *Journal of Mammalogy* 80 (2): 570–588.

Woodroffe, G. L., Lambin, X. and Strachan, R. (2008). Chapter 5 (pp. 110–117), in Harris, S. and Yalden, D. W. (eds.), *Mammals of the British Isles: Handbook*, 4th edition. Mammal Society, London.

Zsebok, S., Czaban, D., Farkas, J., Siemers, B. M. and von Merten, S. (2015). Acoustic species identification of shrews: twittering calls for monitoring. *Ecological Informatics* 27: 1–10.

Index

Page numbers in *italics* denote figures and in **bold** denote tables.